Wind Energy Applications

K.R. Rao, PhD, PE
Editor

ASME
PRESS

Library of Congress Cataloging in Publication Control Number: 2022936849

ISBN: 978-0-7918-8572-7

CONTENTS

DEDICATION

This *Wind Energy Applications* volume in the ASME Press Renewable Energy Series is dedicated to the individuals, associations and nations in the pursuit of the cause, development and promotion to increase the share of Wind Energy power in our global energy supply.

Editor-in-Chief Dr. KR Rao is personally indebted to the late VRP Rao FIE(I) who was the first person who urged me to study and understand the importance of Renewable Energy, as well as indebted to *all of the authors who contributed to ENERGY and POWER GENERATION HANDBOOK of ASME Press* published in 2011.

Dr. K. R. Rao expresses his gratitude to Late Prof Dr. Robert T. Norman, his Doctoral Advisor, who was Officer of USAID, India saw KR Rao's work at Kanpur, India and invited KR Rao for pursuing doctoral studies program at University of Pittsburgh, Pa.

Dr. K.R. Rao
Editor-in-Chief
Renewable Energy Series

Acknowledgements

My appreciation are to all of the original contributors of Chapters 6, 7, 8 and 9 of ASME's Energy and Power Generation Handbook, who for various reasons could not contribute for the current effort. My thanks are for the continuing contributor Ramesh Muthya of Chapter 10 whose dedicated effort is acknowledged.

I invited Dr Weifie Hu after hearing his presentation at an ASME International Conference to contribute for this publication and the resulting contribution for this book is immensely appreciated and acknowledged.

This publication was ably supported by Mary Grace Stefanchik and Tara Smith of ASME Technical Publishing. My appreciation and thanks to them for their cooperation support and their patience is greatly appreciated.

Finally, all of this enduring effort, spread over 15 months, would have never been possible had it not been for the constant help and untiring zeal provided by my wife, Dr. Indira Rao, that included all of the sundry editorial chores associated with this project.

Dr. K.R. Rao
Editor-in-Chief
Renewable Energy Series

Contributor Biographies[1]

BALDWIN, THOMAS L.

Thomas L. Baldwin, PE, PhD, IEEE Fel-low, is a senior engineer at the Idaho National Laboratory. He conducts engineering studies and research in electrical power generation and transmission for the U.S. Department of Energy, U.S. Navy, and EPRI. His research interests are in distribution energy system design, industrial power systems, grounding issues, transformers, and the analysis of power quality problems.

Dr. Baldwin also holds the rank of professor at the FAMU-FSU College of Engineering at florida State University, tallahassee, fl, and has conducted research at the Center for Advanced Power Systems since 1999. he is a registered Professional Engineer in the State of North Carolina.

Thomas l. Baldwin received the BSEE and MSE.E degrees from Clemson University, Clemson, SC, and the Phd degree in electrical engineering from virginia Polytechnic Institute and State University, blacksburg, vA, in 1987, 1989, and 1993, respectively.

Dr. Baldwin is a member of the IEEE Power and Energy Society and the Industrial Applications Society and serves on several committees and working groups including Power System grounding and the IEEE green book.

EECEN, PETER

Peter Eecen is research manager of the group Rotor & Farm Aerodynamics at the Wind Energy department of the Energy research Centre of the Netherlands (ECN). His responsibilities are to establish the research strategy and priorities and manage a group of 20 researchers. In collaboration with the ECN Wind Industrial Support group (EWIS), he is responsible for the business development; an important aspect of ECN is to bring to the market the developed knowledge and technologies in renewable energy.

Peter holds a Master's degree in theoretical physics and did his PhD in the field of nuclear fusion at the FOM Institute for Plasma Physics, after which he worked for 3 years at TNO in the field of underwater acoustics.

Dr. Eecen joined the Wind Energy department of ECN in 2000 as manager of the group "Wind and Waves," concerning the wind and wave descriptions for turbine loading and wind resource assessments. After that, he led the experimental department for a year. He has been working in the field of Operation and Maintenance of large offshore wind farms. He started the project of Operation and Maintenance Cost Estimator (OMCE). Since 4 years, Peter is heading the group Rotor & Farm Aerodynamics. This group aims to optimize the aerodynamic performance of the wind turbine rotor and of the wind farm as a whole and reduce the uncertainties in modeling rotor aerodynamics, wake aerodynamics, boundary layers by development of CFD technology, development of aerodynamic design tools for wind turbines and wind farms. During his career in wind energy, Dr. Eecen performed re-search in a variety of areas,

[1] Biographies of Thomas Baldwin, Peter Eecen, Gary Seifert, Larry Viterna and Jose Zayas who had contributions in the 2011 edition of Energy Generation Handbook are included in this list of contributors since their contributions are retained in the current publication, and they are pertinent even today. These biographies although dated need to be retained "as they are" in the 2011 Handbook. - Editor

which include modeling wind and waves, resource assessments, uncertainties in wind measurements, remote sensing, operation, and maintenance. He was responsible for measurements on full-scale wind farms and the ECN scale wind farm.

Peter is active in international organizations like IEA, MEASNET, TP Wind, and European projects. He is coordinating the subprogram Aerodynamics of EERA-Wind, the European Energy Research Alliance. Ten leading European Research Institutes have founded EERA to accelerate the development of new energy technologies by conceiving and implementing Joint Research Programs in support of the Strategic Energy Technology (SET) plan by pooling and integrating activities and resources.

HU, WEIFEI

Weifei Hu received the B.S. degree in 2008 from Zhejiang University, Hangzhou, China, the M.S. degree in 2010 from Hanyang University, Seoul, South Korea, and the Ph.D. degree in 2015 from the University of Iowa, Iowa city, Iowa, USA, all in mechanical engineering. From February 2016 to September 2018, Dr. Hu was a postdoctoral fellow at Cornell University, Ithaca, New York, USA. He is currently holding a tenure-track ZJU100 Young Professor position at the School of Mechanical Engineering, Zhejiang University.

Dr. Hu's research interests include Digital Twin, Artificial Intelligence, Design Under Uncertainty, and Wind Energy. His work has been published in over 60 peer-reviewed journals and conference proceedings, e.g., Renewable Energy, Wind Energy, Structural and Multidisciplinary Optimization, Journal of Mechanical Design, and many ASME and AIAA proceedings. He is a Technical Committee Member (2016–now) and was the Vice Chair (2017-2019) of the Renewable and Advanced Energy Systems Committee in ASME, and an editorial board member of two international journals (Wind Energy and Journal of Intelligent Manufacturing and Special Equipment).

MUTHYA, RAMESH PRANESH RAO

Mr. Muthya holds a Master's degree in Mechanical Engineering from the Indian Institute of Science from Bengaluru, India, and a Master of Science from Oldenburg University, Germany, in Application of Renewable Energy Technologies. Mr. Muthya started his career in wind energy at the National Aeronautical Laboratory, Bangalore, in 1979. He has many firsts to his credit starting from development of indigenous wind power battery chargers and transferred know-how to industry. He made performance measurements on wind turbines for the first time in India, which resolved many issues of underperformance. In order to extend utility of measured wind information to a larger area, he came up with the idea of localized wind maps superimposed on scaled survey maps. Over a hundred such reports were created and were the basis for wind farming. He holds a patent on passive speed control of windmills with Dr. S. K. Tewari. He introduced small wind chargers in Indian Antarctic station. Mr. Muthya took over the position of the first full time Executive Director, Centre for Wind Energy Technology in 2002. He was instrumental in maximizing the benefits of the DANIDA funded project. He went on to create ground rules for effective, orderly, and sustainable growth of the field by continuous interactions with the industry, the Government, and other stake holders. After a short stint in Canada as the scientific director/advisor to GP CO, Varrenes, Mr. Muthya returned to India to help the Indian wind industry. As the founding Managing Director of Garrad Hassan India, Mr. Muthya brought in some of the most sought-after consulting practices to India that spans across the resource quantification to asset management techniques. One of the least attended to areas in terms of design documentation capabilities in India was effectively addressed by Mr. Muthya. He has served as Chairman, Electro Technical committee #42 (ET-42), set up by Bureau of Indian Standards to interact with International Electro-technical committee on all wind energy-related standardization efforts. He also chaired the committee that brought out draft grid code for wind turbine grid interconnection. Presently, as the President, Wind Resource and Technology, at Vaayu Power Corporation, Mr. Muthya is deeply involved in bringing in the best practices on all aspects of wind energy deployment in India.

RAO, K. R.

K.R. Rao retired as a Senior Staff Engineer with Entergy Operations Inc. and was previously with Westinghouse Electric Corporation at Pittsburgh, PA, and Pullman Swindell Inc., Pittsburgh, PA. KR got his Bachelors in Engineering degree from Banaras University, India, with a Masters Diploma in Planning from School of Planning & Architecture, New Delhi, India. He completed Post Graduate Engineering courses in Seismic Engineering, Finite Element and Stress Analysis, and other engineering subjects at Carnegie Mellon University, Pittsburgh, PA. He earned his PhD, from University of Pittsburgh, PA. He is a Registered Professional Engineer in Pennsylvania and Texas. He is past Member of Operations Research Society of America (ORSA).

KR was Vice President, Southeastern Region of ASME International. He is a Fellow of ASME, active in National, Regional, Section, and Technical Divisions of ASME. He has been the Chair, Director, and Founder of ASME EXPO(s) at Mississippi Section. He was a member of General Awards Committee of ASME International. He was Chair of Codes & Standards Technical Committee, ASME PV&PD. He developed an ASME Tutorial for PVP Division covering select aspects of Code. KR was a member of the Special Working Group on Editing and Review (ASME B&PV Code Section XI) for September 2007 to June 2012 term.

Dr. Rao was Editor of the *Companion Guides to the ASME Boiler & Pressure Vessel Codes* which he initiated in 2002 that ultimately became "classics" and are enumerated below:

- *Companion Guide to the ASME Boiler & Pressure Vessel Codes, First Edition (2002)*
- *Companion Guide to the ASME Boiler & Pressure Vessel Codes, Second Edition (2006) in 3 volumes;*
- *Companion Guide to the ASME Boiler & Pressure Vessel Codes, Third Edition (2009) in 3 volumes;*
- *Energy & Power Generation Handbook – Established and Emerging Technologies (2011)*
- *Companion Guide to the ASME Boiler & Pressure Vessel Codes, Fourth Edition (2012) in 2 volumes;*
- *Global Applications of the ASME Boiler and Pressure Vessel Code by K.R. Rao (2014)*
- *Continuing and Changing Priorities of the ASME Boiler and Pressure Vessel Codes and Standards (2016); and*
- *Companion Guide to the ASME Boiler & Pressure Vessel Codes, Fifth Edition (2018) in 2 volumes;*

Dr. Rao is currently the Editor in Chief of the *Renewable Energy Book Series* being published by ASME Press. The first book of this series was "Solar Energy Applications" published in March 2020 and the second book of the series "Biomass and Waste Energy Applications" published in March 2021. This book is the third of the "*Renewable Energy Book Series*" scheduled to be published in early 2021.

Dr. Rao is a recipient of several Cash, Recognition, and Service Awards from Entergy Operations, Inc. and Westinghouse Electric Corporation. He is also the recipient of several awards, certificates, and plaques from ASME PV&P Division including Outstanding Service Award (2001) and Certificate for "Vision and Leadership" in Mississippi and Dick Duncan Award, Southeastern Region, ASME. Dr. Rao is the recipient of the prestigious ASME Society Level Dedicated Service Award. KR is a member of the Board of ASME District F Professional & Educational Trust Fund for 2008 to 2011.

Dr. Rao is a Fellow of American Society of Mechanical Engineers, Fellow of Institution of Engineers, India, and a Chartered Engineer, India. Dr. Rao was recognized as a "Life Time Member" for inclusion in the Cambridge "Who's Who" registry of executives and professionals. Dr. Rao was listed in the Marquis 25th Silver Anniversary Edition of "Who's Who in the World" as "one of the leading achievers from around the globe."

SEIFERT, GARY D.

Gary D. Seifert, PE, EE, is senior program manager at the Idaho National Laboratory. He has responsibility for multiple technical tasks for the U.S. department of Energy, Department of Homeland Security, the U.S. Air Force, U.S. Navy, and NASA, as well as various power systems upgrades at the Idaho National Laboratory. Renewable projects have included the Ascension Island wind Project and Ascension Island Solar Power projects, which have displaced a significant amount of diesel generation resulting in major financial and emissions savings.

Gary has been involved in multiple projects improving control systems and adding automation. other support tasks include wind Powering America, wind anemometer loan program, wind radar integration, power system distribution upgrades, high reliability power systems, relay system updates, smart substation upgrades, fiber optic communication systems installations, National SCADA Testbed, and the design of process control systems.

Gary is also currently involved in studies for multiple Department of Defense government wind projects and is leading a technical wind radar interaction project for the U.S. DOE and supporting wind prospecting activities in Idaho and surrounding regions.

Gary holds patents in thermal photovoltaic and Electro optical high voltage (EOHV) sensor designs. he was awarded a research and development top 100 award in 1998 for his work on the EOHV sensor and was instrumental in the implementation of the DOD's first island wind farm at Ascension air station.

Gary has a Bachelor of Science Electrical Engineering degree from the University of Idaho in 1981. He is an Adjunct Instructor Department of Engineering Professional Development, University of wisconsin since 1991.

VITERNA, LARRY

Dr. Larry Viterna is a loaned executive to Case Western Reserve University from NASA. At Case, he serves as the Technical Director of the Great Lakes Energy Institute, leading the formation of technology development efforts in renewable energy and energy storage. Most recently at NASA, Dr. Viterna was Lead for Strategic Business Development at the Glenn Research Center, a major federal laboratory with a budget of over $600M and a work-force of 2500. Previously, he was assigned to the NASA Deputy Administrator in Washington, DC, where he coordinated the development of the implementation strategy for Agency-wide change following the Space Shuttle Columbia Accident. Dr. Viterna was on the team that created the world's firs multi-megawatt wind turbines starting in 1979. He is the recipient of NASA's Blue Marble Award for aerodynamic models, now named for him, that are part of international design tools for wind turbines and that helped enable passive aerodynamic power control in the wind energy industry. Dr. Viterna has also been recognized with NASA Glenn's highest Engineering Excellence award for his pioneering work in fuel efficient hybrid vehicles. He is certified for the Senior Executive Service and received his Ph.D. in Engineering from the University of Michigan. He has also completed executive education in business administration at Stanford University, public policy at the Harvard Kennedy School of Government, and international management at the National University of Singapore.

ZAYAS, JOSE

Jose Zayas is the senior manager of the Renewable Energy Technologies group at Sandia National Laboratories. His responsibilities in this role include establishing strategy and priorities, defining technical and programmatic roles, business development, and performing management assurance for the renewable energy-related activities of the laboratory. He manages and develops programs

- Bring together key renewable energy technology capabilities to consistently implement a science-based reliability and systems approach
- Leverage Sandia's broader predictive simulation, testing/evaluation, materials science, and systems engineering capability with expertise in renewable energy technologies
- Expand and accelerate Sandia's role in the innovation, development, and penetration of renewable energy technologies Mr. Zayas joined Sandia National Labs in 1996 and spent the first 10 years of his career supporting the national mission of the labs wind energy portfolio as a senior member of the technical staff. During his technical career, he had responsibilities for several programmatic research activities and new initiatives for the program. Jose's engineering research contributions, innovation, and outreach spanned a variety of areas, which include active aero-dynamic flow control, sensors, dynamic modeling, data acquisition systems, and component testing.

After transition to the position of program manager in 2006, Jose has engaged and supported a variety of national initiatives to promote the expansion of clean energy technologies for the nation. Most recently, Jose has continued to lead the organization's clean energy activities and has coordinated and developed the laboratories crosscutting activities in advanced waterpower systems. This program focuses on developing and supporting an emerging clean energy portfolio (wave, current, tide, and conventional hydro energy sources). Through developed partnerships with key national labs, industry, and academia, Sandia is supporting and leading a variety of activities to accelerate the advancement and viability of both wind energy and the comprehensive marine hydrokinetics industry. Additionally, Jose is currently leading a Federal interagency research program to address barriers affecting the continued deployment and acceptance of wind energy systems across the nation.

Jose holds a bachelor's degree in Mechanical Engineering from the University of New Mexico and a Master's degree in Mechanical and Aeronautical Engineering from the University of California at Davis.

PREFACE

Renewable Energy has emerged as an important source for energy and power generation from renewable resources, naturally replenished on a human timescale, such as sunlight, wind, rain, hydro including tides and waves, geothermal heat, and from traditional and modern biomass. Worldwide investments in renewable energy are surpassing expectations, significantly in Europe (Germany and Spain), the US, in Asia (China and India) and in Australia.

Energy, and Power Generation Handbook: Established and Emerging Technologies, edited by K.R. Rao, and published by ASME Press in 2011 was a comprehensive reference work of 32 chapters authored by 53 expert contributors from around the world, with the authors drawn from different specialties, each an expert in the respective field and with several decades of professional expertise and scores of technical publications. Recognizing the need of treating Renewable Energy and Power Generation as a separate field ASME Press initiated "Renewable Energy Series" to address each entity of Renewable Energy in a separate book, revising pertinent chapters of the 2011 Handbook and bringing the coverage up-to-date.

This book is the third in a series of renewable energy topical books and addresses *WIND ENERGY APPLICATIONS* to update chapters 6, 7, 8, 9 and 10 of the 2011 *Handbook* in which Wind Energy has been covered in chapters 1, 2, 3, 4 and 5, respectively, in this third book of the Renewable Energy Series. In addition, a new chapter numbered Chapter 6 covering "Artificial Intelligence in Wind Energy" is presented by Weifei Hu. Due to various reasons ranging from changing interests, change of jobs and due to lack of interest after a lapse of a decade since the publication of original book, initial authors of original chapters 6, 7, 8 and 9 were unavailable for the current effort.

Dr. KR Rao, who was successful in his 2019 Springer Publication covering "Wind Energy for Power Generation - Meeting the Challenge of Practical Implementation" in two volumes and nearly two thousand pages decided to update chapters 6, 7, 8 and 9 with his knowledge and acumen to update these chapters.

Mr. Ramesh Muthya with his decades of experience in Wind Power Generation in India continues to update Chapter 10 titled "Role of Wind Energy Technology in India and Neighboring Countries", which was authored by him in the 2011 Handbook.

This book is meant to cover the technical discussions relating to Wind Energy source as well as why(s) and wherefore(s) of power generation. A unique aspect of this update is the scholarly discussions and nearly 400 footnotes and 102 references enabling the reader to make "value judgments" regarding wind energy technology. This book has the end user in view from the very beginning to the end. The audience targeted by this publication not only includes libraries, universities for use in their curriculum, utilities, consultants, and regulators, but is also meant to include ASME's global community. ASME's strategic plan includes Energy Technology as a priority.

This book could be of immense use to those looking beyond the conventional discussions contained in similar books that provide the "cost benefit" rationale. Instead of picturing a static view, the contributors portray a futuristic perspective in their depictions, even considering the realities beyond the realm of socio-economic parameters to ramifications of the political climate. These discussions will captivate advocacy planners of global warming and energy conservation. University libraries, the "public-at-large," economists looking for technological answers, practicing engineers who are looking for greener pastures in pursuing their professions, young engineers who are scrutinizing job alternatives, and engineers caught in a limited vision of energy and power generation will find this book informative.

From the update provided by Dr. KR Rao in the footnotes of chapters 6, 7, 8 and 9 from public domain, textbook publications, scholastic literature, professional societies and Governmental publications of Task Forces and Special Study Groups, and ASME's Technical Publications that deal with renewable energy sources including Wind Energy Exploration. This publication will be especially useful for ASME members across most of the Technical Divisions, Associations of the European Union dealing with Wind Energy and Research Laboratories engaged with Wind Energy and Power Generation.

INTRODUCTION

In the current third book of the Renewable Energy Series published by ASME Press and edited by Dr. K. R. Rao, chapters 1, 2, 3, and 4 had been updated by Dr. Rao who authored the 2019 Springer Classic, "Wind Energy for Power Generation - Meeting the Challenge of Practical Implementation", in two volumes and nearly 2,000 pages. Dr. Rao had researched especially for these book chapters tracing the Wind Energy Applications accomplished during the past decade, i.e., since 2020 through now. Based on his research from publications in the public domain Dr. Rao's sum-up of these chapters 1, 2, 3 and 4 are follow the original authors introduction as it appeared in 2011 Handbook.

In the original chapter by Dr. Larry Viterna's introduction that appeared in the 2011 Handbook[1] he stated, "in the decade from 1975 through the 1985 the United States government worked with industry to advance the technology and help enable large commercial wind turbines. The impetus for this effort was two events on the world stage (1) an oil embargo by the Organization of Petroleum Exporting Countries (OPEC) in 1973 followed by (2) the fall of the Shah of Iran in 1979. In response to the first crises, in 1975, the NASA's Glenn Research Center took on the role of leading the technology development of large horizontal-axis wind turbines, the dominant type of wind turbine in use. The following pages document the development of major wind turbine systems and technologies, their impact on the industry and future development areas that could benefit the industry. NASA's Glenn Research Center, Ohio had the facilities and engineering capabilities necessary to advance wind turbine technology. NASA was able to successfully apply its capabilities in areas such as propeller and turbine design, structural analysis, advanced materials, aerodynamics, instrumentation and power system control to the design of wind energy systems. During the 1970's and '80s, a total of 13 experimental wind turbines were put into operation including four major wind turbine designs. This program pioneered many of the multimegawatt turbine technologies including steel tube towers, variable-speed generators, composite blade materials, high lift-to-drag airfoils. In addition, key engineering analysis, design and measurement capabilities were developed such as models for aerodynamic performance, structural dynamics, fatigue life, wind measurement and acoustics. The outstanding success of wind energy can be seen today in the growth of the installed power capacity".

In the current update Editor K. R. Rao contributed in section 1.8 of this chapter "NASA Developments, Potential, and Bibliography Since 2011". Since 2011 Wind energy has been extensively investigated and researched by NASA and cited in many publications covering power generation cited in this section. NASA has been extremely active not only in space exploration for which it is primarily meant but also espousing its efficacy for human habitation by disseminating valuable imagery of wind energy for the current and future urban and regional habitations around the world. Figures 1.33 to 1.37 demonstrate the satellite depiction from the outer space of wind energy around the entire globe. Wind power in USA was largely augmented by NASA's innovative efforts as documented by Dr. Viterina in the original compilation of chapter 6 of classic publication "Energy and Power Generation Handbook -- Established and Emerging Technologies", edited by K. R. Rao and published by ASME in 2011. This update provides additional contributions of NASA since the publication of Viterna's chapter in 2011. Dr. Rao covered Wind Developments Around the Globe by NASA; Potential Wind Capacity in the US 48 Contiguous States; Community-Scale 50-m Wind Maps; Residential-Scale 30-m Wind Maps, Implementation of Residential Wind Towers: Offshore Wind Resource Potential; Installed Wind Capacity as of 2015; Proposed Wind Energy Project Offshore North Carolina; Proposed Wind Projects in the US; Potential Wind Capacity; Wind Generation Developments of NASA since 2010; Capturing Airborne Wind; NASA WorldWind Efforts; Monthly Mean Wind and 10m Wind Speed; and conclusions.

Dr. K. R. Rao, the editor of the current publication, updated this chapter due to the nonavailability of Dr. Peter Viterna to update his original contribution of the 2011 Handbook. As a follow-up of original chapter 6 by Dr. Viterna covering NASA's contributions of Wind Energy in ASME Energy and Power Generation Handbook [1], NASA's contribution for wind energy since 2010 had been captured by Dr. Rao in the current update. Dr. Viterna was a NASA Researcher whose innovation advanced the entire wind energy industry. Despite the advances made by the NASA-led program, there were still significant challenges. Wind turbines, especially in high-wind conditions, can routinely stall, limiting the ability of the turbine to produce electricity. The inability to properly predict stall behavior led to inefficient designs and costly turbine failures. "A Pair of Surprises" were what Viterna developed in a model that took into account three-dimensional effects and predicted stall behavior with far greater accuracy than previous methods, improving wind turbine performance. The model was not well accepted by colleagues in the wind energy field, Dr. Viterna remembers. "I almost got laughed off the stage when I presented it," he says.

Dr. Viterna, however, continued to employ the model for NASA's purposes, even using it to improve Glenn's Icing Research Tunnel, designed to study the effects of ice buildup on aircraft. During a nighttime test, the wind tunnel's pressure release door flew open, sending 400-mph winds ripping through the adjoining work area. This surprised Viterna's coworkers and colleagues inventing this model. He discovered that this initially criticized model had, within a decade of its creation, quietly become the established method of modeling the performance of wind turbine airfoils in stall conditions. "A Method for Success" was what is in the "Viterna Method" which is now routinely used by research institutions and turbine manufacturers alike in the design and certification of modern wind turbines. "The model makes it possible for manufacturers to predict and characterize the performance of airfoils for these turbines," says Paul Gray a turbine developer Aerostar Inc. The method is even used for finding applications beyond horizontal wind turbines; Viterna' once-mocked model is now employed for vertical wind turbines, wind tunnels, and underwater turbines that make use of tidal energy to produce power. The Agency can also expects to see a resurgence of its early work, predicts Sandy Butterfield, wind program chief engineer at the National Wind Technology Center. "As the industry moves forward and becomes more competitive, you will see efforts to explore lower cost, more structurally efficient machines, and that's when people will begin to capitalize again on that good work NASA did in the early 1970s and 1980s," he says. That is good news for the advancement of alternative energy, and for the dedicated researchers at Glenn, so long as this does not involve another office destroyed. Information about this NASA spinoff is in the original article from "Spinoff 2009", see the three figures 1.58, 1.59 and 1.60 all which are attributed to Viterna. Dr. Rao updated Dr. Viterna's chapter with 4 tables, 30 figures, 53 footnotes and 2 references.

Chapter 7 of the 2011 Handbook was authored by Jose Zayas who addressed "Scope of Wind Energy Generation Technologies". Jose Zayas noted the progress of Wind Energy since the early days through time of his contribution in 2011 during which time Wind Energy in the United States has achieved large market installations and continued market acceptance. Through the 3rd quarter of 2010, the United States has approximately 37,000 MW of installed capacity, which represents 2% of the US energy consumption. Since the beginning, Sandia National Laboratories has had a key role in developing innovations in areas such as aerodynamics, materials, design tools, rotor concepts, manufacturing, and sensors, and today, through continued partnerships with industry, academia, and other national labs, Sandia continues to develop and deliver the next set of technology options that will continue to improve the reliability and efficiency of wind systems. It is difficult to predict what the next generation of technologies will bring to this industry, but it continues to mature and leverage technologies from other sectors, the resulting turbines will be smarter, more efficient, and they will represent a significant percentage of the US Energy Mix. the author uses 11 references along with 37 schematics, figures, and pictures to augment the professional and scholastic treatment of the subject.

Dr. K. R. Rao, the Editor of the current publication, up-dated this chapter in Section 2.7, due to the nonavailability of Jose Zayas to update his original contribution in the 2011 Handbook [1]. Dr. Rao in section 2.7 covers the Scope of Wind Energy Generation Technologies since 2011, the earlier period was covered in "Scope of Wind Energy Generation Technologies" by Jose Zayas in "Energy and Power Generation Handbook- Established and Emerging Technologies" edited by K. R. Rao published by ASME, 2011 [1]. The content of Jose Zayas's initial contribution is retained in its entirety since the textual content has relevance even to-day after two decades after the initial publication in 2011.

This section 2.7 has four parts: 2.7.2 dealing with the Next Generation Issues of Wind Energy"; 2.7.3 covers "Next Generation – Analysis"; 2.7.4 addresses "Wind Vision"; 2.7.5 "Wind Map" and lastly 2.7.6 ends with a Conclusion. In 2 7.2 discussion starts with topics which impacted Wind Energy Generation since 2010 to-date, which are issues with impact on next generation issues. The topics are "Bigger the Better", "Hub Height"; "Rotor Diameter"; "Inverters"; "Life Extension and Recycling"; "Circular Economy"; "Name Plate Capacity"; "Transportation Challenges"; "Artificial Intelligence" and "Hybrid" related issues. In 2.7.3 coverage starts with "Market Analysis -On Land"; "Market Analysis - Offshore"; "Distributed Wind"; "Grid Integration"; "Environmental Aspects of Wind Turbine Siting"; "Cyber Security"; "Digital Related Issues"; "High Fidelity – HPC & HPM"; "Compressible and Incompressible flow models" and finally "Nonlinear Structural Flows". Section 2.7.4 picks up information in "Wind Vision" contributed by Jose Zayas as director Wind & Water Technology Office; Section 2.7.5 elaborates the Department of Energies (DOE's) Office of Energy Efficiency and Renewable Energy's Energy Innovation Portal and WETO Projects Wind Map; and finally in 2.7.6 a conclusion is provided elaborating the points discussed in this section and summarizing what the future of wind energy as contemplated by US DOE and NREL.

K. R. Rao contributed this update based on his 'on-line research' and findings from his Classic "Wind Energy for Power Generation - Meeting The Challenge of Practical Implementation" Published by Springer in two volumes in 2019. This update also picks up the essence of a report "WIND VISION - New Era for Wind Power in the United States", by Jose Zayas who was Director, Wind and Water Power Technologies Office U.S. Department of Energy. "WIND VISION - New Era for Wind Power in the United States" dated March 12, 2015. Wind Vison was a collaborative effort of "several hundred stake holders and individuals across the agency, industry, academia, and national labs for their strategic interest in a renewed vision for wind energy", which was a significant comprehensive contribution for Wind Energy Development. The essence of this report has been captured by K. R. Rao spread over in several sections of this current chapter update. Readers are urged to refer to the "Wind Vison" report to capture the message about the growth of the wind energy during the past decade and for the future up to 2050. Dr. Rao updated Jose Zayas chapter with 5 tables, 25 figures, 110 footnotes and 2 references.

Thomas Baldwin and Gary Seifert cover "Wind Energy in the U.S." in Chapter 8 of 2011 Handbook [1] which is retained as Chapter 3 in the current publication. Thomas Baldwin, Gary Seifert used 34 references along with 26 schematics, figures, pictures, and tables to augment the professional and scholastic treatment of the subject. In the 1970s and 1980s, wind turbines were clustered into wind farms and connected to the electric grid in California, which marked the first commercial, utility-scale use of wind energy. The size of those wind turbines was 100 kW and smaller. In the following decades, wind turbine technology progressed quickly, and by 2010, grid-connected wind turbines were typically in the 2-MW range and turbines as large as 6-MW have been deployed. The potential energy in wind is determined as a function of the cube of the speed. Given the cubic relationship between power and wind speed when wind speed doubles there is eight times more power available. In 2009, wind supplied about 1.8% of the electricity in the United States. Wind power currently provides more than 20% of the electricity distributed in northern Spain and in Denmark. A goal is to increase wind generation in the United States to 20% by 2030. Utility transmission lines carry wind-generated electricity from vast and sparsely populated areas where the wind is most abundant, like the Great Plains, to large cities where demand for electricity is high. At the moment, there is insufficient transmission infrastructure connecting the windiest parts of the country with large cities. Enhancing the transmission line power capacity for utility-scale wind plants is a key issue that must be resolved in the coming years. Most regions of the United States are served by "power pools" of utilities that join together to generate electricity and transmit it to where it is needed. The name "power pool" is descriptive of the electric power coming from many different sources (a coal-fired power plant, a hydro plant, and others) flows into a "pool" from which it is distributed to thousands of end users. A power pool can easily absorb the electric power from a wind plant and add to it the generation mixture up to a penetration level of around 20%. Wind penetration greater than 20% requires advancements in transmission capacity, forecasting accuracy, and energy storage as addressed in the Wind Energy Research section. Wind plants could be installed in many locations, providing income, jobs, and electricity for homes and businesses. Wind farms and plants can also revitalize the economy of rural communities, providing steady income through lease or royalty payments to farmers and other landowners. Although leasing arrangements can vary widely, in 2010, a reasonable estimate for income to a landowner from a single utility-scale turbine was ap- proximately $4,000 a year, depending on the wind resource, the size of the turbine, and other factors. For a typical farm, income from wind comes with little interruption in farming activities with about one acre removed from agricultural production per turbine. Farmers can grow crops or raise livestock next to the towers. While wind farms may extend over a large geographical area, their actual "footprint" covers only a very small portion of the land, making wind development a cooperative way for farmers to earn additional income. WIND TURBINE TECHNOLOGIES have been covered by the authors in the initial chapter.

The original authors of chapter 8 that appeared in the "Energy and Power Generation Handbook" published by ASME in 2011 [1] were invited to provide an update for their contributions of the original publication of 2011. It is quite understandable that they were unavailable for updating their original contribution since over a period of a decade their jobs may have taken different directions and their professional interests had shifted to newer and different avenues. Given that perspective and that they were unavailable for the current publication, Dr. K. R. Rao, editor-in-charge of this Renewable Energy Series decided to provide the update reflecting the present scenario of the subject addressed by Thomas Baldwin and Gary Seifert in Chapter 8 titled "WIND ENERGY IN THE U.S". Dr. KR Rao with his classic 2-volume book, nearly 2,000 pages on "Wind Energy for Power Generation- Meeting The Challenge of Practical Implementation" published by Springer in 2019 [2] makes him credible to provide the update for Baldwin/ Seifert Chapter. This chapter update has been renumbered as chapter 3 in the current publication. Most of the content and graphics including references of the original Baldwin / Seifert effort have been retained by the editor considering they are still germane for addressing the subject matter, even today. However, considering several advances which have occurred over a period of one decade, editor who provided this update retained the topical outline which is still pertinent today as initially published and the updates are provided against each of the topics mentioned in the 2011, ASME Publication [1]. Wind power in the United States has expanded rapidly over the past few decades, and January through December 2020, 337.5 terawatt-hours were generated by wind power, or 8.42% of all generated electrical energy in the United States. In 2019, wind power surpassed hydroelectric power as the largest renewable energy source generated in the U.S. As of January 2021, the total installed wind power nameplate generating capacity in the United States was 122,478 megawatts (MW). This capacity is exceeded only by China and the European Union. Thus far, wind power's largest growth in capacity was in 2020, when 16,913 MW of wind power was installed and surpassed by leaps and bounds what was achieved in 2012, which saw the addition of 11,895 MW, representing 26.5% of the new power capacity installed during that year. By September 2019, 19 states had over 1,000 MW of installed capacity with 5 states (Texas, Iowa, Oklahoma, Kansas, and California) generating over half of all wind energy in the nation. Texas, with 28,843 MW of capacity, about 16.8% of the state's electricity usage, had the most installed wind power capacity of any U.S. state at the end of 2019. The state generating the highest percentage of energy from wind power is Iowa at over 57% of total energy production, while North Dakota has the most per capita wind generation. The Alta Wind Energy Center in California is the largest wind farm in the United States with a capacity of 1,548 MW. GE Power is the largest domestic wind turbine manufacturer. In this update closely following the topical outline of the original authors this coverage in Section 3.15 "Scope of Wind Energy in the US Since 2011", after a brief introduction in section 3.15.1 has in the following sections in section 3.15.2 Wind Turbine Technologies, in 3.15.3 Wind Resources in The U.S., in section 3.15.4 covers Wind Plant Economics, in section 3.15.5 addresses Technical Issues, in section 3.15.6 coverage is Environmental Issues, in section 3.15.7 Radar Impacts, in section 3.15.8 Local Impacts, in section 3.15.9 Addressing Needs For Wind To Reach Its Full Potential In The U.S., in section 3.15.10, Wind Farm Development, in section 3.15.11 Wind Resource Assessment, in section 3.15.12 Materials Required for The Land-Based Wind Turbine Industry, and a conclusion in 3.15.13. The sum up of this chapter update states wind capacity additions continued at a robust pace during the past decade. Wind power has been a significant source of new electric generation capacity in the US in recent years. Supply chain is diverse and multifaceted, with strong domestic content for nacelle assembly, towers and blades. Turbine scaling is significantly boosting wind project performances, while the installed cost of wind projects has declined. Wind power sales prices and levelized cost of energy are at all-time lows, enabling economic competitiveness despite low gas prices. Growth beyond current cycle remains uncertain and could be blunted by declining federal tax support, expectations for low natural gas prices and solar costs, and modest electricity demand growth. The outlook for future wind energy growth will depend upon phaseout of federal tax incentives; continued low natural gas and wholesale electricity prices; potential decline in market value as wind penetration increases; modest electricity demand growth; limited near-term demand from based on state policies; and finally growing competition from solar in some regions. Dr. Rao updated Baldwin & Seifert chapter with 4 tables, 37 figures, 82 footnotes and 2 references.

Chapter 9 "Wind Energy Research in The Netherlands" by Dr. Peter Eecen was published in the 2011 Handbook [1]. Since Dr. Peter Eecen was unavailable to up-date his original write-up Editor Dr. K. R. Rao with his credentials as author of the classic 2-volume 2,000 page-book on "Wind Energy for Power Generation-Meeting The Challenge of Practical Implementation" published by Springer in 2019 [2] is credible to up-date this chapter. In the current book it is renumbered as chapter 4 in which the original write-up is retained in "to-to" since it is still applicable for the current times, as well. The essence of Dr Peter Eecen's original introduction as it appeared in the 2011 Handbook is retained in this introduction. The research in wind energy is concentrated at the wind energy department of the Energy Research Centre of the Netherlands, ECN and the interfaculty wind energy department DUWIND at Delft University of Technology. Both institutes have been involved in wind energy research from the start in the 1970s and closely match their research programs. In the Netherlands, the wind energy research is supported by an extensive experimental infrastructure. Interest in the application of modern wind energy grew in the Netherlands in the 1970s when the limit of fossil fuels became clear. The current wind energy research and associated industrial activities are taking place in an international context, mostly the European context, therefore the research activities not only take account of the long-term energy research program of the Dutch government, such as the long-term energy research program, but also of the R&D priorities defined in the international context, such as the Strategic Research Agenda (SRA) of the wind energy sector. The three wind energy research organizations are well represented in international bodies such as European Wind Energy Association (EWEA), European Academy of Wind Energy (EAWE), International Energy Agency (IEA), International Electrotechnical Commission (IEC), International Network for Harmonized and Recognized Measurements in Wind Energy (MEASNET), European Wind Energy Technology Platform (TPWind) and the European Energy Research Alliance (EERA). Several of wind energy programs with technical details have been documented in Chapter 9 by Eecen and are retained in chapter4 of the current publication. This Chapter 9 by Peter Eecen describes the developments within the Netherlands with regard to the wind energy research since funding by the National Wind Energy Research Program during 1976 to 1985. Wind energy research activities in the Netherlands have been predominantly performed at the Wind Energy Department of The Energy Research Centre of the Netherlands (ECN) and the interfaculty wind energy department Duwind at Delft University of Technology. Both institutes are involved in wind energy research since the start of the modern wind turbines. These institutes match their research programs with each other so that a consistent research program in the Netherlands is in place. In the Netherlands the wind energy research is supported by an extensive experimental infrastructure at The Knowledge Centre WMC with is a research institute for materials, components, and structures performing blade tests for large wind turbines up to 60 m in length. ECN has a research wind farm where prototype wind turbines are tested with five full-scale turbines research activities The wind farm research include aerodynamics of wind with a large selection of experimental facilities for wind energy applications. The most prominent facilities are the wind tunnels, of which the open Jet facility is the most recent addition. A historic overview of the wind energy research activities in the Netherlands is written from the perspective of the research community and provides alternative insights with a focus on the implementation of wind energy and the development of support mechanisms. The description of research activities, the developed advanced design tools, developed knowledge, and intellectual property provide an alternative source for further activities to reduce the cost of energy of wind power. Dr. Eecen used 22 references along with ten schematics, figures, pictures, and tables to augment the professional and scholastic treatment of the subject.

"Wind Energy in Netherlands" documented in Chapter 9 by Eecen in the ASME's 2011 classic publication "Energy and Power Generation Handbook - Established and Emerging Technologies" edited by K. R. Rao [1], has been updated by Editor-in-Chief of the current Renewable Energy Series, K. R. Rao. This update "WIND ENERGY IN THE NETHERLANDS Since 2011 – INTRODUCTION" has been provided in section 4.10 and follows the same topical outline initially contributed by Eecen in the 2011 publication. Eecen with his professional and scholastic expertise provided information which is applicable even to-day, however the topical discussions covering what happened since 2011 to date need to be up-dated. Even though the topics are retained as much as possible, a continuation of the same is not appropriate. Thus, discussions are as applicable for the topics listed in this update and cannot be attributed do the original write-up of Eecen. All of the material in this update is from information available in the online public domain and is documented in the footnotes pertinent to the discussions. Material in this update can be easily researched from the footnotes provided and readers have an opportunity to even look up to the latest positions of the topic at the time of browsing the manuscript. As of October 2020, wind power in the Netherlands has an installed capacity of 4,990 MW, 19% of which is based offshore. In 2019, the wind turbines provided the country with 12% of its electricity demand during the year. The Netherlands reached an Energy Agreement for Sustainable Growth for offshore wind turbines to supply 3.3% of the Netherlands' total energy needs. The Coalition Agreement and the Climate Agreement (2019) include a commitment to continue the successful offshore wind energy policy. This results in a total offshore wind power capacity of 11 GW by 2030, which is enough to supply 8.5% of all the energy in the Netherlands. The total installed capacity of offshore wind power in the Netherlands is around 2.5 gigawatts (GW) in 2021 and targets to increase to at least 4.5 GW by 2023 which is enshrined in the Energy Agreement for Sustainable Growth.

The Netherlands is working on a transition to a sustainable, reliable, and affordable energy supply. The Netherlands wind energy market is expected to grow at more than 9.5% during the forecast period 2020-2025. Wind energy in the Netherlands powers about 5.7 million homes and cuts the CO2 footprint of electricity by 12%. The Netherlands is expected to build the world's largest offshore wind farm by 2027, along with a 2.3-square-mile artificial island to support it. The wind farm is expected to be capable of producing 30 GW of power. Onshore farms in The Netherland's wind farms made up 80% of the new installations with 11.8 GW. This was 22% lower than the government's pre-COVID forecast. The Netherlands installed the most wind power capacity in 2020 (1.98 GW), 75% of that was offshore wind. The Netherlands installed the most wind capacity use due to strong offshore additions. Offshore wind made up 20% of new installations in Europe with 2.9 GW of new capacity connected to the grid in 2020 of which The Netherlands installed half of that capacity. Offshore installations were driven by the completion of the Borssele Wind Farm zones, the three wind farms were awarded to different consortia in 2016 and 2017. 220 GW of wind power capacity are now installed in Europe, 11% of this figure is offshore, and Netherlands has 5 GW. In the Netherlands an increase of 1.6 GW of wind energy installations are expected in 2021. In 2021-2025 new onshore wind installations in The Netherlands will be 20 GW. According to Realistic Expectations Scenario, between 2021 and 2025, offshore installations in The Netherlands were 4.4 GW. In cumulative terms of total capacity, Europe would reach 318 GW of installed capacity by the end of 2025 of which The Netherlands will be above the 10 GW.

The update has been covered in Section 4.10 "WIND ENERGY IN THE NETHERLANDS – INTRODUCTION", with several discussions contained in this section sequentially in section 4.10.1 "Wind Energy in The Netherlands Since 2011"covering in sections 4.10.1.1 Upwind, 4.10.1.2 Smart Rotors, 4.10.1.3 Delft University of Technology (DU) Airfoil Concepts followed by section 4.10.2 RESEARCH CENTERS and PROGRAMS IN THE NETHERLANDS SINCE 2011 which contain in sections 4.10.2.1 ECN (Energy Research Center, The Netherlands), 4.10.2.2 WMC (Knowledge Center), 4.10.2.3 Blade Testing and Material Research at WMC, 4.10.2.4 An Advanced Method for Wind Turbine Wake Modeling, 4.10.2.5 DUWIND, DELFT NL, 4.10.2.6 Environmental Research Impacting Urban Areas 4.10.2.7 European Wind Atlas and the Netherlands Wind Energy Research 4.10.2.8 Netherlands Wind Energy Association (NWEA) Research Policy Plan, and in section 4.10.3 PROGRAMS and PROGRESS of WIND ENERGY IN NETHERLANDS SINCE 2011 with sections 4.10.3.1 RWE, NL Wind Turbines on Dutch Dams, 4.10.3.2 Infrastructure for Wind Energy in The Netherland, 4.10.3.3 Trains by Wind Power in The Netherland, 4.10.3.4, The German-Dutch Wind Tunnels DNW, 4.10.3.5 Recent Developments of Wind Farm Design 4.10.3.6 Recent Developments of Offshore Wind Farm Design, 4.10.3.7 Climate Control in Netherlands with Impact on Wind Energy and in Section 4.10.4 SUMMARY with 4.10.4.1 addressing Wind Farms in Netherlands During 2020-through-2026, and in 4.10.4.2 Futuristic Wind Farms in Netherlands.

In section 4.10.4.2 "Futuristic Wind Farms in Netherlands" update covers The Netherlands wind energy market - growth, trends, Covid-19 impact, and forecasts (2021 - 2026), which have been studied with 2020 as the base year for the study. The Netherlands is working on a transition to a sustainable, reliable, and affordable energy supply. The Netherlands wind energy market is expected to grow at a CAGR of more than 9.5% during the forecast period 2020-2025. Wind energy in the Netherlands powers about 5.7 million homes and also cuts the CO2 footprint of electricity by 12%. Factors like increasing demand for renewable energy, rising investments in wind farms, and reducing CO2 emissions are driving the wind energy market in the country. Offshore wind is expected to witness significant growth in the wind energy market in the Netherlands during the forecast period. The Netherlands has an overly ambitious renewable-energy plan in the works. The country is expected to build the world's largest offshore wind farm by 2027, along with a 2.3-square-mile artificial island to support it. The wind farm is expected to be capable of producing 30 GW of power, which is likely to provide an opportunity to a growth in the deployment of wind energy in coming future. The Netherlands is making strides toward the country's renewable energy targets i.e., 16% renewable energy sources (RES) by 2023 of total energy demand. The factors like shallow waters, good wind resources, good harbor facilities, experienced industry, and a robust support system are driving the offshore wind energy market. The active offshore wind farms in the North Sea are Gemini (600 MW), Luchterduinen (129 MW), Prinses Amalia (120 MW), and Egmond aan Zee (OWEZ) (108 MW). Offshore wind farms which are in the Dutch part of the North Sea. The upcoming offshore wind farm zones for the 3,500 MW new offshore wind capacity are Borssele (1,400 MW), South Holland coast wind farm zone (1,400 MW) and North Holland coast wind farm zone (700 MW). Netherlands had 4463 MW of wind energy installed in 2019 and the wind-generated electricity in Netherlands accounted for 10.5 TWh in 2018. Key players in The Netherlands wind energy market include ENERCON GmbH, Siemens Gamesa Renewable Energy SA, Mitsubishi Corp, General Electric Company, and Lagerwey Wind BV. The notable features for the foreseeable future for wind energy are Airborne Wind Energy, Offshore Floating Concepts, Smart Rotors, WindInduced Energy-Harvesting Devices, Blade Tip-Mounted Rotors, Unconventional Power Transmission Systems, Multi-Rotor Turbines, Alternative Support Structures, Modular High Voltage Direct Current Generators, Innovative Blade Manufacturing Techniques, and Diffuser-Augmented Turbines, and Small Turbine Technologies. The optimization of floating offshore wind platforms are through the integrated design of the platform and the wind turbine, including downwind rotors, high tip speed ratio operation and two bladed rotors. These could have an impact on the cost of the platform and the whole floating system. Dr. Rao updated Dr. Eecen's chapter with 7 tables, 36 figures, 134 footnotes and 2 references.

In the current publication the sole author from the original ASME Classical 2011 Handbook is Mr. Ramesh Muthya to whom the Editor Dr. K. R. Rao expresses his immense appreciation for participating in this publication in spite of Mr. Muthya's several professional consulting services, (see his Bio). Mr. Ramesh Muthya authored the last chapter of Section II covering Wind Energy in the 2011 Handbook [1]. Mr. Ramesh Muthya covered in Chapter 8 "Role of Wind Energy Technology in India and Neighboring Countries". Mr. Ramesh Muthya with his long association with the field has treated the subject of wind energy development as a source of power in the Indian sub-continent with a thorough understanding of development cycle. India engaged on wind power as early as the late 1950s but was mostly for nonelectric applications. Origins of usage of wind for grid connection started only in the mid-1980s with few small turbines installed at known windy areas. Approach to resource development in India was approached cautiously except for a few biogas plants. China had also during that time frame been slow in taking to large-scale development in renewable energy. In 2007 China claimed about 4 GW of wind power installations and by 2009, the installations touched an astounding 26 GW, with scores of companies undertaking development on a war footing. A spate of wind turbine designs had been developed, and prototypes are being built and tested. Sri Lanka has good wind resource however it cannot be expected to reach the market size potential that India or China have reached. Ramesh Muthya had attempted to capture some of the salient aspects of Wind Energy Technology Development and deployment in India in the context of power supply systems management. The author uses 14 references along with 16 schematics, figures, pictures, and tables to augment the professional and scholastic treatment of the subject. Mr. Muthya in the current contribution stated that the expectations for the next decade onshore wind power in Asia would follow the trends of West. New ways of handling the variable nature of the wind power availability will cease to be an issue with the grid managers. Infrastructure development although will continue to be a source of concern, increased revenue allocation will be made for wind power development. In a country like India grid extension will need considerable effort and time because of right of way issues and network congestion which at certain times could adversely affect the stability of grid. The engineering problems associated with such issues have solutions and India as well as rest of the world are grappling with such issues. This field being highly capital intensive, it will face financing problems to make it viable. It can only be assumed that increased cost of energy would result in making wind more attractive. Forecasting for wind power has caught imagination of the grid managers and in India there is a commercial and policy push to target increase in wind energy potential. Mr. Ramesh Muthya updated his earlier 2011 submission with 6 tables and 10 figures.

Perhaps the most interesting chapter of this book is authored in the last Chapter 6 by Dr. Weifei Hu titled "Artificial Intelligence in Wind Energy" which dwells on a futuristic aspect of wind energy, perhaps still in the exploratory stage. Despite the increasing development of wind energy (WE) which has notably become the leading renewable energy now, there are still persistent barriers to wider implementation related to technology, cost, and environmental and social impacts, e.g., super large-scale ($\geq$10 megawatt) wind turbine (WT) technology, wind turbine deployment at low wind speed and/or residential- closed areas, and offshore wind farm under severe wind-wave conditions. As one of the emerging technologies, artificial intelligence (AI) could take some of these challenges and assist in achieving the goal of further lowering the levelized cost of energy (LCOE) from wind. Dr. Hu discusses these issues with the help of bibliographical references, schematics and scholastic equations.

Recently AI has been tremendously investigated in an extremely wide range from financial data analytics and industrial production design to novel biometric and forensic applications and various renewable energy development. This chapter will provides the fundamental theories of four categories of state-of-the-art AI methods including (i) neural learning methods, (ii) statistical learning methods, (iii) evolutionary learning methods, and (iv) hybrid learning methods in a lucid way, then carry out a comprehensive survey of various AI methods used in WE including (i) wind speed prediction, (ii) wind power estimation, (iii) wind turbine condition monitoring and diagnosis, (iv) wind turbine systems design, and (v) wind farm operation and maintenance. In addition, realistic case studies of AI in WE are also explained along with results and discussion replete with details. Concluding remarks summarize some research goals of AI in WE, for example, accurate WE forecasting, optimal WE efficiency and improved WE accessibility.

The significant development of WE brings challenges, e.g., exploring new available wind resources on land and offshore, designing reliable and cost effective titanic WTs (e.g., 10 MW WTs), and developing new monitoring approaches that can reduce the operational and maintenance (O&M) cost. Artificial intelligence (AI) techniques, emerging from computer science are becoming promising tools to overcome these challenges and boosting the further development of WE. AI is a kind of intelligence demonstrated by machines in contrast to the natural intelligence displayed by humans. It has been performing more complex tasks than straightforward automation in different domains and applications, such as robotics, cyber-physical systems, and renewable energy. It consists of many branches including artificial neural network (ANN), genetic algorithm (GA), expert system (ES), fuzzy logic (FL), and various hybrid systems, which are combinations of two or more of the branches, to produce efficient and effective computing solutions. Even though some current AI techniques still require interaction with human intelligence, the appropriate use of AI techniques leads to smart systems with improved performance that cannot be achieved by traditional methods. A large number of AI techniques have been proposed and applied in WE. This chapter distinguishes various AI techniques into four categories, neural learning methods, statistical learning methods, evolutionary learning methods, and hybrid learning methods, which are further separated into different specific AI methods as explained in Section 6.2. Section 6.3 provides surveys of various AI methods used in WE focusing on wind speed prediction, wind power prediction, WT design and optimization, and WT condition monitoring (WTCM). Concluding remarks are provided in Section 6.4. This chapter aims to offer non-exhaustive surveys due to the extremely wide topic (the combination of AI and WE) and present readers an overall but clear picture of AI in WE. The content of chapter 6 weaving through Introduction in 6.1,to section 6.2 deals with Wind Energy, then Section 6.3 deals with Artificial Intelligence, 6.4 with Artificial Intelligence in Wind Energy, section 6.5 with the "Theories of State-of-the-art Artificial Intelligence", Section 6.2 with "Neural Learning Methods", Section 6.3 covers "Statistical Learning Methods", Section 6.4 covers "Evolutionary Learning Methods" and 6.2.4 "Hybrid Learning Methods", Applications of Artificial Intelligence in Wind Energy, Wind Speed Prediction, Wind Power Estimation, Wind Turbine Condition Monitoring and Diagnosis, Wind Turbine Systems Design, Wind Farm Operation and Maintenance and the author provides conclusions summarizing his research with Acknowledgments, Acronyms and References. Dr. Hu's contribution had 3 tables, 5 figures, 8 equations and 100 references.

Update contained in this book is rich with 28 tables, 143 figures, 379 footnotes, 8 equations and 102 references, in addition to several of the tables, figures and references of 2011 Handbook [1]. This provides the readers in addition to consulting scripts of the professional write-up of initial authors, immense scholastic and professional material for the readers to use.

1. Energy and Power Generation Handbook - Established and Emerging Technologies, edited by K. R. Rao, ASME Press, New York, 2011.

2. Wind Energy for Power Generation - Meeting the Challenge of Practical Implementation by K. R. Rao, Volumes 1 &2, This Springer book is published by Springer International Publishing AG part of Springer Nature, Gewerbestrasse 11, 6330 Cham, Switzerland.

CHAPTER

1

ROLE OF NASA IN WIND ENERGY

Larry A. Viterna and K. R. Rao[1]

1.1 INTRODUCTION

Since the beginning of NASA over 50 years ago there has been a strong link between the energy and environmental skills developed by NASA for the space environment and the needs of the terrestrial energy program. The technologies that served dual uses included solar, nuclear, biofuels and biomass, wind, geothermal, large-scale energy storage and distribution, efficiency and heat utilization, carbon mitigation and utilization, aviation and ground transportation systems, hydrogen utilization and infrastructure and advance energy technologies such as high-altitude wind, wave and hydro, space solar power (from space to earth) and nanotech photovoltaics. NASA, in particular with wind and solar energy, had extensive experience dating back to the 1970's and 1980's and continues today to have skills appropriate for solving our nation's energy and environmental issues that mimic in fact those needed for space flight. NASA has long been interested in maturing new laboratory level technologies into industry products and has a well-founded system of interactions with Universities for research leading to roof-of-Concept and then to prototype demonstrations and mission applications with industry. The capability to assess and then test the feasibility for future commercial development is a well-honed NASA skill. This has certainly promoted new businesses and job growth as a result. That is particularly evident in the case of photovoltaic energy but also relevant for wind energy as well.

1.2 WIND ENERGY

In the decade from 1975 through the 1985 the United States government worked with industry to advance the technology and help enable large commercial wind turbines. The impetus for this effort was two events on the world stage (1) an oil embargo by the Organization of Petroleum Exporting Countries (OPEC) in 1973 followed by (2) the fall of the Shah of Iran in 1979. In response to the first crises, in 1975, the NASA Lewis Research Center in Ohio (now named the Glenn Research Center) took on the role of leading the technology development of large horizontal-axis wind turbines, the dominant type of wind turbine in use today. The following pages document the development of major wind turbine systems and technologies, their impact on the industry and future development areas that could benefit the industry.

As a major national laboratory, NASA's Lewis Research Center (now named the Glenn Research Center) in Ohio had the facilities and engineering capabilities necessary to advance wind turbine technology. At that time, as it is still true today, NASA was able to successfully apply its capabilities in areas such as propeller and turbine design, structural analysis, advanced materials, aero-dynamics, instrumentation and power system control to the design of wind energy systems. During the 1970's and '80s, with funding from the National Science Foundation, the Department of Energy (DOE), and the Department of Interior (DOI) a total of 13 experimental wind turbines were put into operation including four major wind turbine designs. This program pioneered many of the multi-megawatt turbine technologies in use today, including: steel tube towers, variable-speed generators, composite blade materials, high lift-to-drag airfoils. In addition, key engineering analysis, design and measurement capabilities were developed such as models for aerodynamic performance, structural dynamics, fatigue life, wind measurement and acoustics. The outstanding success of wind energy can be seen today in the growth of the installed power capacity. In 2008 and 2009 the U.S. increased its installed capability for wind energy by 45% and achieved the largest installed capacity of any country in the world. Nearly all of this increased capacity was from horizontal-axis wind turbines that generate electrical energy. Large horizontal-axis wind turbines are in fact the dominant type in use throughout the world and owe much of their characteristics to the NASA technologies and the DOE investment.

The large wind turbines developed under the NASA-led technology development effort set several world records for diameter and power output. Figure 1.1 shows the size and power output of the designs developed.

All turbines shown were built and operated in single or multiple units with the exception of the Mod-5A that was a competitive design to the Mod-5B, the winner in a down selection process. Each of the remaining turbines will be discussed below.

[1]Viterna was the original author of this chapter that appeared as Chapter 6 in the "Energy and Power Generation Handbook Established and Emerging Technologies", published in ASME in 2011. This chapter has been updated by Editor K. R. Rao in Section 1.7 who has provided relevant write-ups from on-line sources in the public domain for "Wind Energy Applications Pertaining to Role of NASA from 2009 to date".

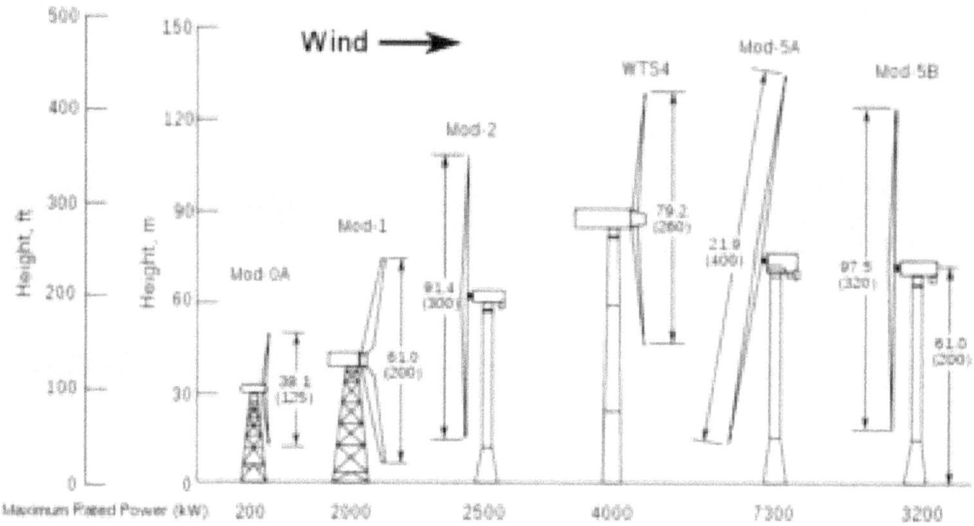

FIG. 1.1: SIZE AND RATED POWER OUTPUT OF WIND TURBINES DEVELOPED UNDER THE NASA-LED PROGRAM

NASA designed, constructed and operated its first experimental wind turbine, the Mod-0, in Sandusky Ohio (see Figure 1.2).

The Mod-0 had a rotor diameter of 38 meters and a power rating of 100 kilowatt. The size of the Mod-0 was nearly ideal because it was large enough to be representative of electric utility wind turbines but not too large to be overly costly to make changes to the hardware. As further explanation, a critical parameter for aerodynamic performance is the Reynolds number. The size scale of the Mod-0 resulted in the airfoils operating above the critical value of 1 million for modern airfoils. Alternatively, the relatively small size of the turbine allowed NASA to select off the shelf commercial products for key components such as the gearbox and generator. This allowed for many technologies to be tested and developed on the Mod-0 at much lower cost than on a larger scale turbine.

Given that it was the first wind turbine design for NASA, the Mod-0 operated as mostly expected with regard to most design goals. Power /speed control were excellent and synchronization with the utility grid's line frequency was found to be easier than anticipated. One important exception to the Mod-0's excellent performance was the blade stress and life. Within months of beginning operation, cracks were found in the aluminum blades near the hub of the rotor. Even though this was a significant challenge, it fostered the development of reliable wind turbine blades and improved system design. It took two years and the best of NASA's capabilities in materials, structures and aerodynamics testing and analysis to resolve the problems. The causes were several. First, analytical methods to model the coupled aeroelastic structural behavior needed to be improved. The rotor used a rigid hub causing blade loads to be resolved primarily by internal stress rather than dissipated into acceleration of blade mass. The cyclic bending load out of the rotor plane was higher than expected. The problem was further defined through NASA wind tunnel testing that showed that the truss tower and access stairs in the center blocked much of the wind passage. The Mod-0 was designed with the rotor downwind of the tower. As the blades passed by the tower each revolution, the aerodynamic forces changed significantly resulting in damaging fatigue stresses. Removal of the stairs resolved much of this factor. Finally, the riveted aluminum blades were found to not be a good

FIG. 1.2: NASA MOD-0 WIND TURBINE, INSTALLED IN SANDUSKY, OHIO IN 1975 BECAME THE TEST BED FOR DEVELOPING NEW TECHNOLOGIES FOR LARGE HORIZONTAL AXIS WIND TURBINES

choice for a fatigue driven machine such as a wind turbine. A significant alternative blade development program began and over the next year successful blades were developed that were made from wood or fiberglass.

With the Mod-0 design proven reliable, four other turbines of a more powerful version of the Mod-0 were built and operated throughout the US. The generator was doubled in size to 200kw, but most all other characteristics remained unchanged. These wind turbines were designated the Mod-0As were built in partnership with Westinghouse. They were installed and operated at Culebra Island, Puerto Rico; Clayton, New Mexico; Block Island, Rhode Island and Oahu, Hawaii. These turbines were instrumental in de-fining the grid interaction and electrical characteristics for small utilities (soft grids) and large utilities (hard grids).

The next step was increasing the size of wind turbines. Under the technical management of NASA in Cleveland, the General Electric Company (Space Division/Valley Forge) scaled up the Mod-0 wind turbine in Sandusky Ohio to achieve 2 megawatts of power output from a 61 meter diameter rotor [65]. The Mod-1 wind turbine shown in Figure 1.3 was the world's first multi-megawatt wind turbine under partnership with commercial industry.

It was installed and operated in Boone North Carolina and helped develop the transportation and erection methods for large wind turbines. The Mod-1 location next to residential housing however uncovered two unexpected environmental impacts – low-frequency acoustic emissions and electromagnetic television interference. The same truss tower and downwind rotor configuration as the Mod-0 gave the Mod-1 a similar "tower shadow" effect on the blade aerodynamics. This manifested itself as a sound wave that was propagated and focused towards local residential housing in the mountainous site. NASA developed a new analytical model [66] to predict the low-frequency sound waves due to the rotation of the blades through the wake of the tower. Studies using this analysis tool showed that decreasing the rotational speed could reduce the sound levels by 10dB. This change was implemented and noise complaints from residents came to an end. The electromagnetic interference proved more difficult to solve at this mountainous site. Residents were provided cable television as a local solution. Challenging as the world's first multi-megawatt wind turbine was, it was instrumental in developing location standards for wind turbines near populated areas.

In an effort to further build the wind industry, NASA and DOE awarded the design and construction of the next generation of lower cost wind turbines to Boeing Engineering and Construction using new technology proven out at the NASA Mod-0 wind turbine in Sandusky, designated the Mod-2 (Figure 1.4), the 300 foot diameter rotor produced 2.5 megawatts.

One of the new technologies was the steel monopole tube tower and the upwind rotor configuration. The steel tube was unproven at these large sizes and a primary concern was the possibility of a mechanical vibration occurring at a resonant natural frequency that could destroy the wind turbine. The tube tower, referred to by

FIG. 1.3: THE NASA/GENERAL ELECTRIC MOD-1 WIND TURBINE WAS THE WORLD'S FIRST MULTI-MEGAWATT COMMERCIAL PROTOTYPE WIND TURBINE (DERIVED FROM THE MOD-0 IN SANDUSKY, OHIO)

FIG. 1.4: THE NASA / DOE/ BOEING MOD-2 (2.5 MEGAWATT) WIND TURBINE WAS A SECOND GENERATION DESIGN THAT PIONEERED THE LOWER COST "SOFT" STEEL TUBE TOWERS USED IN THE INDUSTRY TODAY

NASA as the "soft tower" had a natural frequency that would be excited by the rotor as it increased in speed during startup.

Development of the steel tube tower required NASA to advance the structural analysis models and create electronic control systems to quickly increase the rotational speeds before resonance vibrations damage the wind turbines. This "soft" steel tube tower is used throughout the industry today. Five Mod-2 wind turbines were put into operation. Another technology introduced on the Mod-2 was partial span (tip) pitch control as seen in Figure 1.4. This technology reduced the weight and cost of the aerodynamic power control compared to full span pitch control used on the Mod-1. The cluster of three Mod-2 wind turbines at Goodnoe Hills in the state of Washington produced a total of 7.5 megawatt of power in 1981. Two other Mod-2 turbines were built and operated in California and Wyoming.

Success in moving from the early rigid towers to lighter and "softer" towers provided a basis for applying this "compliant" design approach throughout the wind turbine system. The 4 megawatt WTS-4 wind turbine shown in Figure 1.5 is an example of this compliant design approach.

Designed by Hamilton Standard, under the technical management of NASA and with funding this time from the Department of Interior, the WTS-4 was placed into operation in Wyoming in 1982. It featured a "soft" steel tube tower, fiberglass blades, torsional springs and dashpots in the drivetrain, and a flexible teetered hub (to be discussed later). To this day, the WTS-4 is the most powerful wind turbine to have operated in the US and it held the world record for power output for over 20 years. A second commercial prototype with a smaller generator (3 megawatts) designated the WTS-3 was constructed and operated in Sweden.

The last generation of wind turbines developed under the NASA-led program was the Mod-5. Two competing designs were developed as a parallel effort, the 7.3 megawatt Mod-5A design in partnership with the General Electric Company, and the 3.2 megawatt Mod-5B design in partnership with Boeing Engineering and Construction. Government funding limitations allowed only one to be funded the Mod-5B. With a rotor diameter of nearly 100 meters, the Mod-5B (Figure 1.6) was the world's largest operating wind turbine into the 1990's.

It demonstrated an availability of 95 percent, an unparalleled level for a new first-unit wind turbine. The Mod-5B had the first large-scale variable speed drive train using a double fed induction generator, the industry standard today. It also featured a sectioned, two-blade rotor that enabled easy transport of the very large blades.

The significant government and industry effort led by NASA begs the question of what happened to these large commercial prototypes in terms of the development of the utility scale wind turbine market in the 1990's. The story is an important case study of public policy and business decisions that set the course of the emerging

FIG. 1.5: THE NASA / DOI / HAMILTON STANDARD WTS-4 (4 MEGAWATT) WIND TURBINE IN WYOMING TYPIFIED THE "COMPLIANT" DESIGN APPROACH AND HELD THE WORLD RECORD FOR POWER OUTPUT FOR OVER 20 YEARS

FIG. 1.6: THE NASA/DOE/BOEING MOD-5 IN HAWAII PIONEERED VARIABLE SPEED GENERATORS AND WAS THE WORLD'S LARGEST OPERATING WIND TURBINE INTO THE 1990s

wind industry in the US. In 1982 the price of oil decreased by a factor of 3. By 1986 the US government began to phase out large wind turbine development and support. The original orders of 100 commercial units of the Mod-5B for Hawaii were cancelled and the industry partners left the US wind turbine market. European countries, in particular Denmark, on the other hand developed a long-term public policy and technology program to address their nation's oil dependency. The effect of the change in US policy in 1986 can be seen in Figure 1.7 [67] where the Europeans came from behind to take the lead in installed wind turbine capacity.

Denmark specifically went on to capture the dominant position in a market that exceeded $15 billion a year by 2004. Fortunately, US wind industry manufacturers such as General Electric are now back in the market and have recovered substantial position. In 2010, the US took the lead again in the world for total installed wind energy capacity.

The success of both the US and European manufacturers owe much to the NASA program and the technologies already discussed including the "soft" steel tube towers, variable speed generators, structural dynamic and acoustic emission modeling. Other technologies not discussed include the NACA (predecessor to NASA) 6 series airfoils, used widely throughout the industry and pioneered on the NASA Mod-0. Also, more accurate modeling of stall aerodynamics [68] that helped enable the early industry's dominant use of passive fixed pitch power control. These are just a few of the examples of impact on this important industry.

Some technologies developed under the NASA program have yet to be fully duplicated by industry. One example is the ability to produce multi-megawatt power using a flexible two blade rotor. Two blades can capture essentially the same energy in the wind as three blade turbines but offer several cost and reliability advantages. First, there is a direct cost reduction of not having to buy one blade, a very expensive component. Second, the reduction in weight at the rotor permeates throughout the entire system and usually yields about a 25% reduction in system weight. Weight is a surrogate for cost because it affects material, manufacturing and transportation costs. Third it is easier to assemble – some two blade turbines have been assembled by lifting the rotor by the tower. Fourth, two blade rotors require a smaller gearbox ratio because the typically operate at higher tip speed ratios. This also reduces the torque in the drive train for the same amount of power. Fifth, two blades easily allow for flexibility in the rotor hub that can effectively uncouple the rotor plane from the motion of the tower. The simplest is the teetered hub used on the NASA Mod-2, WTS-4 and Mod-5B turbines (Figure 1.8).

The uncoupling of the cyclic loads into the drivetrain also greatly increases reliability of key components in the drivetrain. Today's wind turbine industry uses a rigid hub and has experienced a broad and critical problem of gearbox failures in the drivetrain. None of the multi megawatt wind turbines NASA developed had gearbox problems, and some ran for over a decade. The WTS-4 (4MW) and WTS-3 (3 MW) still hold records for operating time and total energy produced – and they were prototypes. The ability of NASA to successfully design and construct these flexible rotor systems was the result of its expertise in helicopter and rotorcraft technology, something very few other countries in the world possess.

Another technology that could be adopted by today's industry is automated blade manufacturing. Today all large wind turbine blades in the world are manufactured using hand layup of the com-

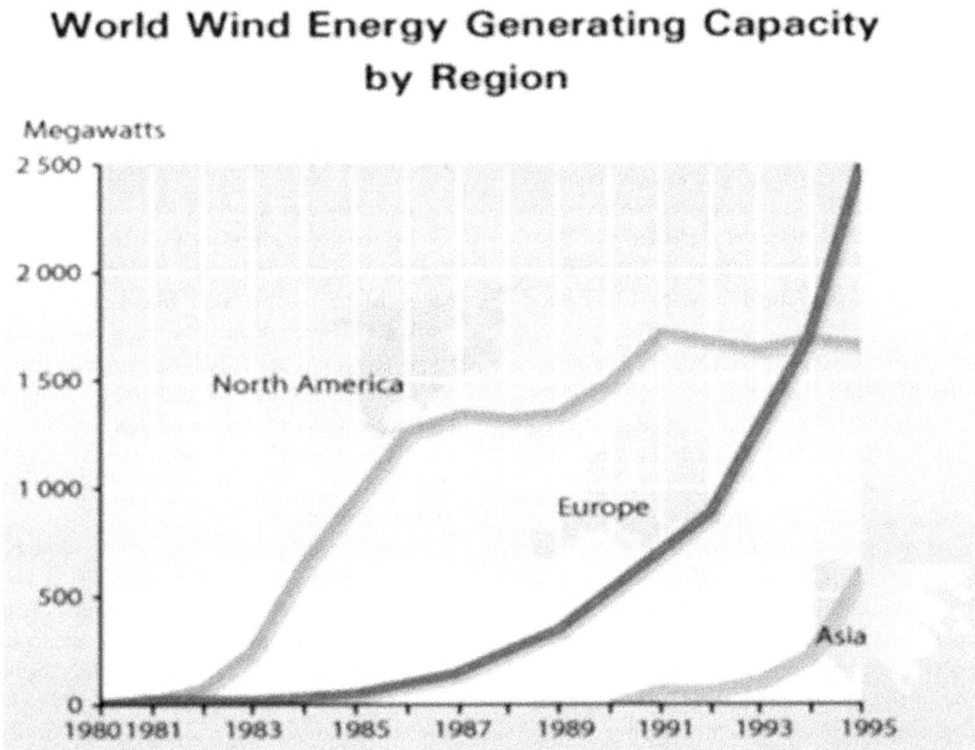

FIG. 1.7 A US PUBLIC POLICY CHANGE IN 1986 POSTPONED US LEADERSHIP IN WIND ENERGY

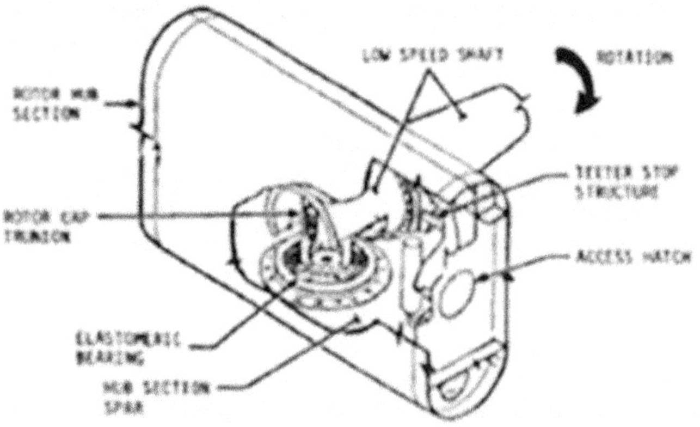

FIG. 1.8: FLEXIBLE ROTOR HUBS SUCH AS THE MOD-5B TEETERED HUBS COULD REDUCE COSTS AND INCREASE THE RELIABILITY OF TODAY'S DRIVETRAINS

posite materials, a very labor intensive process. Thus most blades are imported from countries where labor is relatively inexpensive. Figure 1.9 shows an automated blade manufacturing system developed on the NASA-led program that could be implemented in to enhance the US wind industry manufacturing capabilities.

Several automated designs were developed and blades were fabricated and tested in lengths up to 50 meters.

Details of the all the efforts and results by NASA and its partners are too broad to do justice within this chapter. An extensive list of reports on the technology development conducted during the Large Horizontal Axis Wind turbine program can be found in the references [69]. It should assist those active in the field of wind energy in locating the technical information they need on wind power planning, wind loads, turbine design and analysis, fabrication and installation, laboratory and field testing, and operations and maintenance. That bibliography contains approximately 620 citations of publications by over 520 authors and coauthors. Sources are: (1) NASA reports authored by government, grantee, and contractor personnel, (2) papers presented by attendees at NASA-sponsored workshops and conferences, (3) papers presented by NASA personnel at outside workshops and conferences, and (4) outside publications related to research performed at NASA/DOE wind turbine sites.

1.2.1 The Future of Wind Energy

Today and into the future, NASA will continue to develop new technologies for the aeronautics and space industry that will find application in the future wind industry as well. As a vision for this future, perhaps the emerging offshore wind industry is a case in point [70]. Currently no offshore wind turbines exist in the US, but there are a number of wind farms in the shallow waters off of Europe. A primary challenge to this new industry is getting the weight and cost of deployment reduced. Deeper waters will require floating wind turbines where weight and reliability will be even more important. Some of the design characteristics of the compliant wind turbines pioneered by NASA such as flexible towers, articulated blades and drive trains as well as downwind rotors and

FIG. 1.9: NASA'S BLADE TECHNOLOGY PROGRAM DEVELOPED 50 METER BLADES USING AUTOMATED MANUFACTURING TECHNIQUES THAT COULD BE ADOPTED TODAY

free yaw (Figure 1.10) may offer significant advantage to this new offshore industry (U.S. Offshore Wind Collaborative) for many years into the future.

1.3 CONCLUSIONS[2]

In the wake of Apollo, GRC, then named Lewis Research Center, helped the predecessor to DOE by making major foundational contributions in wind energy, solar power, batteries, and fuel cells. Many of the people who spearheaded these achievements still work at GRC, augmented with new talent in a broad range of aerospace technologies. GRC also currently serves as the lead center for power and propulsion technology within NASA. The technologies that it has developed for aeronautical and space applications has given GRC a comprehensive perspective for applying NASA's skills and experience in energy on a problem with national and global implications.

Perhaps what is needed is an "Apollo-type" effort to solve our nation's energy crises. One could envision a national effort directed at renewable energy and the environment that would combine not only the DOE's program in renewable energy but also would encompass contributions from multiple other agencies (NASA, DOD, EPA, Commerce, Transportation, etc.) NASA GRC could provide a lead role in such an endeavor based on its history of projects, ranging from concept studies and fundamental research efforts, to

[2, 3, 4] Footnotes refer to contributions by the original author Larry A. Viterna

FIG. 1.10: NASA'S LIGHT WEIGHT COMPLIANT DESIGN APPROACH AND TECHNOLOGIES MAY ENABLE LOW COST AND RELIABLE OFFSHORE WIND TURBINES IN THE FUTURE

proof-of-concept experiments and commercial prototype demonstrations. NASA's ability to work cooperatively with other agencies in developing technologies, it's long-standing interest in alternative/ renewable energy research, demonstrating technologies, and infusing these products into commercial applications and it's capability to lead and coordinate projects makes it an ideal choice to lead the way to developing a sustainable energy future for our nation.

1.4 ACKNOWLEDGMENTS[3]

Dr. Bailey would like to acknowledge the significant contributions of Dr. Geoffrey Landis of the Photovoltaic and Power Technologies Branch, George Schmidt, Deputy Director of Research and Technology Directorate, and Dr. Ryne Raffaelle, Director of the National Center for Photovoltaics, DOE.

1.5 ACRONYMS[4]

AFM	Atomic Force Microscopy
AFRL	Air Force Research Laboratory
ALSEP	Apollo Lunar Surface Experiments Package
AMO	Air Mass Zero
APSA	Advance Photovoltaic Solar Array
AU	Astronomical Unit
CIGS	Copper Indium Gallium Diselenide
CLEFT	A Peeled Film Technique
D3100	Model Number
DDF	Director's Discretionary Fund
DOD	Department of Defense
DOE	Department of Energy
DOI	Department of the Interior
ELO	Epitaxial Liff-Off
EPA	Environmental Protection Agency
ERDA	Energy Research and Development Administration
ETDD	Enabling Technology Development and Demonstration
GaAs	Gallium Arsenide
GRC	Glenn Research Center
GSFC	Goddard Space Flight Center H&RT Human and Robotic Technology
IEEE	Institute of Electrical and Electronics Engineers
InP	Indium Phosphide
ISS	International Space Station
ITO	Indium Tin Oxide
JPL	Jet Propulsion Laboratory
KSC	Kennedy Space Center
LED	Light Emitting Diode
LeRC	Lewis Research Center
MAE	Materials Adherence Experiment
MER	Mars Exploration Rovers
MOD	Module
MSFC	Marshall Space Flight Center
NACA	National Advisory Committee for Aeronautics
NASA	National Aeronautics and Space Administration
NRC	National Research Council
OPEC	Organization of Petroleum Exporting Countries
P30T	Poly 3-Octyl-Thiophene
PET	Polyethylene Terephthalate
PV	Photovoltaics
PVSC	Photovoltaic Specialists Conference
RIT	Rochester Institute of Technology
SAA	Space Act Agreement
SEP	Solar Electric Propulsion
SPRAT	Space Photovoltaic Research and Technology
SSP	Space Solar Power
STF	Systems Test Facility
STM	Scanning Tunneling Microscopy
STORM	Scanning Tunneling Optical Resonance Microscopy
TPV	Thermophotovoltaic
UTJ	Ultra Triple Junction
WTS	Wind Turbine System
xPRM	Exploration Precursor Robotic Missions
xPRP	Exploration Precursor Robotic Program
xSCOUT	Exploration Scout Missions

1.6 REFERENCES[5]

1) Chapin D.M., Fuller C.S., Pearson, G.L. (1954), "A new silicon photocell for converting solar radiation into electrical power". J. Appl Phys. 25, 676.

2) Reynolds D.C., Leies G., Antes L., Marburger R.E. ,"*Photovoltaic effect in cadmium sulfide*", Phys. Rev. 96, 1954, p533.

[5] References are by Viterna

3) Jenny D.A., Loverski J.J., Rappaport P., *Photovoltaic effect in GaAs p-m-junctions and solar energy conversion,* Phys. Rev. 101, 1956, p 1208.
4) Easton R.L., Votaw M.J., *Vanguard I IGY Satellite,* Rev. of Scientific Instruments, 30, 2, 1958, pp. 70-75.
5) *The Telstar Experiment,* Bell Syst. Tech. J., 1963, p. 42.
6) Loeferski J., *Theoretical considerations governing the choice of the optimum semi-conductor for the photovoltaic solar energy conversion,* J. Appl. Phys. 27, 1956, p. 777.
7) Jackson E.D., *Areas for improvement of the semiconductor solar energy converter.* Trans. Of the Conf. On the Use of Solar energy, Tucson, Arizona, 5, 1955, p. 122.
8) Perlin J., *From Space to Earth,* aacet publications, 1999, p.50.
9) Terrestrial Photovoltaic Measurments, *Workshop Proceedings,* March 19-21, Cleveland, OH, 1975.
10) I Weinberg, H. Curtis and A. Forestieri, *Effects of Outdoor Exposure on Solar Cell Modules in the ERDA/NASA LewisS Research Center Systems TestT Facility,* NASA TM X-73657, *1977.*
11) A. Ratajczak, W. Bifano, J. Martz, and P. O'Donnell, *NASA Lewis Research Center Photovoltaic Application Experiments,* AIAA/ASERC Conference on Solar Energy: Technology Status, Phoenix, Ariz./ Nov. 27-29, 1978.
12) W. Bifano, R. DeLobard, A. Ratajczak and L. Scudder, *Status of DOE and AID Stand-Alone Photovoltaic System Field Tests*, Proc. Of the 17th Photovoltaic Specialist Conference, pp1159-1167, 1984.
13) A. Forestieri and C. Baraona, *Space Station Power System*, Proc. Of the 17th Photovoltaic Specialist Conference, pp7-11, 1984.
14) J. Mullin, J. Loria, and H. Brandhorst, *The NASA Photovoltaic Terchnology Program,* Proc. Of the 17th Photovoltaic Specialist Conference, pp12-16, 1984.
15) J. Fan, R. McClelland, and B. King, *GaAs CLEFT Solar Cells for Space Applications,* Proc. of the 17th Photovoltaic Specialist Conference, 1984, pp32-35.
16) T. Trumble and F. Betz, *Evaluation of a Gallium Arsenide Solar Panel on the LIPS-II Satellite*, Proc. Of the 17th Photovoltaic Specialist Conference, 1984, pp12-16.
17) P. Stella and R. Kurland, "Thin Film GaAs for Space – Moving Out of the Laboratory", Proc. Of the 23rd IEEE Photovoltaic Specialist Conference, 1993, pp 21-26.
18) P. A. Jones, *et al.*, "The SCARLET Light Concentrating Solar Array," *25th IEEE Photovoltaic Specialists Conference*, Washington DC, 1996.
19) J. Stubstad, *et al.*, "SCARLET and Deep Space 1: Successfully Validating Advanced Solar Array Technology," *AIAA paper AIAA99-4487*, Space Technology Conference & Exposition, Albuquerque NM September 28-30 1999.
20) G. Landis, P. Jenkins, and D. Scheiman, "Photovoltaic Cell Operation on Mars," *19th European Photovoltaic Science and Engineering Conf.*, Paris, France, June 7-11 2004, 3674-3677.
21) R. Ewell and D.R. Burger, "Solar Array Model Corrections from Mars Pathfinder Lander Data," 26th IEEE Photovoltaic Specialists Conference, Sept.-Oct. 1997. 1019-1022.
22) G. Landis and P. Jenkins, "Measurement of the settling rate of atmospheric dust on Mars by the MAE instrument on Mars Pathfinder, *Journal of Geophysical Research, 105, No. El*, 2000, pp. 1855-1857.
23) P. Stella *et al.*, "Managing PV Power on Mars – MER Rovers", 35th IEEE Photovoltaic Specialists Conf., 2009.
24) G. Landis, "Exploring Mars with Solar-Powered Rovers," *31st. IEEE Photovoltaic Specialists Conference,* Orlando FL, Jan 3-7 2005, 858861.
25) Gasner, *et al.*, "The Stardust Solar Array," World Conf. Photovoltaic Energy Conversion, Osaka Japan, May 12-18, 2003.
26) P. Stella and S. DiStefano, "Early Mission Power Assessment of the Dawn Solar Array," *34th IEEE Photovoltaic Specialists Conf.*, Philadelphia PA, June 7-9, 2007.
27) M. Piszczor, *et al.*, "Advanced Solar Cell and Array Technology for NASA Deep Space Missions," 33rd IEEE Photovoltaic Specialists Conf., San Diego CA, May 12-16 2008.
28) J. Dunne and E. Burgess, *The Voyage of Mariner 10,* NASA Special Publication SP-424, 1978.
29) G. Dakermanji, J. Jenkins, and C. J. Ercol, "The MESSENGER Spacecraft Solar Array Design and Early Mission Performance," *4th World Conf. Photovoltaic Energy Conversion*, Waikoloa, Hawaii, May 7-12, 2006, 1919-1922.
30) A. F. Hepp, J. S. McNatt, S. G. Bailey, and J. E. Dickman, R. P. Raffaelle, B. J. Landi, A. Anctil, and R. DiLeoM. H.-C. Jin, Chun-Young Lee and T.J. Friske, S.-S. Sun, C. Zhang, S. Choi, A. Ledbetter, K. Seo, C. E. BonnerK. K. Banger, S. L. Castro, David Rauh "Ultra-Lightweight Hybrid Thin-Film Solar Cells: A Survey of Enabling Technologies for Space Power Applications", Proceedings of the 5th IECEC, 2007, pp 179-200.
31) A.N. MacInnes, M.B. Power, A.R. Barron, A.F. Hepp, and P.P. Jenkins, "Chemical Vapor Deposition from Single Organometallic Precursors", *United States Patent* 5,300,320, April 5, 1994. United States Patent.
32) S.G.Bailey, G.A.Landis, and D.M.Wilt, "Effect of Crystal Orientation on Anisotropic Etching and MOCVD Growth of Grooves on GaAs," *J. Electrochemical Soc.*, 136, No.11, 1989, pp 3444-3449.
33) "Textured InP for Device Applications", S. Bailey, N. Fatimi, and G. Landis, *Chemical Surface Preparation, Passivation and Cleaning for Semiconductor Growth and Processing,* Vol. 259, 341-346 (1992).
34) S. Bailey, D. Wilt, F. DeAngelo, and E. Clark, "Preferentially Etched Epitaxial Liftoff of InP", , *Proceedings of the 23rd IEEE PVSC*, 1993, pp. 783-785.
35) S. Bailey, D. Wilt and F. DeAngelo, U.S. Patent No. 5,641,381, "Preferentially Etched Epitaxial Liftoff of InP Material", issued on June 24, 1997.
36) J. Carlin, M. Hudait, S. Ringet, D. Wilt, E. Clark, C. Leitz, M. Currie, T. Langdo, E. Fitzgerald, "High Efficiency GaAs on Si Solar Cells", *Proceedings of the 28th IEEE PVSC*, pp. 1006-10011(2000).
37) S. Bailey, D.Wilt, R. Raffaelle, S. Hubbard, "Thin Film Poly III-V Space Solar Cells", C. Bailey *Proceedings of the 33rd IEEE PVSC,* 2008.
38) S. Bailey, R. Raffaelle, P. Nuedeck, S. Hubbard, "Optical and Electrical Characterization of SiC Devices", *Proceedings of the 28th IEEE PVSC*, 2000, pp. 1257-1260.
39) R. Raffaelle, S. Kurinec, D. Scheiman, P. Jenkins, S. Bailey, "Silicon Carbide Cells for Space Applications", *Proceedings of the the 17th European PVSEC, Volume 1, 2001, pp 281283.*
40) S. Bailey, H. Curtis, M. Piszczor, R. Surampudi, T. Hamilton, D. Rapp, P. Stella, N. Mardesich, J. Mondt, R. Bunker, B. Nesmith, E. Gaddy, D. Marvin, and L. Kazmerski, "Photovoltaic Cell and Array Technology Development for Future Unique NASA Missions", *Proceedings of the 29th IEEE PVSC 2002,,* pp, 799-803.
41) S. Bailey, R. Raffaelle, and Keith Emery, "Space and Terrestrial Photovoltaics – Synergy and Diversity", *Progress in Photovoltaics*, Vol 10 Number 6, 2002, pp. 399-406.
42) S.G. Bailey, R.P. Raffaelle, J. E. Lau, P. Jenkins, S.L. Castro, P. Tin, D.M. Wilt, A.M. Pal, and S. Fahey. "Scanning Tunneling Optical Resonance Spectroscopy", *Proceedings of the 1st International Energy Conversion Engineering Conference,* 2003.
43) S.G. Bailey, R.P. Raffaelle, J. E. Lau, P. Jenkins, S.L. Castro, D. Scheiman, P. Tin, D.M. Wilt, "Scanning Tunneling Optical Resonance Spectroscopy", *Photonics Tech Briefs,* 2a-6a, October 2003.
44) S.L. Castro, S.G. Bailey, R.P. Raffaelle, K.K. Banger, A.F. Hepp, "Nanocrystalline Chalcopyrite Materials (CuInS2 and CuInSe2) via Low-Temperature Pyrolysis of Molecular Single-Source Precursors", *Chem. Mater.,* American Chemical Society, August 2003.
45) S. Bailey, S. Castro, R. Raffaelle, "Nano-structured Materials for Solar Cells", *Proceedings of the 3rd World Conference on Photovoltaic Energy Conversion,* (2003).

46) R. Raffaelle, T. Gennett, S. Bailey, D. Wilt, P. Tin, J. Lau, P. Jenkins, D. Scheiman, "Scanning Tunneling Optical Resonance Spectroscopy (STORM)", 3/28/02, LEW-17344-1.
47) R. Rafaelle, T. Gennett, G. Landis, S. Bailey, J. Lau, "Carbon Nanotube Thermal Power Converter", 8/9/02 . LEW-17413-1.
48) A. Hepp, S. Castro, K. Banger, S. Bailey, "Preparation of Ternary Nanostructured Crystallites, Via Ternary Single source Precursors", 10/31/02, LEW-17446-1.
49) Chubb, D., "Fundamentals of Thermophotovotlaic Energy Conversion", Elsevier Science, 2007, pp. 1-530.
50) NASA Space Exploration workshop, May 24 2010; see http://www.nasa.gov/exploration/new_space_enterprise/home/workshop_home.html.
51) "Ubiquitous Access to Abundant Power," *Technology Frontiers* workshop, hosted by NASA Directorate Interrogation Office, (Washington, DC: June 16, 2010.
52) C. Cress, C. Bailey, S. Hubbard, D. Wilt, S. Bailey, R. Raffaelle, "Radiation Effects on Strain Compensated Quantum Dot Solar Cells", *Proceedings of the 33rd IEEE PVSC,* 2008.
53) Stephen Markley, "MIT Prints Solar Cells on Paper; Could It Work as Car Paint?," *Kicking Tires*, May 6, 2010, accessed September 24, 2010, http://blogs.cars.com/kickingtires/2010/05/mit-prints-solar-cellsonpaper-could-it-work-as-car-paint.html; U.S. Department of Energy, Solar Energy Technologies Program, "Paint-On Solar Cell Captures Infrared Radiation," news release, January 19, 2005, accessed September 24, 2010, http://www1.eere.energy.gov/solar/news_detail.html? news_id=8785.
54) S. Bailey, B. Landi, R. Raffaelle, "Carbon Nanotubes Synthesized and Assessed for Space Photovoltaics", *GRC 2005 Research and Technology,* 2006, pp78-80.
55) Sheila Bailey, Dave Wilt, Stephanie Castro & Ryne Raffaelle, "Alpha-Voltaic Power Source Designs Investigated", *GRC 2005 Research and Technology 2006*, pp 76-77.
56) P. E. Glaser, "Power from the Sun: Its Future," *Science Vol. 162, 1968, pp.* 957-961.
57) "The Final Proceedings of the Solar Power Satellite Program Review", NASA TM-94183, July, 1980, pp. 1-330.
58) NASA, *Satellite Power System Concept Development and Evaluation Program System Definition Technical Assessment Report*, U.S. Department of Energy, Office of Energy Research, Dec. 1980, Report DOE/ER/10035-03.
59) U. S. Office of Technology Assessment, *Solar Power Satellites,* 1981.
60) J. Mankins, "A Fresh Look At Space Solar Power: New Architecture, Concepts, and Technologies," paper IAF-97-R.2.03, 48th International Astronautical Conference, October 6-10 1997, Turin, Italy.
61) J. Mankins, J. Howell, "Overview of the space solar power exploratory research and technology program", AIAA 2000-3060, 35th Intersociety Energy Conversion Engineering Conference, July, 2000, pp. 24-28.
62) G. Landis, Re-evaluating Satellite SolarPower Systems for Earth", IEEE 4th World Conference on Photovoltaic Energy Conversion, Waikoloa, HI, May 7-12, 2006.
63) Landis, G., "Solar Power from Space: separating speculation from reality", XXIth Space Photovoltaic Research and Technology Conference, Cleveland, OH, October 6-8, 2009.
64) Mankins, J., "New Directions for Space Solar Power", Acta Astronautica 65, 2009, pp. 146-156.
65) Spera, D.A., Viterna, L.A., Richards, T.R. and Neustadter, H. E. (1979) *Preliminary analysis of performance and loads data from the 2-megawatt Mod-1 wind turbine generator.* Cleveland OH: NASA Lewis Research Center., NASA TM-81408 http://ntrs.nasa.gov/archive/nasa/casi.ntrs.nasa.gov/19800008234_1980008234.pdf.
66) Viterna, L. A., 1981, The NASA-LeRC Wind Turbine Sound Prediction Code, NASA TM-81737, Cleveland, Ohio, NASA Lewis Research Center, http://gltrs.grc.nasa.gov/reports/1981/TM-81737.pdf.
67) UNEP *United Nations Environment Programme.* http://maps.grida.no/go/graphic/wind_power_generation_per_region.
68) Viterna. Larry A., 1981, *Fixed Pitch Rotor Performance of Large Horizontal Axis Wind Turbines*, NASA CR195462, http://ntrs.nasa.gov/search.jsp?R=194076&id=9&as=false&or=false&qs=Ne%3D35%26Ns%3DHarvestDate%257c1%26N%3D245%2B4294888887%2 B33.
69) Spera, David A., (1995) *Bibliography of NASA-Related Publications on Wind Turbine Technology 1973-1995*, NASA CR-195462, http://ntrs.nasa.gov/archive/nasa/casi.ntrs.nasa.gov/19950020314_ 1995120314.pdf.
70) "U.S. Offshore Wind Energy: A Path Forward", U.S. Offshore Wind Collaborative, 2009-10-16. http://www.usowc.org/pdfs/PathForward-final.pdf.

1.7 NASA DEVELOPMENTS, POTENTIAL AND BIBLIOGRAPHY SINCE 2010[6]

Since 2011 Wind energy has been extensively investigated and researched by NASA and cited in many publications covering power generation cited in this section. NASA has been extremely active not only in space exploration for which it is primarily meant but also espousing its efficacy for human habitation by disseminating valuable imagery of wind energy for the current and future urban and regional habitations around the world. Figures 1.11 to 1.14 demonstrate the satellite depiction from the outer space of wind energy around the entire globe. Wind power in USA was largely augmented by NASA's innovative efforts as documented by Viterina in the original compilation of chapter 6 of classic publication "Energy and Power Generation Handbook -Established and Emerging Technologies", edited by K. R. Rao and published by ASME in 2011. [1] This update provides additional contributions of NASA since the publication of Viterna chapter in 2011. [1]

1.7.1 Wind Developments Around the Globe by NASA

NASA's contributions for wind energy in USA has been extensively covered in K. R. Rao's book published by ASME Press in 2011 on pages 168, 359, 982, 1090, 1124, 1128, 1139, 1203, 1204, and indexed on pages 1393, 1394, 1407, and 1433 [. Several other NASA publications since 2011, covering wind energy have been similarly recorded which are noted in the following paragraphs. ASME's Classic publication of 2011 covered NASA's global efforts for wind energy in countries such as Australia, China, Mongolia, India, Sri Lanka, Nepal, Bhutan, Sikkim, Europe, Africa, South and Central America, Caribbean, USA, Canada, Puerto Rico and Virgin Islands just to name few of the places around the globe which are covered. ASME's 2011 edition also addressed wind energy impact on global warming (GW), climate change (CC), and environment. NASA's contribution about energy efficiency in wind energy has been elaborately covered in several chapters of ASME publication of 2011. Conceptually, on an average a single nuclear power plant can generate 2 gigabytes (2000 megabytes) of electrical power nuclear power whereas a single wind power tower can generate up to 4 mb of electrical power. However, what is interest-

[6] NASA Developments, Potential, and Bibliography since 2010 are contributed by Editor-in-Chief K.R Rao

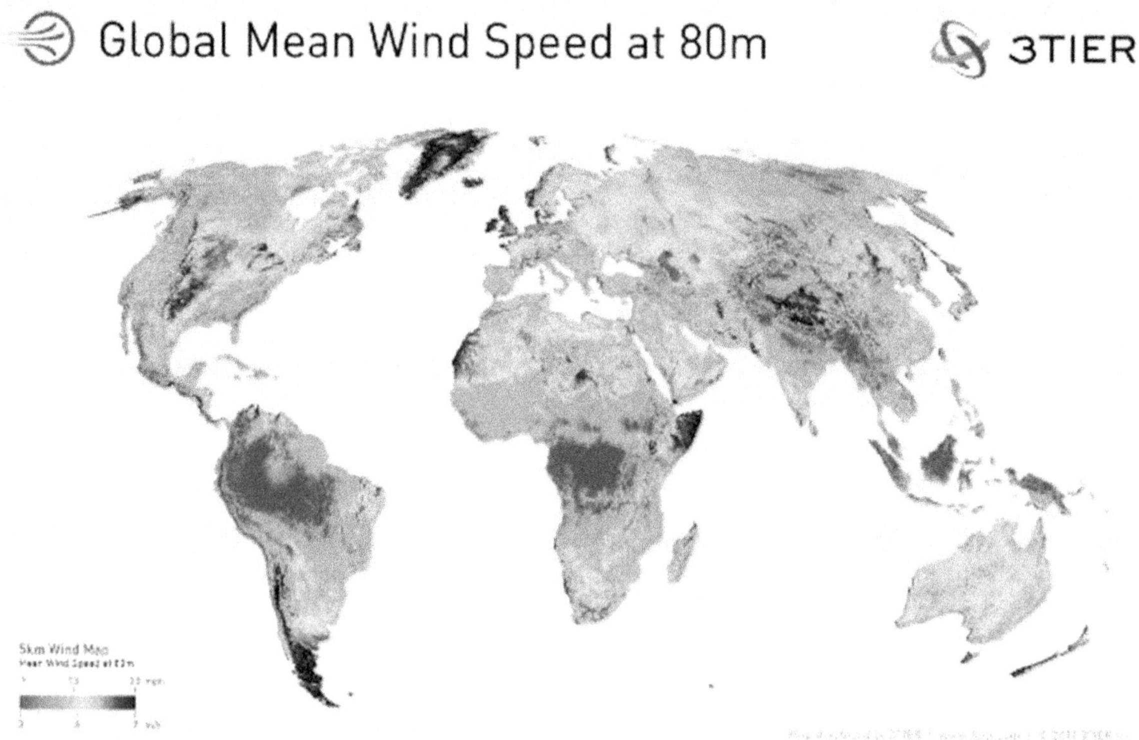

FIG. 1.11: NASA GLOBAL WIND SPEED MAP (SOURCE: [7])

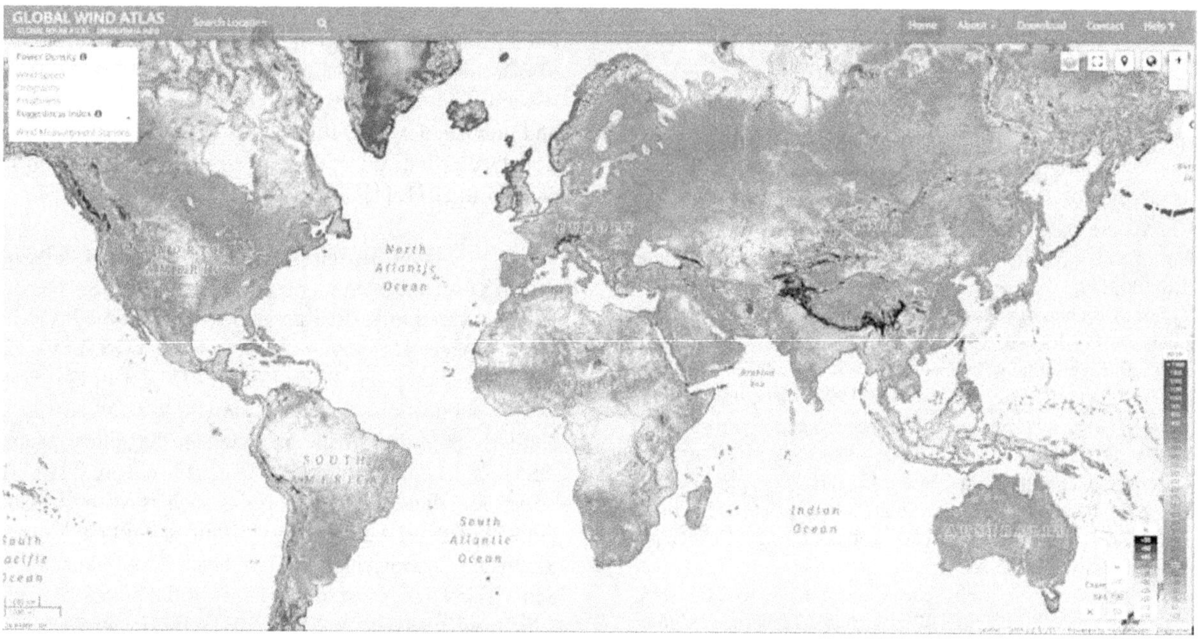

FIG. 1.12: GLOBAL WIND ATLAS_JAN 2018: (SOURCE: [8])

---

[7]https://www.bing.com/images/search?view=detailV2&ccid=kgDwuXAE&id=10D28197B4A93FB941741585A362F39E1840 11BC&thid=OIP.kgDwuXAENrfenRx5pKWzogHaEu&mediaurl=https%3a%2f%2fi.pinimg.com%2foriginals%2f99%2fd0%2 fd3%2f99d0d3e4311a5f581c1af-1114f6ace89.jpg&cdnurl=https%3a%2f%2fth.bing.com%2fth%2fid%2fR.9200f0b9700436b7 de9d1c79a4a5b3a2%3frik%3dvBFAGJ7zY-qOFFQ%26pid%3dImgRaw%26r%3d0&exph=2200&expw=3443&q=World+Win d+Speed+Map&simid=607988999261981594&FORM=IR-PRST&ck=672CA4AF8F5E2C0C1AFF3191BB7A6B9E&selectedIndex=0&ajaxhist=0&ajaxserp=0 _8-11-21

[8]https://images.search.yahoo.com/yhs/search?p=global+wind+atlas&fr=yhs-omr- 001&type=813167290&hspart=omr&hsimp=yhs-001&imgurl=http%3A%2F%2Fen.wikipedia.org%2Fwiki%2FSpecial%3AFilePath%2FGlobal_Wind_Atlas_2.2.png#id=0&iu rl=https%3A%2F%2Fwww.esa.int%2Fvar%2Fesa%2Fstorage%2Fimages%2Fesa_multimedia%2Fimages%2F2018%2F01%2 Fglobal_wind_atlas%2F17331266-1-eng-GB%2F-Global_wind_atlas.jpg&action=click _8-7-21

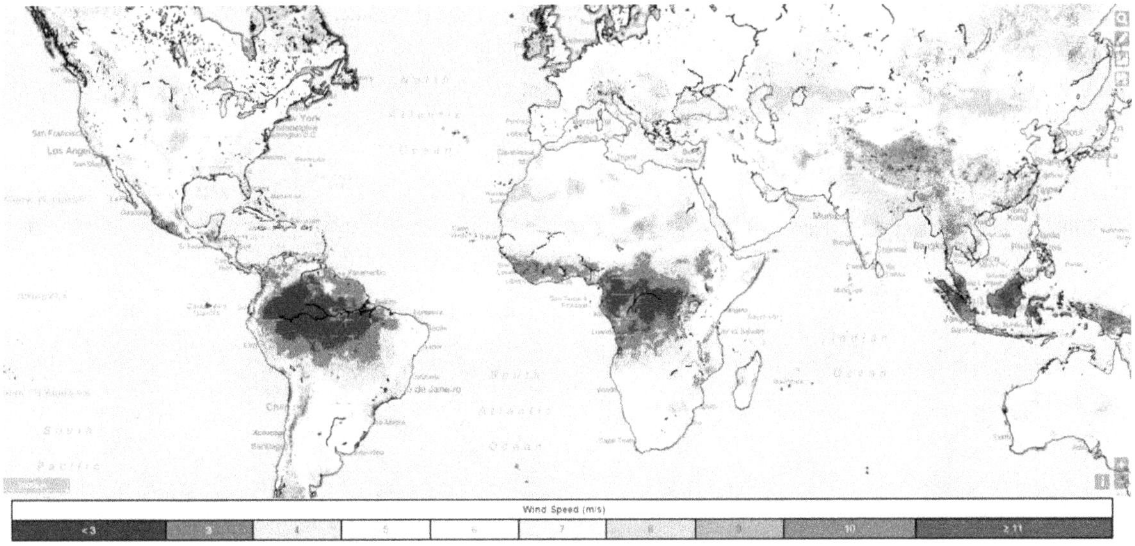

FIG. 1.13: GLOBAL WIND SPEED MAP (SOURCE: [9])

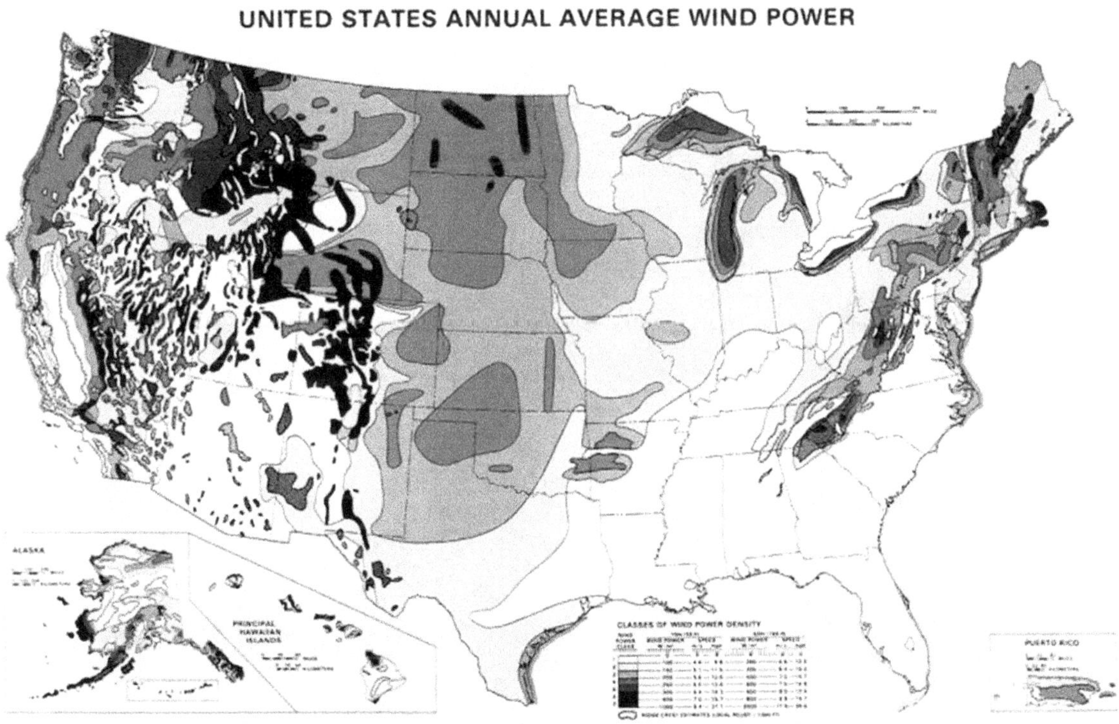

FIG. 1.14: US WIND MAP (SOURCE: [10])

---

[9]https://www.bing.com/images/search?view=detailV2&ccid=bme1hXqp&id=F60A33E61C87D2AB8A0881D9C685E13E3707 19EF&thid=OIP.bme1hXqpVNcp-jhfrUBTk6gHaDb&mediaurl=https%3A%2F%2Fwww.wasp.dk%2F-%2Fmedia%2FSites%2FWASP%2FNews%2FGWA.ashx%3Fmw%3D700%26hash%3D-06C2B82A0798C3CD8DD54E5576808770754D8A14&cdnurl=https%3A%2F%2Fth.bing.com%2Fth%2Fid%2FR.6e67b5857aa954d7298e17eb5014e4ea%3Frik%3D7xkHNz7hhcbZgQ%26pid%3DImgRaw%26r%3D0&exph=611&expw=1322&q=World+Wind+Speed+Map&simid=608 048939814579860&form=IR-PRST&ck=A394FB47F3606943AC136D27D9019FE5&selectedindex=8&ajaxhist=0&ajaxserp= 0&vt=0&sim=11

[10]https://www.bing.com/images/search?view=detailV2&ccid=wTjvesAK&id=E74F132E3A83F99CAE5BB5E6D2BB4CF50AC 1F174&thid=OIP.wTjvesAKbXEy0yT- qQ3rdwHaEr&mediaurl=https%3a%2f%2fth.bing.com%2fth%2fid%2fR.c138ef7ac00a6d7132d324fea90deb77%3frik%3ddP HBCv-VMu9LmtQ%26riu%3dhttp%253a%252f%252fwww.nooutage.com%252fimages%252fmap-wind-annual-avg- us.gif%26ehk%3dcu2%252ft16qad1A3pVAn-ghnnZYD8%252b%252bc1gGcNmiseRe9dB0%253d%26risl%3d%26pid%3dIm gRaw%26r%3d0&exph=713&expw=1130&q=Us+Wind+Map&sim-id=607999161145114221&FORM=IRPRST&ck=42C074 EA0169FECF83D617973DF4DABA&selectedIndex=0&ajaxhist=0&ajaxserp=0 _8-12-21

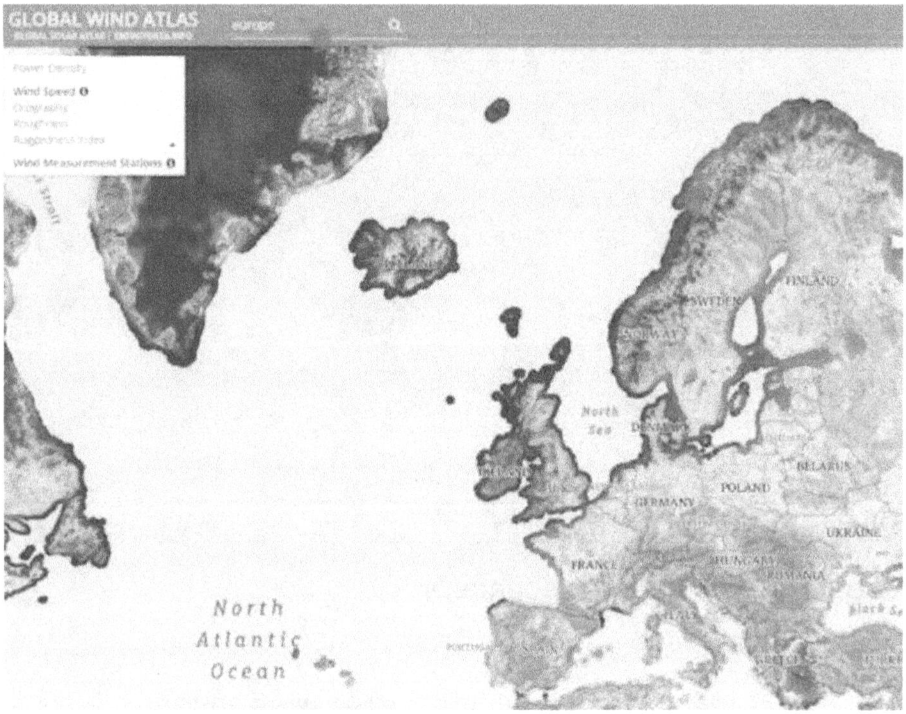

FIG. 1.15: WIND SPEED MAP OF EUROPE (SOURCE: [11])

FIG. 1.16: SOUTH ASIA WIND SPEED MAP (SOURCE: [12])

[11]https://www.esa.int/var/esa/storage/images/esa_multimedia/images/2018/01/wind_speed_in_europe/17331303-1-eng- GB/Wind_speed_in_Europe_pillars.jpg _8-12-21

[12]https://www.bing.com/images/search?view=detailV2&ccid=k59NmPEI&id=6360E51AAFFE770BBB245B6CE631BD98E0E 86B01&thid=OIP. k59NmPEI_QsvDegFYHmUgwHaFj&mediaurl=https%3a%2f%2fmaps.aerisapi.com%2fguVtifS8CxSbIJsv 6zU5p_20NQQMjd2dz6GQoqP8vP-WMACE99r63nj2oPzGwMY%2fwind-speeds%2cwind-speeds-text%2cadmin- dk%2f640x480%2f- 11. 45%2c59.59%2c55.97%2c171.04%2f-current.png&cdnurl=https%3a%2f%2fth.bing.com%2fth%2fid%2fR.939f4d98f108fd 0b2f0de80560799483%3frik%3dAWvo-4Ji9MeZsWw%26pid%3dImgRaw%26r%3d0&exph=480&expw=640&q=nasa+wind+speed+map+asia&simid=608035719905955218&FORM=IRPRS-T&ck=45D4C22D64F2AC4247C56DCB4C10A278&select edIndex=0&ajaxhist=0&ajaxserp=0 _8-12-21

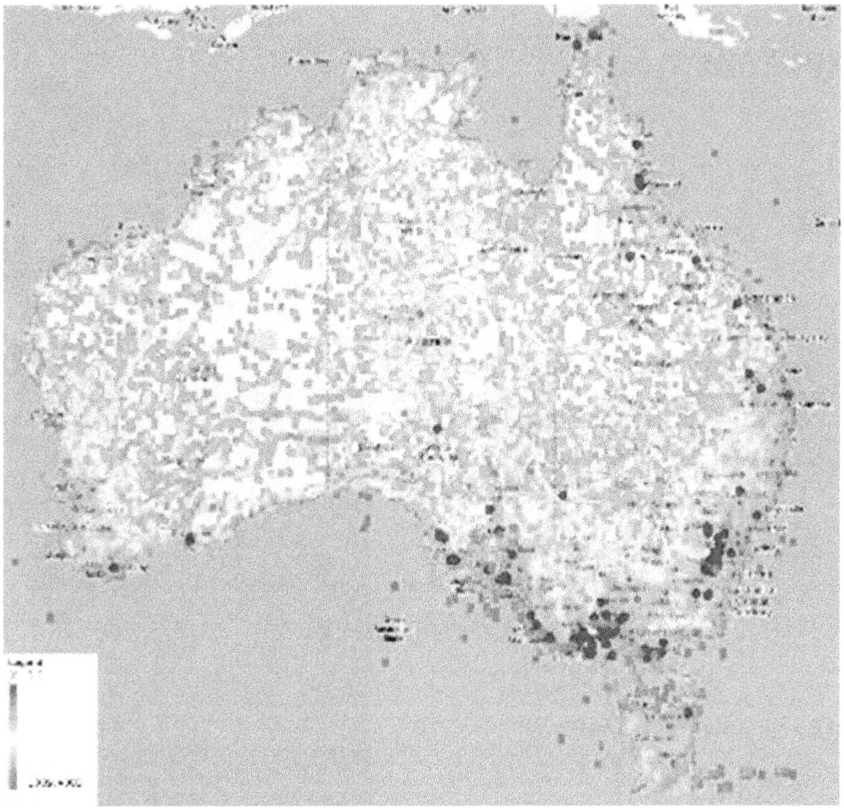

FIG. 1.17: WIND, WIND-FARMS, BIRDS AND BATS MAP OF AUSTRALIA (SOURCE: [13])

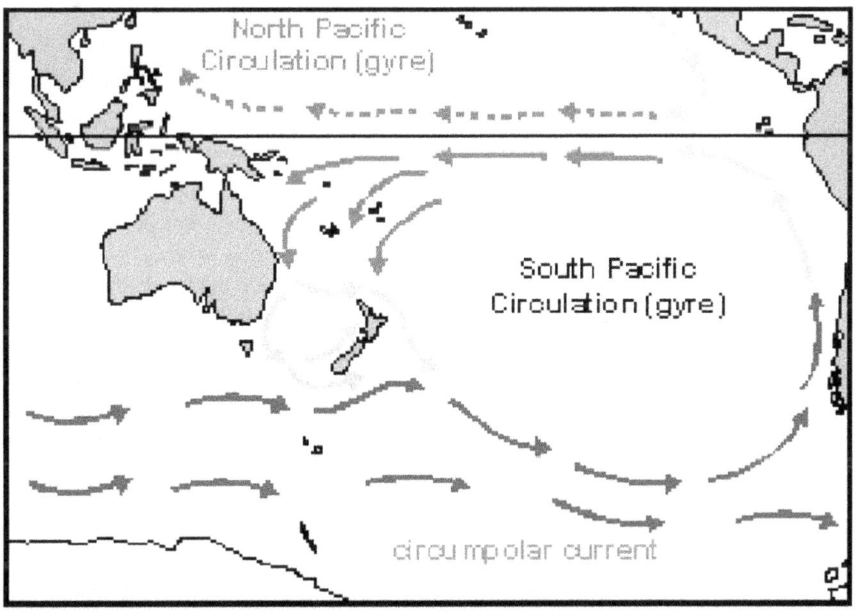

FIG. 1.18: SOUTH PACIFIC WINDS (SOURCE: [14])

[13]https://www.ala.org.au/wp-content/uploads/2012/07/Pic3.png_8-12-21

[14]https://www.bing.com/images/search?view=detailV2&ccid=9Lf9ivNT&id=49DA53AC30969879F54BC1A2D20FD2E9DF15 4ED-D&thid=OIP.9Lf9ivNTEeXwwhbjTeVCAgHaFD&mediaurl=https%3A%2F%2Fth.bing.com%2Fth%2Fid%2FR.f4b7fd8 af35311e5f-0c216e34de54202%3Frik%3D3U4V3%252bnSD9KiwQ%26riu%3Dhttp%253a%252f%252fwww.seafriends.org.n z%252fniue%252fworld16.gif%26ehk%3DU8i8KbE4aEebvm7YA4sYFydIHhrpyMjSSiEayRk%252fvM8%253d%26risl%3D%26pid%3DImgRaw%26r%3D0&exph=205&expw=300&q=australia%2c+newzealand+and+pacific+wind+map&form=IRPR ST&ck=AA8140FE55AE9E79877264A15942AFCC&selectedindex=1&ajaxhist=0&ajaxserp=0&vt=0&sim=11&pivotparams=insightsToken%3Dccid_bqBjrI7m*cp_46A0E82D7CF1F8E0D3AE795EA3E0AFA3*mid_BAAA-0C4AE0F66ACF0AB1183 E4A0B7A15E966774E*simid_607996962119506694*thid_OIP.bqBjrI7maOoNKN!_C0aUDPwHaFE&iss=VSI _8-12-21

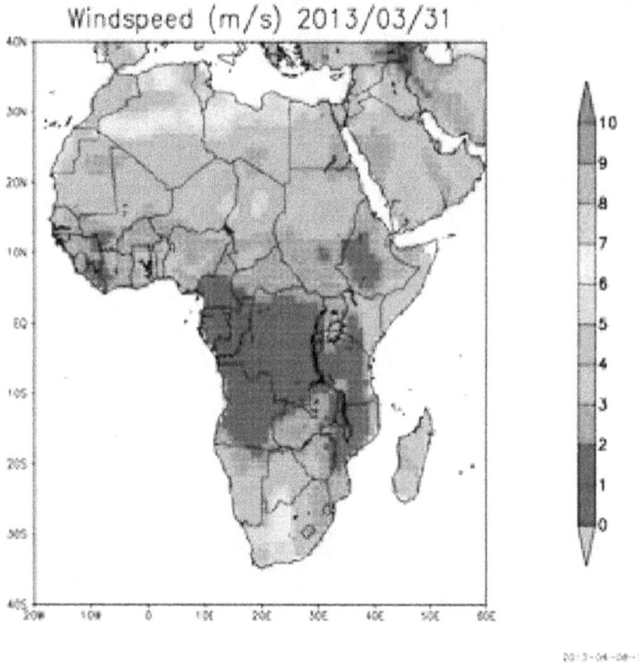

FIG. 1.19: WIND SPEED MAP OF AFRICA (SOURCE: [15])

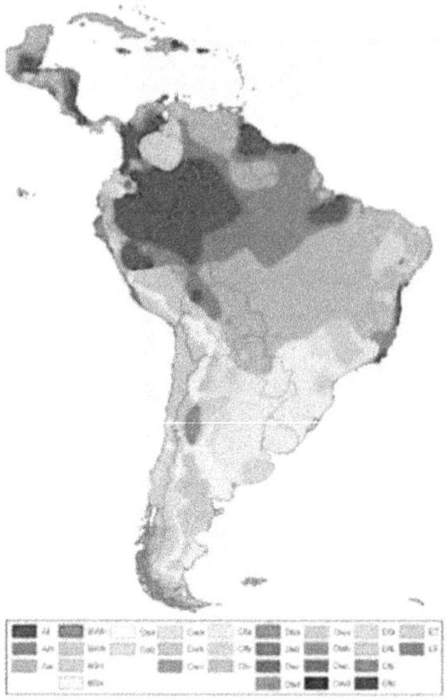

FIG. 1.20: WIND SPEED MAP OF CENTRAL AND SOUTH AMERICA (SOURCE: [16])

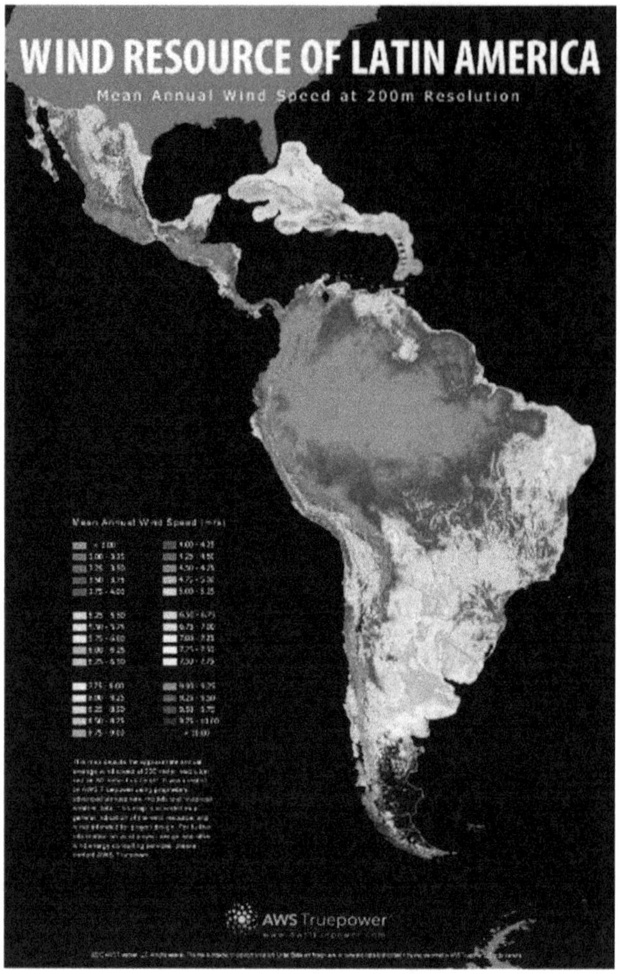

FIG. 1.21: NASA'S WIND SPEED MAP OF CENTRAL AND SOUTH AMERICA (SOURCE: [17])

ing is that for generation of nuclear energy we need plutonium as energy which may not be indigenously available at any location around the globe, whereas wind energy would be available on this globe for which the local authorities need not have to pay a price!

According to Paul Dimotakis, chief technologist at NASA's Jet Propulsion Laboratory Wind energy has the potential to provide 10 to 15 percent of the world's future energy. The following figures show the wind potential around the globe and in several countries around the world generated by NASA's efforts.

---

[15]http://hydrology.princeton.edu/~justin/research/project_global_monitor/figures/2013/20130331/figure_map_africa_wind.gif_8-12-21

[16] https://www.bing.com/images/search?view=detailV2&ccid=UoRNzU1R&id=7DB46209EDD33A00E9BF-753B16A17D309D 295EC8&thid=OIP.UoRNzU1RXlJ4ObK-sKmn- lAHaGw&mediaurl=https%3a%2f%2fwww.nicepng.com%2fpng%2fdetail%2f201-2014135_map-of-south-america- according-to-kppen- climate.png&cdnurl=https%3a%2f%2fth.bing.com%2fth%2fid%2fR.52844dcd4d515e527839b2ac2a69fe94%-3frik%3dyF4pn TB9oRY7dQ%26pid%3dImgRaw%26r%3d0&exph=748&expw=820&q=wind+speed+map+of+south+america&simid=608022031844993736&FORM=IRPRST&ck=170D06A56629350AE523EB62AD7E403F&selectedIndex=71&ajaxhist=0&ajaxserp=0_ 8-12-21

[17]https://www.bing.com/images/search?view=detailV2&ccid=YJQ6R7%2bV&id=FF65D4DD423AE8587161300D67A6988C9 B6EF4C5&thid=OIP.YJQ6R7-VkoB6-a5qxXEmUgHaLW&mediaurl=https%3a%2f%2fi.pinimg.com%2foriginals%2faf%2f78%2fe8%2faf78e8e5860e36c6137c9225ac9e5013.jpg&cdnurl=https%3a%2f%2fth.bing.com%2fth%2fid%2fR.60943a47bf9592807af9ae6ac5712652%3frik%3dxfRu m4yYpmcN-MA%26pid%3dImgRaw%26r%3d0&exph=750&expw=489&q=aws+wind+speed+map+of+central+and+south+a merica&simid=608022749103797268&-FORM=IRPRST&ck=851786D0C9D05F467B6207F592C6AE8B&selectedIndex=0&i dpp=overlayview&ajaxhist=0&ajaxserp=0 _8-12-21

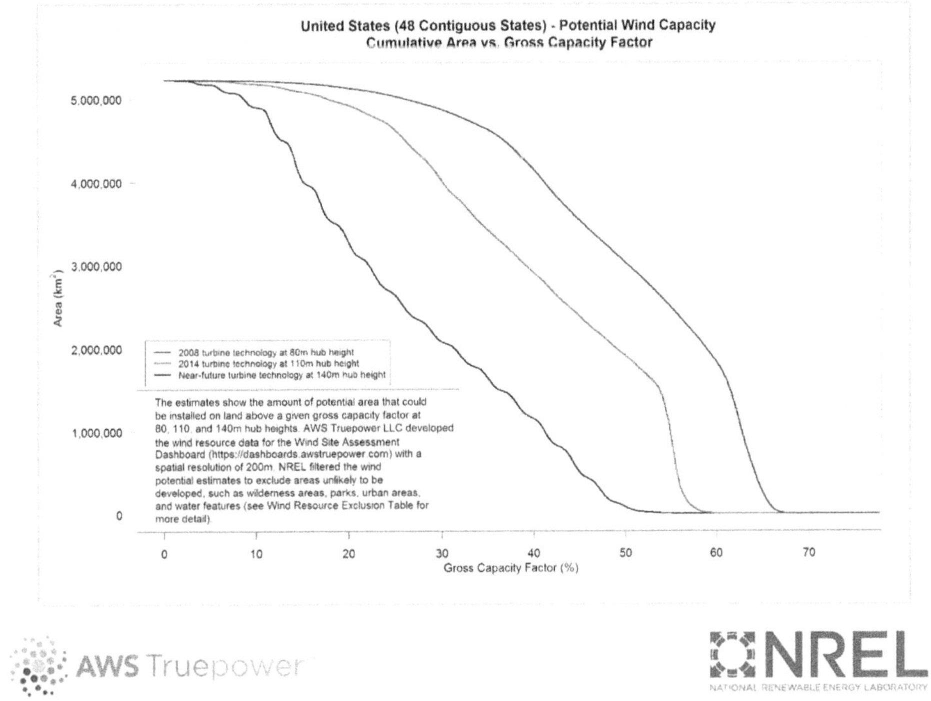

FIG. 1.22: UNITED STATES (48 CONTIGUOUS STATES) - POTENTIAL WIND CAPACITY CUMULATIVE AREA VS. GROSS CAPACITY FACTOR (SOURCE: [20])

1.7.2 US 48 Contiguous States vs. Potential Wind Capacity

The 48 contiguous states chart (Fig. 1.22) shows the potential gigawatts of rated capacity above a given gross capacity factor (without losses) at heights of 80-m, 110-m, and 140-m above ground. The potential wind capacity maps show land area with a gross capacity factor of 414 35% and higher, which may be suitable for wind energy development. AWS Truepower LLC produced the wind resource data with a spatial resolution of 200 m, which was binned into 20-km grid cells. Map shading shows the amount of area with the potential to be developed within each 20-km cell: the darker the color, the larger the potentially developable area within each cell. Areas that are excluded from development by law, such as wilderness areas and national parks, and other areas unlikely to be developed, such as urban areas and water bodies, have also been excluded. Site evaluation for wind projects should be coordinated with appropriate authorities and should consider potential effects on local resources and land uses, including but not limited to impacts on wildlife, sound, visual, radar, aviation, safety, and local priorities. Various federal agencies have authorities relating to the siting and permitting of wind plants, depending on the location and resources that might be affected. For example, potential impacts to air traffic and military missions are evaluated by the Federal Aviation Administration (FAA) and the US Department of Defense Siting Clearinghouse through the FAA's airspace obstruction evaluation process. Potential impacts to migratory birds or endangered species are evaluated with the US Fish and Wildlife Service. Other federal agencies that may have a role include, but are not limited to, the US Army Corps of Engineers for projects impacting wetlands and waterways and the Bureau of Land Management or US Forest Service for projects on federal lands. Similar state, tribal, and local agencies may also have jurisdiction. In addition to resource maps, NREL produced charts showing the potential wind capacity above a given gross capacity factor at various heights as another way to express how new technologies affect the potential developable land area. This shows the wind potential for the 48 contiguous United States in both map and chart format; additional wind potential charts and maps have been developed for states (on the 110-m map above, click on a state). NREL also estimated the land area and potential wind capacity in various capacity factor ranges for each state using the Wind Site Assessment Dashboard (https://dashboards.awstruepower.com[18]) from AWS Truepower LLC.[19]

1.7.3 Community-Scale 50-m Wind Maps[21]

Though not directly linked with NASA the US Department of Energy provides 50-m height, high-resolution wind resource maps for most of the states and territories of Puerto Rico and the Virgin Islands in the United States. Counties, towns, utilities, and schools use community scale wind resource maps to locate and quantify the wind resource, identifying potentially windy sites determining a potential site's economic and technical viability. These resource maps show estimates of wind power density at 50

[18]https://dashboards.awstruepower.com

[19]Wind Energy for Power Generation Meeting the Challenge of Practical Implementation by K. R. Rao, Volumes 1 & 2, This Springer book is published by Springer International Publishing AG part of Springer Nature, Gewerbestrasse 11, 6330 Cham, Switzerland

[20]https://windexchange.energy.gov/files/u/visualization/pdf/us_contiguous_wind_potential_chart.pdf

[21]Wind Energy for Power Generation Meeting the Challenge of Practical Implementation by K. R. Rao, Op. Cit.

m above the ground and depicts the resource that could be used for community-scale wind development using wind turbines at 50-60-m hub heights. As a renewable resource, wind was classified according to wind power classes, which were based on wind speed frequency distributions and air density. These classes ranged from Class 1 (the lowest) to Class 7 (the highest). In general, at a 50-m height, wind power Class 4 or higher could have been useful for generating wind power with turbines in the 250-kW to 750-kW rating. Given the advances in technology, resources below Class 4 may now be suitable for the new midsize wind turbines. In recognition of these continuing advancements in wind energy technologies and the ability for the current generation of wind turbines to extract cost competitive wind energy from lower wind speeds the Energy Department has moved away from the wind power classification system and now reports wind speeds only. These 50-m height maps still provide useful insight for the identification of energetic sites for community-scale projects.[22]

1.7.4 Residential-Scale 30-m Wind Maps[23]

These maps although attributable to WINDExchange which is a resource of the Department of Energy's Wind Program DOE with help from NASA to show at the 30-m height, high-resolution wind resource maps for the United States. Businesses, farms, and homeowners use residential-scale wind resource maps to identify wind sites that may be appropriate for small-scale wind projects. This map shows the predicted mean annual wind speeds at a 30-m height, presented at a spatial resolution of 2 km that is interpolated to a finer scale. Areas with good exposure to prevailing winds and annual average wind speeds around 4 m per second and greater at a 30-m height are generally considered to have a suitable wind resource for small wind projects. Small wind turbines are typically installed between 15 and 40 m high. The average wind speeds indicated on this map are model derived estimates that may not represent the true wind resource at any given location. Small terrain features, vegetation, buildings, and atmospheric effects may cause the wind speed to depart from the map estimates.

Figure 1.23 shows land area with a gross capacity factor of 35% and higher, which may be suitable for wind energy development. AWS Truepower LLC produced the wind resource data with a spatial resolution of 200 m, which was binned into 20-km grid cells. Map shading shows the amount of area with the potential to be developed within each 20-km cell: the darker the color, the larger the potentially developable area within each cell. Thus this information for all of the US States, Hawaii, Puerto Rico and Virgin Islands can be extremely useful for installing land based wind turbines.

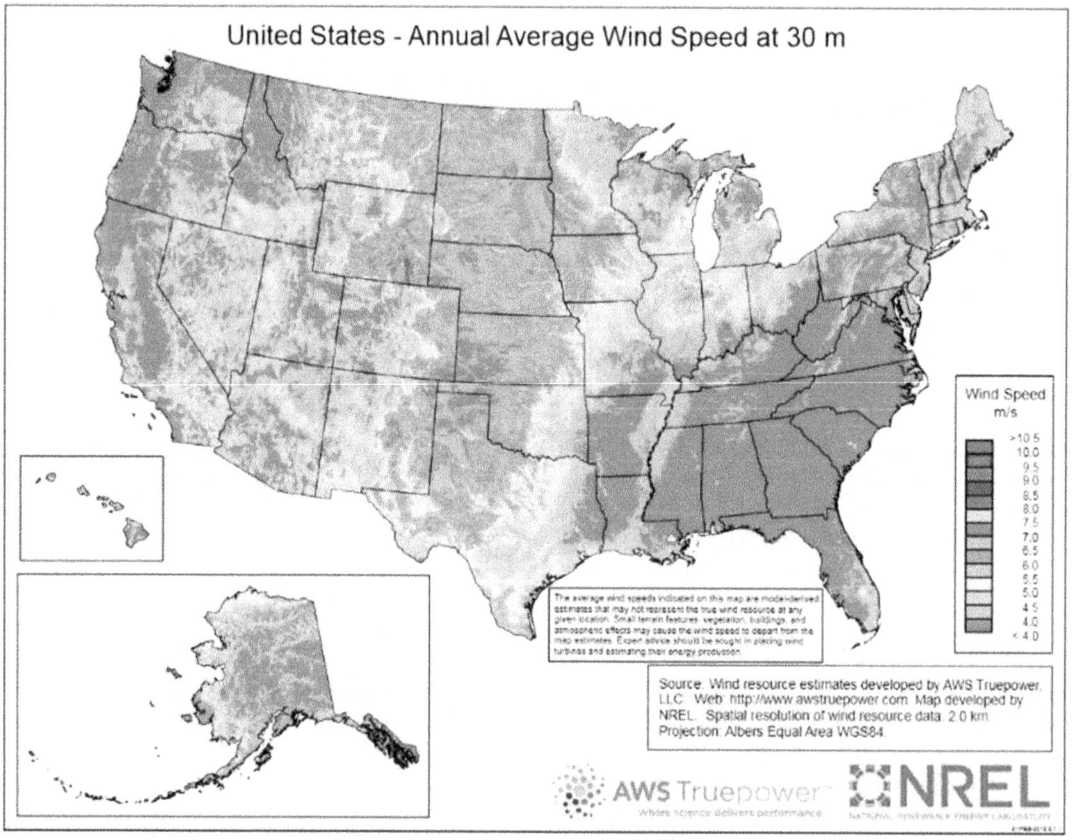

FIG. 1.23: RESIDENTIAL SCALE WIND MAPS (SOURCE: [24])

[22] https://windexchange.energy.gov/maps-data/96 _8-20-21

[23] Wind Energy for Power Generation Meeting the Challenge of Practical Implementation by K. R. Rao, Op. Cit.

[24] https://windexchange.energy.gov/maps-data/325 _8-20-21

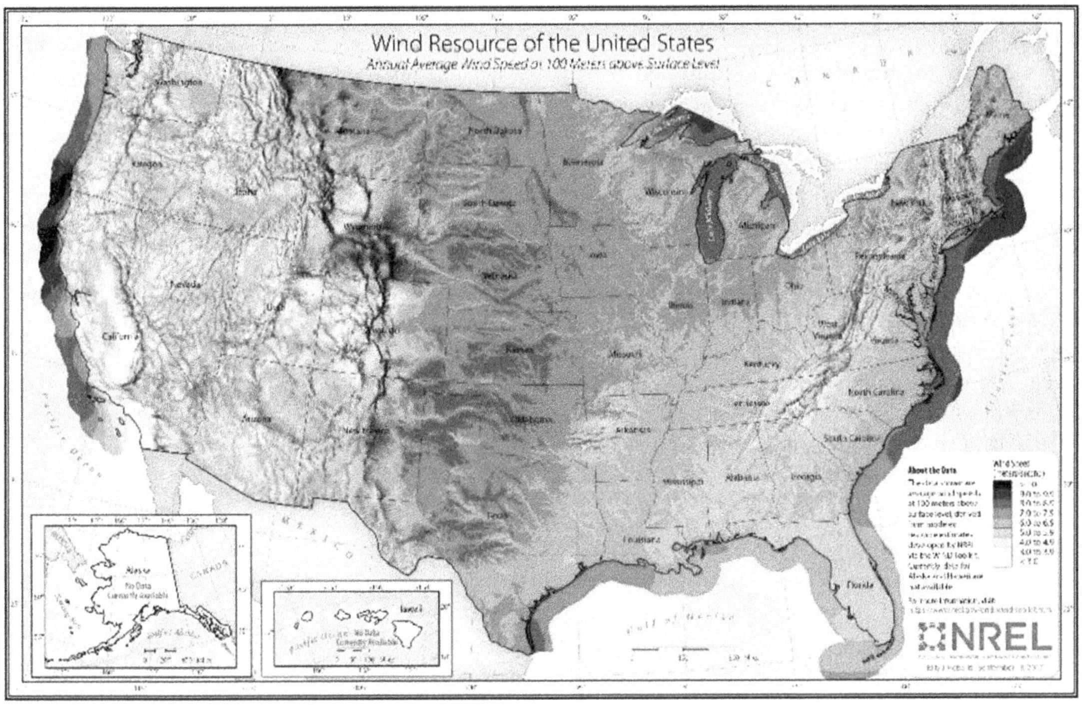

FIG. 1.24: U.S. WIND POWER RESOURCE AT 100-METER HUB HEIGHT (SOURCE: [25])

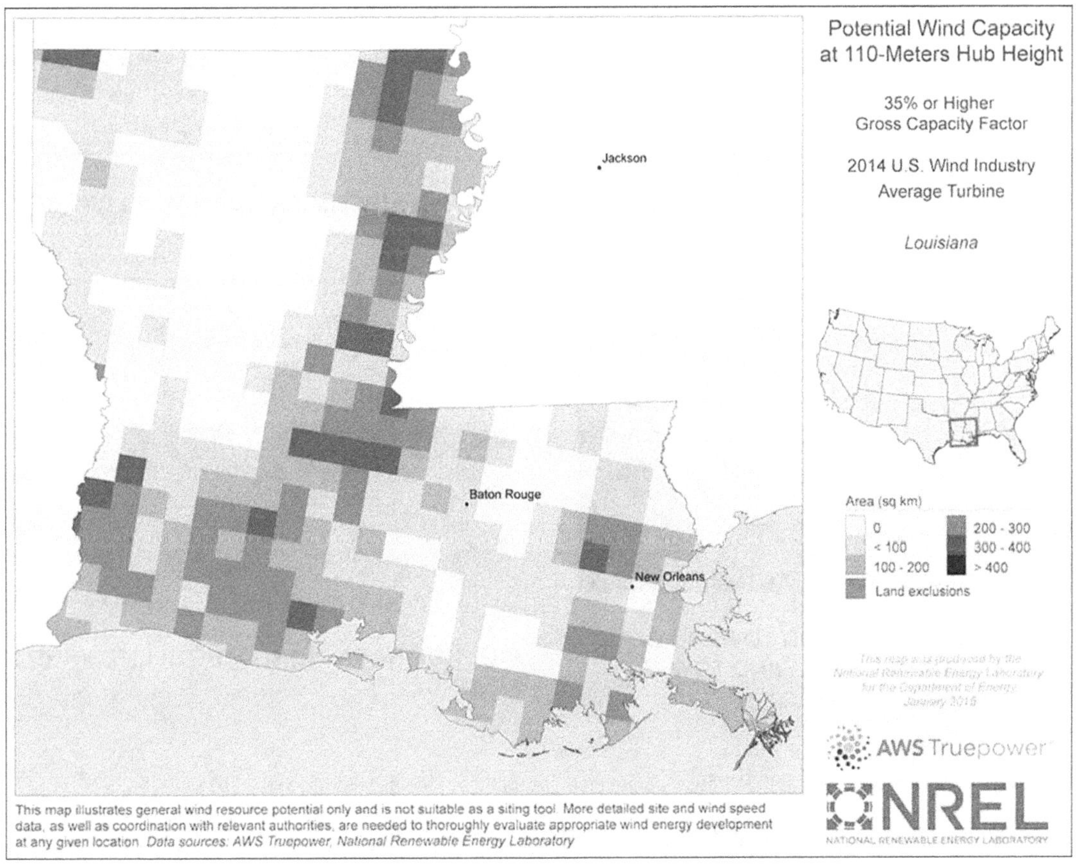

FIG. 1.25: US WIND MAP OF LOUISIANA AT 110 M HUB HEIGHT (SOURCE: [26])

[25]https://windexchange.energy.gov/maps-data/324 _8-20-21
[26]https://windexchange.energy.gov/maps-data/173 _8-20-21

1.7.5 Implementation of Residential Wind Towers[27]

A 25-kilowatt wind turbine was installed in 2018 on a 140-foot tower, which separates the turbine from treetops and other potential barriers to winds. High energy-use by home owner was motivated by personal values and the desire for a net-zero energy home (meaning renewable energy is generating enough to cover all electricity, heating, and cooling needs. Edison Utility: Company (Met-Ed) is part of FirstEnergy, which is one of the nation's largest investor owned utilities.

1.7.6 Offshore Wind Resource Potential

Once windmills are installed, wind can be converted to electricity inexpensively. But not everyone likes wind farms. The giant collection of whirling blades mars scenic views and can kill birds and bats, particularly if located in a high-traffic flyway. To minimize these risks, one solution may be to place wind farms in the ocean. Wind tends to blow stronger over the ocean than over land. The ocean presents a smooth surface over which wind can glide without interruption, while hills, mountains, and forests tend to slow or channel wind over land.

Wind over the ocean isn't consistent, in some places, the air is still, while in others, the wind blows fiercely. To identify potential

FIGS 1.26A: 140M RESIDENTIAL WIND TOWER ABOVE ALL OBSTACLES; 1.26B: 140M HIGH WIND TOWER (SOURCE [28])

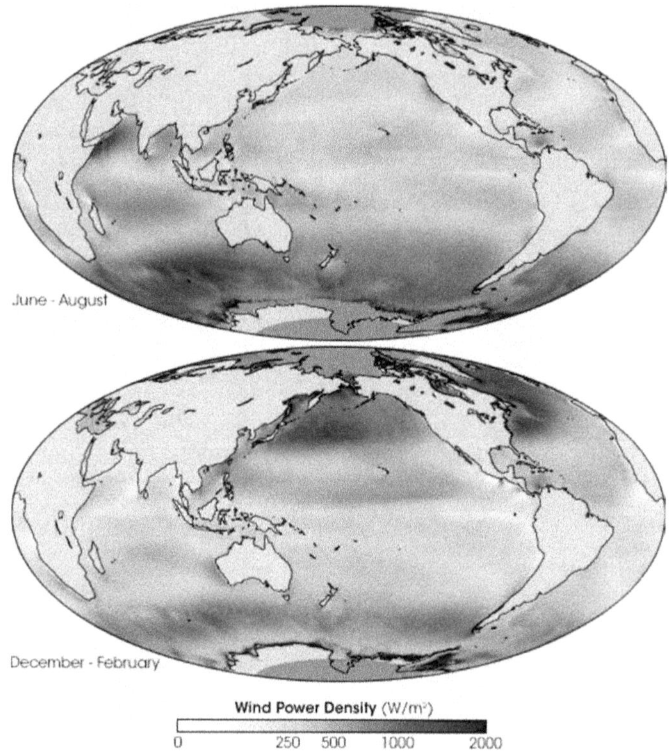

FIG. 1.27: GLOBAL OCEAN WIND ENERGY POTENTIAL (SOURCE: [29])

FIG. 1.28: NASA WORLD SATELLITE WIND MAP (SOURCE: [30])

---

[27] https://www.energy.gov/sites/prod/files/2018/08/f54/dwic_deployment_example_res_eocycle_draft_8_03_2018.pdf_8-20-21

[28] Ibid _8-20-21

[29] https://earthobservatory.nasa.gov/images/8916/global-ocean-wind-energy-potential _8-11-21

[30] https://www.bing.com/images/search?view=detailV2&ccid=jfYotPBD&id=0382CB03B9381596FBC6819FBAABBEE-B815E 321E&thid=OIP.jfYotPBDPteYpdG09x3mqAHaDt&mediaurl=https%3a%2f%2fth.bing.com%2fth%2fid%2fR.8df628b-4f0433ed798a5d1b4f71de6a8%3frik%3dHjJegeu%252bq7qfgQ%26riu%3dhttp%253a%252f%252f173.254.228.79%252f%257eo pp idanlibrary%252fwp-content%252fuploads%252f2017%252f05%252fWorld-Map-Satellite- India.jpg%26ehk%3d2oaidwtqvErWB0d-U6Uf%252fT68rmS3TPsaLCFGvrgFghhE%253d%26risl%3d%26pid%3dImgRaw% 26r%3d0&exph=2000&expw=4000&q=World+Map+Satellite&-simid=608030656140151746&FORM=IRPRST&ck=988A7E 613264D11E83533E235CB631D5&selectedIndex=0&ajaxhist=0&ajaxserp=0 _8-11-21

wind farm locations, NASA scientists Tim Liu, Wenqing Tang, and Xiaosu Xie, all at the Jet Propulsion Laboratory, mapped out average wind intensity over the ocean between 2000 and 2007, shown in Figure 1.28. They created their maps from data collected by NASA's Quick Scatterometer (QuikSCAT), which measures wind speed and direction over the world's oceans. The satellite sends pulses of microwave energy through the atmosphere to the ocean surface and measures the energy that bounces back from the wind-roughened surface. The energy of the microwave pulses changes depending on wind speed and direction. The scientists averaged QuikSCAT's measured wind speeds by season, and then calculated the wind power density, the amount of energy that could be derived from a wind turbine in a given location. Their maps for the winter and summer seasons are shown in figure 1.29.

Wind strength is influenced by seasonal patterns, land-ocean interactions, land topography, and ocean temperatures. All of these interactions are evident in this pair of images. Areas of high wind power density, where winds are strongest, are purple, while low

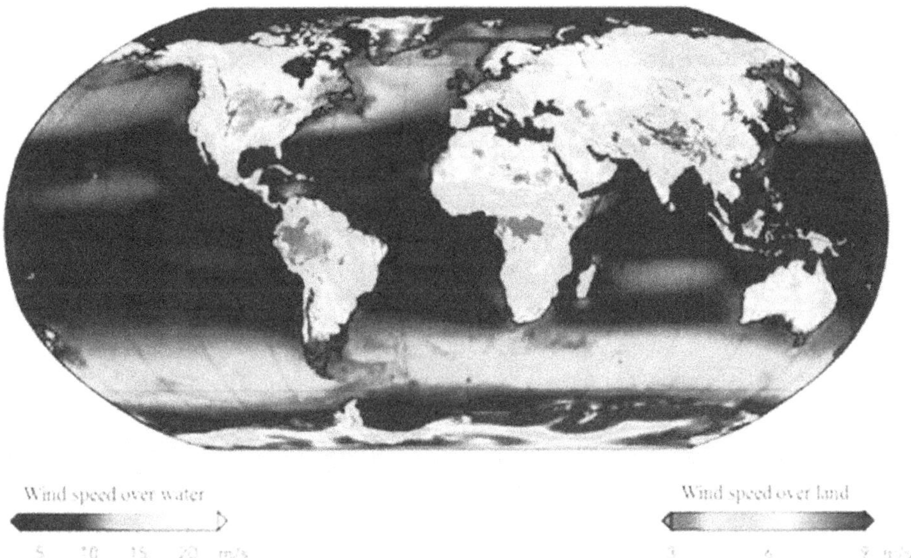

FIG. 1.29: NASA GLOBAL WIND SPEED MAP (SOURCE: [31])

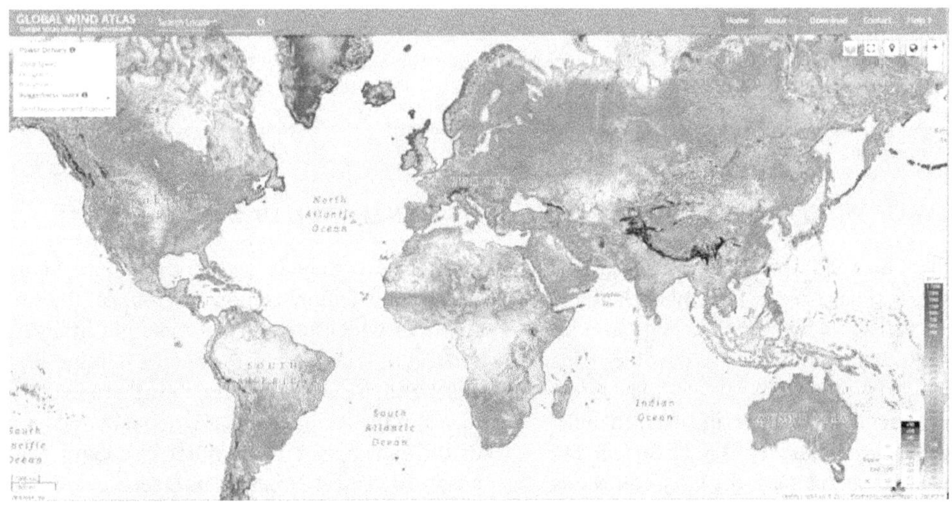

FIG. 1.30: GLOBAL WIND ATLAS_JAN 2018: (SOURCE: [32])

[31]https://www.bing.com/images/search?view=detailV2&ccid=owBNL5aK&id=AB25DB882CB94523E5A219D4D3AE0E1F11 3C6705&thid=OIP.owBNL5aKIx3iF4CIMieHKwHaEc&mediaurl=https%3a%2f%2fwww.researchgate.net%2fprofile%2fJua n_M_Ortiz%2fpublication%2f311857024%2f-figure%2ffig3%2fAS%3a442555537072128%401482524870140%2fWorldwide-average-wind-speed-map-created-based-on-wind-speed-measurements-at-theheight.png&cdnurl=https%3a%2f%2fth.bing.com%2fth%2fid%2fR.a3004d2f968a231de21780883227872b%3frik%3dBWc8E R8OrtPUGQ%26pid%3dImgRaw%26r%3d0&exph=464&expw=773&q=nasa+global+wind+speed+map&simid=6080428967 99135927&FORM=IR-PRST&ck=B13C3781D4B0DC6D1BF0B0D3A5F70E7D&selectedIndex=136&ajaxhist=0&ajaxserp=0_8-11-21

[32]https://images.search.yahoo.com/yhs/search?p=global+wind+atlas&fr=yhs-omr- 001&type=813167290&hspart=omr&hsimp=yhs-001&imgurl=http%3A%2F%2Fen.wikipedia.org%2Fwiki%2FSpecial%3AFilePath%2FGlobal_Wind_Atlas_2.2.png#id=0&iu rl=https%3A%2F%2Fww.esa.int%2Fvar%2Fesa%2Fstorage%2Fimages%2Fesa_multimedia%2Fimages%2F2018%2F01%2 Fglobal_wind_atlas%2F17331266-1-eng-GB%2F-Global_wind_atlas.jpg&action=click _8-7-21

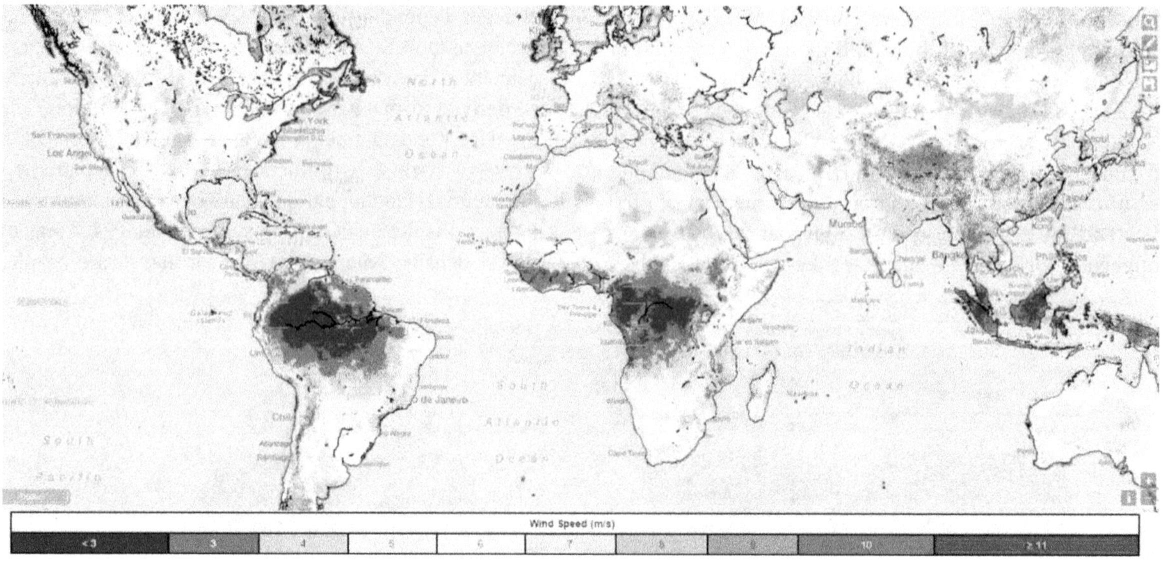

FIG. 1.31: GLOBAL WIND SPEED MAP (SOURCE: [33])

FIG. 1.32: VIEW OF WIND MILL FARM THAT USES NASA GENERATED DATA (SOURCE: [34])

power density regions are light blue and white. The largest patterns shown in the images are seasonal patterns. In December, January, and February, winter storms fuel strong winds in the mid-latitudes of the Northern Hemisphere. In June, July, and August, winter reins in the Southern Hemisphere, and the pattern is reversed. The Asian monsoon also controls the seasonal distribution of wind. In June, July, and August, strong winds gust across the Bay of Bengal and the Arabian Sea. From December to February, the monsoon winds blow over the East China Sea. Finally, the trade winds trace their way across the tropics, stronger in the winter than in the summer.

In places where the land and the ocean interact to generate strong winds, the wind power density is more consistent. Winds accelerate as they deflect around a bit of land sticking into the ocean. This feature is most obvious off the southern tip of South America, which has high wind potential in both summer and winter, see figure 1.29. The effect occurs in other locations, such as off Tasmania, New Zealand, and northern California, and is easiest to see in the summer when seasonal winds are calmer. The spot of brilliant purple in an otherwise calm ocean off the Pacific coast of Central America in December, January, and February is an example of strong winds created by the topography of the land. In Central America, as in other places, mountains channel the wind, creating a natural wind tunnel that sends strong gusts over the ocean. The final wind pattern shown in this

[33]https://www.bing.com/images/search?view=detailV2&ccid=bme1hXqp&id=F60A33E61C87D2AB8A0881D9C685E13E3707 19EF&thid=OIP.bme1hXqpVNcpjhfrUBTk6gHaDb&mediaurl=https%3A%2F%2Fwww.wasp.dk%2F-%2Fmedia%2FSites%2FWASP%2FNews%2FGWA.ashx-%3Fmw%3D700%26hash%3D06C2B82A0798C3CD8DD54E5576808770754D8A14&cdnurl=https%3A%2F%2Fth.bing.com%2Fth%2Fid%-2FR.6e67b5857aa954d7298e17eb5014e4ea%3Frik%3D7xkHNz7hhcbZgQ%26pid%3DImgRaw%26r%3D0&exph=611&expw=1322&q=World+Wind+-Speed+Map&simid=608 048939814579860&form=IRPRST&ck=A394FB47F3606943AC136D27D9019FE5&selectedindex=8&ajaxhist=0&ajaxserp=0&vt=0&sim=11 _8-25-21

[34]https//ghrc.nsstc.nasa.gov/home/measurements/winds _8-21-21

image is the wind created by the ocean itself. Warm ocean currents naturally warm the air above them. When ocean currents carry warm water into an area of cool water, or vice versa, the temperature difference in the air generates wind. The clearest example in these images is the Gulf Stream, which snakes up the east coast of the United States and east over the North Atlantic. In both winter and summer, winds are stronger over the Gulf Stream than the surrounding ocean.[35]

Global Hydrology Resource Center (GHRC), NASA conducts measurements of ocean surface wind speed and direction from satellite microwave radiometers, and vertical wind profiles from airborne instruments and dropsondes, as well as surface wind observations. The GHRC DISCOVER's Distributed Information Services (DIS) funded by NASA's MEASUREs program, generates a suite of inter-related atmospheric parameters including ocean surface winds and water vapor derived from passive microwave instruments aboard TRMM, NOAA-15, NOAA-16, and DMSP satellites. Ground-based and airborne instruments also measure surface and upper level winds using NASA Hurricane Science field campaigns. These field campaigns provide for intensive observation of specific phenomena using a variety of instruments on aircraft, satellites and surface networks[36] (see figure 1.33).

1.7.7 Installed Wind Capacity as of 2015

1.7.8 Proposed Wind Energy Project Offshore North Carolina[38]

As part of the Biden-Harris administration's commitment to creating nearly 80,000 jobs through developing 30 gigawatts of offshore wind energy by 2030, the Department of the Interior announced on 30 July 2021 that the Bureau of Ocean Energy Management (BOEM) will conduct an environmental review of a proposed wind energy project offshore North Carolina. This project would be the first to operate offshore North Carolina, would make landfall in Virginia Beach and could eventually provide power to Virginia jump starting country's transition to a cleaner energy future and working together to lift up the local industry, see figure 1.34. BOEM is to review a Construction and Operations Plan (COP) submitted by Kitty Hawk Wind scale, offshore wind energy project consisting of up to 69 total wind turbine generators, one offshore substation, inter-array cables, and up to two transmission cables that will make landfall in Virginia Beach. North Carolina has set ambitious goals to develop 2.8 GW of offshore wind energy off of the state's coast by 2030 and 8 GW by 2040. The state estimates that meeting that goal would power 2.3 million homes by 2040. The Commonwealth of Virginia enacted the Virginia Clean Economy Act in 2020, which sets a target to produce its electricity from 100% renewable sources by 2045, with 5.2 GW of offshore wind energy by 2034. Additionally, the

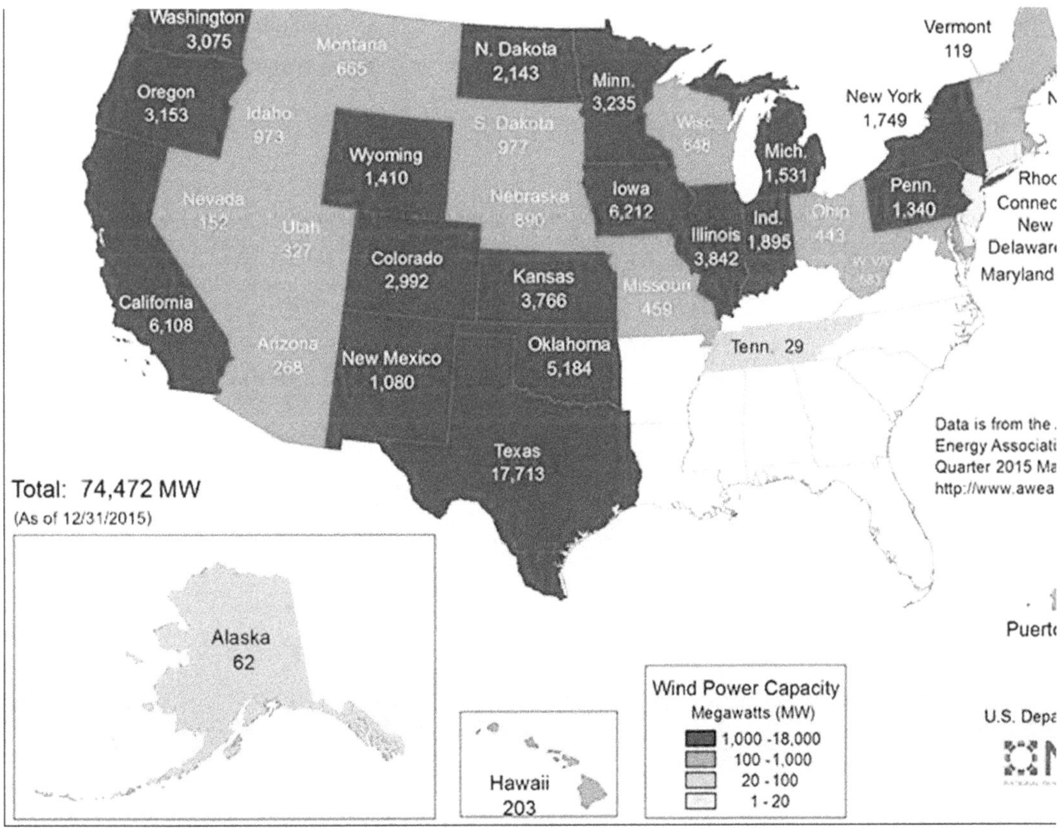

FIG. 1.33: INSTALLED WIND CAPACITY, 2015 (SOURCE: [37])

[35]https://earthobservatory.nasa.gov/images/8916/global-ocean-wind-energy-potential _8-11-21

[36]https//ghrc.nsstc.nasa.gov/home/measurements/winds_8-21-21

[37]https://www.nrel.gov/docs/fy16osti/62942.pdf _8-22-21

[38]https://www.renewableenergymagazine.com/wind/next-steps-for-first-proposed-wind-energy-20210730_8-21-21

FIG. 1.34: FOR AN OFFSHORE WIND FARM, NORTH CAROLINA (SOURCE: [39])

Commonwealth of Virginia, State of North Carolina, and State of Maryland have established the Southeast and Mid-Atlantic Regional Transformative Partnership for Offshore Wind Energy Resources (SMART-POWER) to "promote, develop, and expand offshore wind energy generation and the accompanying industry supply chain and workforce" in the region. As part of BOEM's environmental review, the agency must first identify what should be considered in the EIS, such as important resources and issues, potential impacts to the environment, reasonable alternatives, and mitigation measures.[40]

1.7.9 Proposed Wind Projects in the US[41]

Not entirely with the help and collaboration with NASA proposed wind projects have been mapped shown in Figure 1.30. The U.S. Department of Energy's Wind Powering America program supported a collaborative effort between the National Renewable Energy Laboratory (NREL) and AWS Truepower (formerly AWS Truewind) to update the U.S. wind potential for the first time in almost two decades. The new wind energy potential estimates at 80-m and 100-m heights are based on wind resource data at 200-m resolution and estimated capacity factors. The major products developed for the contiguous United States include: – State and National maps of mean wind speed.

1.7.10 Potential Wind Capacity

Wind power capacity potential reflects the amount of wind power that is technologically possible to have installed in a given region. We would not expect this potential to be fully utilized, but it is indicative of the physical limits of the assumed technology. For example, between 2016 installed capacity that is close to 75,000 megawatts and potential capacity what is technologically possible at 80 meters is 10,600,000 megawatts. To learn more about your location, compare a state's current installed capacity to what is technologically possible. Wind power capacity potential is estimated assuming that all available land area, after legal and technical exclusions are applied, is populated with wind turbines at a density of 3 megawatts per square kilometer. For additional detail on exclusions and methods for determining resource potential, see Figure 1.36.[43]

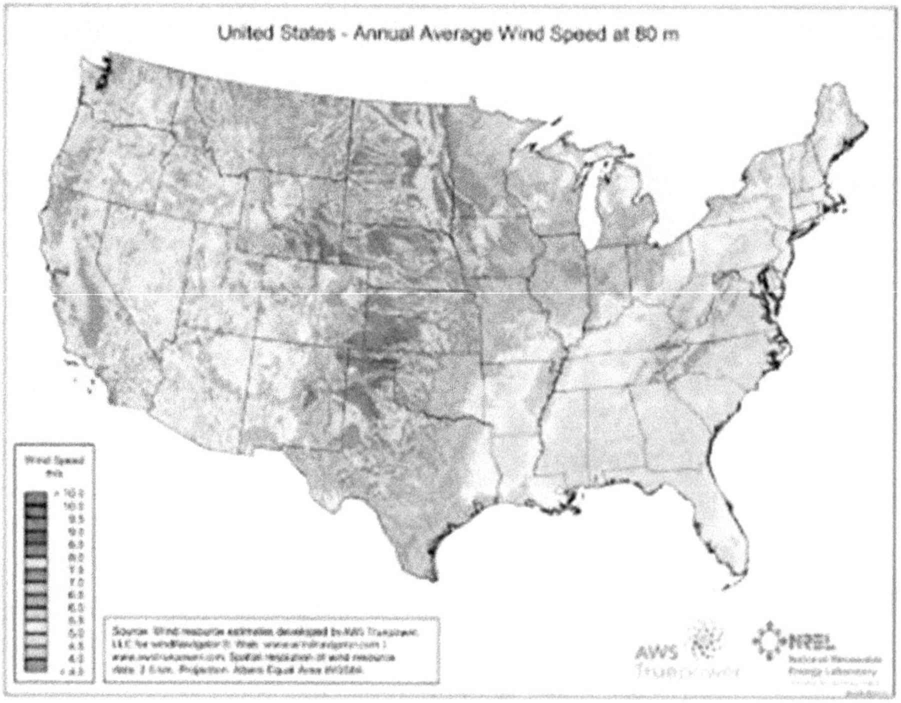

FIG. 1.35: ANNUAL AVERAGE WIND SPEED AT 80M (SOURCE: [42])

[39]Ibid.

[40]Op. cit.

[41]https://windexchange.energy.gov/maps-data/321 _8-22-31

[42]https://www.nrel.gov/docs/fy10osti/48036.pdf _8-22-21

[43]https://www.nrel.gov/docs/fy10osti/48036.pdf_8-22-21

1.7.11 Wind Generation Developments of NASA since 2010

1.7.11.1 Capturing Airborne Wind

NASA aerospace engineer Mark Moore speculated how long nanotubes could reach into the clouds to tether a turbine-flying at 2,000 feet or 10,000 to 30,000 feet to harvest the power from the wind back to Earth. Moore is part of the first federally-funded research effort to examine airborne wind capturing platforms. He's developing methods to evaluate how to capture wind energy efficiently. Aloft might be a funnel-shaped blimp with a turbine at its back; or a balloon with vanes that rotate; a truss-braced wing; a parachute; a kite or all of them are ideas being considered by renewable energy industry. Tethers for airborne wind generation assets don't require a lot of ground space, nor are they labor intensive. And they don't pollute. "They could stay up a year, then come down for a maintenance check and then go back up, or they could be reeled in in case of a storm. A single operator could watch over 100 of these air borne platforms. At 2,000 feet (610 m), there is two to three times the wind velocity compared to ground level," Moore said. "The power goes up with the cube of that wind velocity, so it's eight to 27 times the power production just by getting 2,000 feet up, and the wind velocity is more consistent." Send turbines farther aloft, into the 150 mph jet stream at 30,000 feet, and "instead of 500 watts per meter for ground-based wind turbines, 20,000 to 40,000 watts per square meter," Moore said. "That's very high energy density and potentially lower cost wind energy because of the 50-fold increase in energy density. "So why isn't it being done? Or at least, why isn't it being researched more expansively? One answer involves the vehicle to be flown. Another involves where to fly it. "All you have right now are small companies doing the research, the ultimate answer could be the federal government itself. There are two aspects. One involves the technology and geography. The other involves the interaction between those elements and other competitors for airspace. That means dealing with current Federal Aviation Administration regulations and with those that might be necessary to accommodate an airspace that includes manned aircraft, the unmanned aircraft in the future, plus wind-borne energy turbines. But first things first, help the FAA in its decision-making. "Offshore deployment of these airborne systems probably makes sense in terms of both airspace, and land use, because there is no demand for low altitude flight over oceans 12 miles offshore. "Also, unlike ground-based turbines, there is almost no additional cost for airborne systems offshore because huge platforms are not required to support the structure or resist large tower bending moments. "NASA could

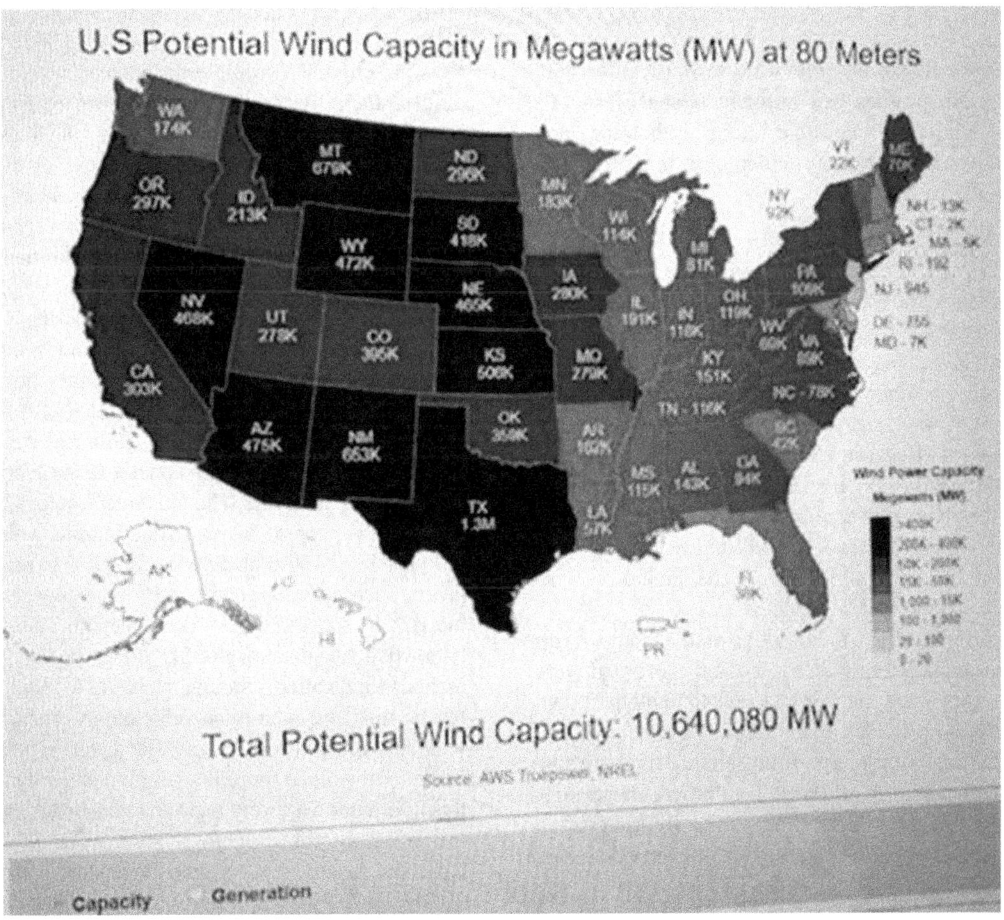

FIG. 1.36: WINDEXCHANGE: U.S. INSTALLED AND POTENTIAL WIND POWER CAPACITY AND GENERATION (SOURCE: [44])

[44]https://windexchange.energy.gov/maps-data/321 _8-22-21

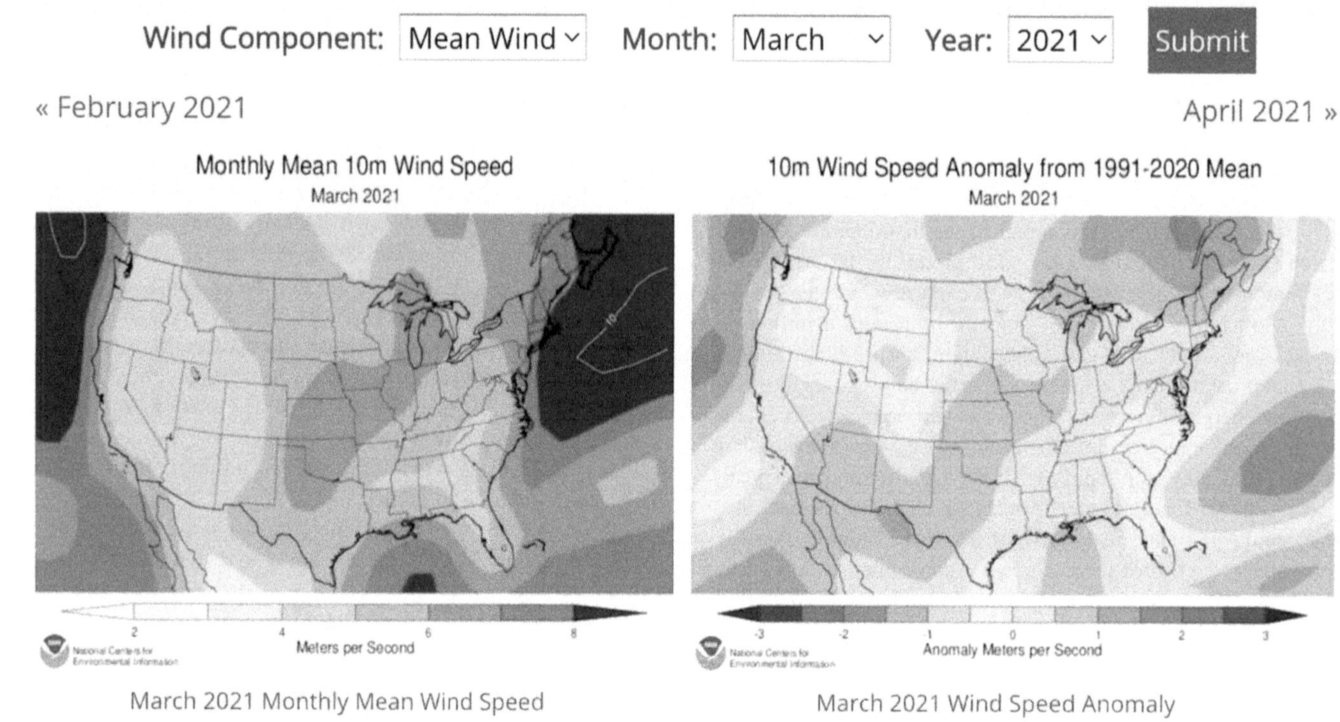

FIG. 1.37: NASA MICRO WIND SPEED MAP (SOURCE: FN [49])

have an important role as an airborne wind testing center with access to offshore wind profiles in controlled airspace NASA's expertise, FAA and National Renewable Energy Laboratory and Department of Energy should explore undergoing this project to harvest power from the wind".[45]

1.7.11.2 NASA WorldWind Efforts[46]

WorldWind is an open source virtual globe application programming interface (API). WorldWind allows developers to quickly and easily create interactive visualizations of 3D globe, map and geographical information. Organizations around the world use WorldWind to monitor weather patterns, visualize cities and terrain, track vehicle movement, analyze geospatial data and educate humanity about the Earth. Because WorldWind is completely open source, extending the API is simple and easy to do. This creates a powerful platform for giving any application the means to express, manipulate and analyze spatial data. WorldWind technology can be incorporated into a wide range of applications, including Windows, Mac, Linux, web, and mobile devices.

WorldWind is different from a 3D globe like Google Earth because it is not an application. Instead, it is an SDK (software development kit) that software engineers can use to build their own applications. WorldWind provides a geographic rendering engine for powering a wide range of projects, from satellite tracking systems to flight simulators. With WorldWind handling the hard work of visualizing geographic data (generating terrain from elevation models, selecting and displaying images from imagery servers), software engineers are free to focus on solving the problems specific to their own domains and more easily building whatever geospatial applications they choose, such as wind availability over states, counties, residential blocks or even over individual building lots. To get right into the code, check out the Web WorldWind GitHub repository. WorldWind's source code, issue tracking and releases are available on GitHub and managed by the NASA WorldWind development team.[47]

NASA World Wind is a graphically rich 3D virtual globe for use on desktop computers running Windows. It combines NASA imagery generated from satellites that have produced Blue Marble, Landsat 7, SRTM, MODIS and more. NASA renews commitment to WorldWind from NASA for continued development and support with continuous release updates to the WorldWind and make it accessible for all Users. The first layer is a locally sourced, very low resolution version of NASA's Blue Marble, which it may be used as a 'fallback' if Web WorldWind is not able to request online imagery layers, such as the second one (Landsat imagery with higher resolution). WorldWind allows developers to quickly and easily create interactive visualizations of 3D globe, map and geographical information, wind patterns around the world to monitor and visualize cities, terrains and even track vehicle movement. World Wind comes with data you can zoom into. After a certain point, World Wind will begin to download more images automatically. The servers hosting this information are very busy and may be unresponsive until we get the necessary infrastructure. NASA World Wind leverages Microsoft .NET technology for rapid development and to easily access[48]

---

[45]Jim Hodges, The Researcher News, NASA angley Research Center, https://www.nasa.gov/topics/technology/features/capturingwind.html_8-25-21
[46]https://worldwind.arc.nasa.gov/_Aug. 12, 2020
[47]https://worldwind.arc.nasa.gov/web/get-started _8-27-21
[48]https://worldwind.arc.nasa.gov/ _8-28-21
[49]https://www.climate.gov/maps-data/dataset/average-wind-speeds-map-viewer _8-28-21

1.7.11.3 Monthly Mean Wind and 10m Wind Speed[50]

Figure 1.27 shows maps of average monthly wind speed and direction for the contiguous United States from 1950 to the present. Maps show the average (mean) wind speed as well as two components of wind direction: U-wind represents the east-west component of wind and V-wind represents the north-south component. For each month and parameter, the tool shows the climatological mean wind (average over the previous three decades), observed winds, and wind anomaly (how much faster or slower wind blew compared to the long-term average for that month). Where did these data come from? The National Centers for Environmental Prediction (NCEP) and the National Center for Atmospheric Research (NCAR) used a computer model of Earth's atmosphere and a database of atmospheric pressure measurements from around the world to produce a picture of how Earth's atmosphere has moved since 1948. This dataset is called the NCEP/NCAR Reanalysis. Knowing the difference in atmospheric pressure from one location to another enabled scientists to program the model to compute wind speeds. Colors on these maps indicate the estimate monthly average wind speeds calculated from the Reanalysis data.[51]

1.7.12 Conclusions

As a follow-up of original chapter 6 by Viterna covering NASA's contributions of Wind Energy in ASME Energy and Power Generation Handbook [1], the above excerpts of NASA's contribution for wind energy since 2010 had been captured. Viterna is a NASA Researcher whose innovation advanced the entire wind energy industry. Despite the advances made by the NASA-led program, there were still significant challenges. Wind turbines, especially in high-wind conditions, can routinely stall, limiting the ability of the turbine to produce electricity. The inability to properly predict stall behavior led to inefficient designs and costly turbine failures. "A Pair of Surprises" were what Viterna developed in a model that took into account three-dimensional effects and predicted stall behavior with far greater accuracy than previous methods, improving wind turbine performance. The model was not well accepted

FIG. 1.38: BY SHAVING THE ICING RESEARCH TUNNEL'S WOODEN FAN BLADES IN ACCORDANCE WITH THE RESULTS OF LARRY VITERNA'S MODEL, GLENN ENGINEERS BOOSTED THE TUNNEL'S TOP WIND SPEED BY OVER 130 MILES PER HOUR (SOURCE:[52])

FIG. 1.39: FOR OVER 20 YEARS, NASA'S 4-MEGAWATT WTS-4 WIND TURBINE HELD THE WORLD RECORD FOR MAXIMUM POWER OUTPUT (SOURCE: [53])

FIG. 1.40: WIND TURBINES LIKE THIS AEROSTAR MODEL ARE STALL REGULATED, A FEATURE ENABLED BY THE VITERNA MODEL. (SOURCE: [54])

[50] https://www.climate.gov/maps-data/dataset/average-wind-speeds-map-viewer _8-28-21
[51] https://www.climate.gov/maps-data/dataset/average-wind-speeds-map-viewer_8-31-21
[52] https://www.nasa.gov/offices/oct/home/tech_life_viterna.html_8-28-21
[53] https://www.nasa.gov/offices/oct/home/tech_life_viterna.html_8-28-21
[54] https://www.nasa.gov/offices/oct/home/tech_life_viterna.html_8-28-21

by colleagues in the wind energy field, Viterna remembers, see figure 1.38. "I almost got laughed off the stage when I presented it," he says. Viterna, however, continued to employ the model for NASA's purposes, even using it to improve Glenn's Icing Research Tunnel, designed to study the effects of ice buildup on aircraft. During a nighttime test, the wind tunnel's pressure release door flew open, sending 400-mph winds ripping through the adjoining work area. This surprized Viterna's coworkers for inventing his model. He discovered his initially criticized model had, within a decade of its creation, quietly became the established method of modeling the performance of wind turbine airfoils in stall conditions. "A Method for Success" was what is in the "Viterna Method" which is now routinely used by research institutions and turbine manufacturers alike in the design and certification of modern wind turbines. "The model makes it possible for manufacturers to predict and characterize the performance of airfoils for these turbines," says Paul Gay of turbine developer Aerostar Inc. The method is even used for finding applications beyond horizontal wind turbines; Viterna' once-mocked model is now employed for vertical wind turbines, wind tunnels, and underwater turbines that make use of tidal energy to produce power. The Agency can also expect to see a resurgence of its early work, predicts Sandy Butterfield, wind program chief engineer at the National Wind Technology Center. "As the industry moves forward and becomes more competitive, you will see efforts to explore lower cost, more structurally efficient machines, and that's when people will begin to capitalize again on that good work NASA did in the early 1970s and 1980s," he says. That is good news for the advancement of alternative energy, and for the dedicated researchers at Glenn, so long as this does not involve another office destroyed. Information about this NASA spinoff is in the original article from *"Spinoff* 2009",[55] see the three figures 1.39 and 1.40 which are attributed to vitena.[56]

1.7.13 Additional Bibliographical References

[1] Energy and Power Generation Handbook Established and Emerging Technologies, edited by K. R. Rao, ASME Press, New York, 2011.

[2] Wind Energy for Power Generation Meeting the Challenge of Practical Implementation by K. R. Rao, Volumes 1 & 2, This Springer book is published by Springer International Publishing AG part of Springer Nature, Gewerbestrasse 11, 6330 Cham, Switzerland.

[55] http://www.sti.nasa.gov/tto/Spinoff2009/_8-28-21

[56] https://www.nasa.gov/vision/earth/technologies/wind_turbines.html_8-28-21

CHAPTER

2

Scope of Wind Energy Generation Technologies

Jose Zayas and K. R. Rao[1]

The author wishes to acknowledge many figures in this chapter that were provided by Sandia National Laboratories, and which are used here with permission.

2.1 INTRODUCTION: WIND ENERGY TREND AND CURRENT STATUS

The energy from the wind has been harnessed since early recorded history all across the world. There are proofs that wind energy propelled boats along the Nile River around 5000 B.C. The use of wind to provide mechanical power came somewhat later in time — by 200 B.C. simple windmills started pumping water in China, and vertical-axis windmills with woven reed sails were grinding grain in the Middle East. The Europeans got the idea of using wind power from the Persians who introduced it to the Roman Empire by 250 A.D. By the 11th century, a strong focus on technical improvements enabled wind power to be leveraged by the people in the Middle East extensively for food production. Returning merchants and crusaders carried this idea back to Europe where the Dutch refined the windmill and adapted it for draining lakes and marshes in 1300s.

In the late 19th century settlers in America began using windmills to pump water for farms and ranches, and later, to generate electricity for homes and industry applications. Although the industrial revolution influenced the propagation of wind energy, larger wind turbines generating electricity continued to appear. The first one was built in Scotland in 1887 by prof. James Blyth from Glasgow. Blyth's 33-ft. tall, cloth-sailed wind turbine was installed in the garden of his holiday home and was used to charge accumulators that powered the lights, thus making it the first house in the world to have its wind power supplied electricity. At the same time across the Atlantic, in Cleveland, Ohio, a larger and heavily engineered machine was constructed in 1888 by Charles F. Brush. His wind turbine had a rotor 17 m in diameter and was mounted on an 18-m tower. Although relatively large, the machine was only rated at 12 kW. The connected dynamo had the ability to charge a bank of batteries or to operate up to 100 incandescent light bulbs, three arc lamps, and various motors in Brush's laboratory. The machine was decomissioned soon after the turn of the century. In the 1940s, the largest wind turbine of the time began operating on a Vermont hilltop known as Grandpa's Knob. This turbine, rated at 1.25 MW fed electric power to the local utility network for several months during World War II.

In Denmark, wind power has played an important role since the first quarter of the 20th century, partly because of Poul la Cour who constructed wind turbines. In 1956, a 24-m diameter wind turbine had been installed at Gedser, where it ran until 1967. This was a three-bladed, horizontal-axis, upwind, stall-regulated turbine similar to those used through the 1980s and into the 1990s for commercial wind energy development, see Fig. 2.1. The popularity of using the wind energy has always fluctuated with the price of fossil fuels. When fuel prices fell in the late 1940s, interest in wind

FIG. 2.1: EARLY WIND FARM IN TEHACHAPI, CA (SOURCE: WEBCOIST.COM)

[1] Jose Zayas was the original author of this chapter that appeared as Chapter 7 in the "Energy and Power Generation Handbook - Established and Emerging Technologies", published in ASME in 2011. This chapter has been updated by Editor K. R. Rao in Section 2.7. K.R. Rao has provided relevant write-ups from on-line sources in the public domain covering "Scope of Wind Energy Generation Technologies since 2009 to date".

FIG. 2.2: NASA MOD-0 WIND TURBINE

turbines decreased, but when the price of oil skyrocketed in the 1970s, so did worldwide interest in wind turbine generator

The sudden increase in the price of oil stimulated a number of substantial government-funded programs of research, development, and demonstration. In the USA, this led to the construction of a series of prototype turbines starting with the 38-m diameter 100-kW Mod-0 in 1975 (Fig. 2.2) and culminating in the 97.5-m diameter 2.5-MW Mod-5B in 1987.

Similar programs were pursued in the UK, Germany, and Sweden. There was considerable uncertainty as to which architecture might prove most cost-effective, and several innovative concepts were investigated at full scale. In Canada, a 4-MW vertical-axis Darrieus wind turbine was constructed and this concept was also investigated by one of the Department of Energy's (DOE) National labs, Sandia National Laboratories (Sandia). The 34-m diameter Sandia Vertical Axis Test Bed was rated at 500 kW and was tested at the USDA-ARS site in Bushland, TX, see Fig. 2.3.

In the UK, an alternative vertical-axis design using straight blades to give an 'H' type rotor was proposed by Dr Peter Musgrove and a 500-kW prototype was constructed. In 1981, an innovative horizontal-axis 3 MW wind turbine was built and tested in the USA. This machine used a hydraulic transmission and, as an alternative to a yaw drive, the entire structure was orientated into the wind. The best choice for the number of blades remained unclear for some time and the industry and research entities experimented with large turbines constructed with one, two, or three blades, eventually converging with three.

Since the early 1980s through today, wind farms and wind power plants have been built throughout the country, and now wind energy appears to be the world's fastest-growing energy source that will power our industry as well as homes with clean, renewable electricity, see Fig. 2.4.

2.2 SANDIA'S HISTORY IN WIND ENERGY

Sandia National Laboratories' (Sandia) roots lie in World War II's Manhattan Project and its history reflects the changing national security needs of postwar America. Sandia's original emphasis on ordnance engineering — turning the nuclear physics packages created by Los Alamos and Lawrence Livermore National Laboratories into deployable weapons — expanded into new areas as national security requirements changed. In addition to ensuring the safety and reliability of the stockpile, Sandia applied the expertise it acquired in weapons work to a variety of related areas, such as energy research, supercomputing, treaty verification, and nonproliferation.

That expertise both in terms of capabilities and facilities was applied to wind energy during the mid-1970s, when the price of oil rose to unprecedented levels, and the nation began a commitment in identifying alternative, clean, and affordable energy generation. For the last 35 years, the laboratory has been committed to this mission and has contributed key technology advancements targeted at reducing the cost of delivered wind energy, while improving the reliability and efficiency of wind system. Historical contributions are captured below and are represented in time by the SNL Vertical Axis Wind Turbine (VAWT) Program, Rotor Innovations, Material and Manufacturing Program, to today's diverse wind research portfolio structured to meet the industry's needs and develop the next generation of components that will continue to improve the efficiency, reliability, and cost effectiveness of wind turbines.

FIG. 2.3: SNL 34-M VAWT

FIG. 2.4: MODERN WIND FARM IN NEW MEXICO (GE 1.5 MW)

2.2.1 Sandia's VAWT Program

2.2.1.1 History: Transition to The Modern Vertical Axis Wind Turbine

French inventor Georges Jean Marie Darrieus filed the first patent for a modern type of vertical axis wind turbine (VAWT) in France in 1925, then in the U.S. in 1931. His idea received little attention at that time, so little in fact that two Canadian researchers re-invented his concept in the late 1960s for the National Aeronautics Establishment of Canada without knowing Darrieus' patent. They later learned of the French inventor, and today's VAWT is known as Darrieus-type wind turbine.

In 1973, the Atomic Energy Commission, a predecessor to the current DOE, asked Sandia National Laboratories, a national laboratory devoted to engineering research and development, to investigate and develop alternative energy sources. Using their extensive experience in aerodynamics and structural dynamics from years of work with delivery systems for weapons, Sandia's engineers began to look into the feasibility of developing an efficient wind turbine that the industry could manufacture. During this time, the Canadians shared their re-invention with Sandia, and interest in the VAWT concept began in earnest.

2.2.1.2 R&D Beginning: From Desktop to Rooftop

The first Darrieus-type VAWT in America was actually only 12 in. tall and was constructed on top of an engineer's desk. To demonstrate that the VAWT concept worked, Sandia's engineers used a fan to create wind for the miniature turbine and a blackboard to perform their calculations — using these simple means, they converted non-believers.

Darrieus' concept appeals to engineers because it works on the principle of aerodynamic lift. Lift is what keeps an airplane in the sky — the wind actually pulls the blades along. In contrast, the traditional Holland-type windmill operates on the principle of drag, meaning the wind has to push a manmade barrier, such as a blade. Modern vertical- and horizontal-axis wind machines both use lift, which makes them more efficient compared with traditional windmills.

Sandia's original modal VAWT combined Darrieus' design with another concept for a wind turbine, called Savonius after its inventor, a Swede. Because the Darrieus VAWT could not start itself, some researchers thought it might be at a disadvantage. The turbine Savonius design used some lift, but its theoretical advantage was in using cups or vanes to trap thc wind — employing the principle of drag — and so it was able to catch the wind and start itself in motion. However, the Savonius element was soon abandoned because of the blade size required for it to work: instead, engineers opted to start up the turbine manually and to use the Darrieus design.

To test the aerodynamics of the turbine, a larger working model was built on the rooftop of the main administration building at Sandia. This model measured 5 m across the outer edges of the two bowed blades, each constructed out of a shank of steel covered by foam and fiberglass, then molded into the characteristic teardrop airfoil shape commonly used in the aircraft industry. Putting the test turbine in motion was no easy feat — researchers patiently waited for the wind to begin to blow, strapped themselves onto the roof of the building and spun the blades by hand. Whenever a thunderstorm, with its accompanying high winds, would blow into Albuquerque — night or day — the engineers rushed to the laboratory, climbed to the roof, and began turning the blades.

Starting the blades was not the only problem, however. To sustain their rotation, the blades had to be turning at least two to four times faster than the wind so that lift could work properly. At this early stage in the turbine's development, the blades required certain wind conditions, which did not occur on a daily basis. In the spring of 1974, however, the winds cooperated, the VAWT blades rotated smoothly on their own, and the demonstration phase began.

Another factor engineers had to consider was that under certain conditions, wind turbines can literally spin apart; they go into what is known as a runaway condition. Researchers knew that if their VAWT had a load to power — a generator for example — the load would act as a brake against runaway, but at that time, there was no load in the test system. For this reason, they built a disc brake consisted of a commercially available automobile disk caliper clamped onto a machined disk.

2.2.1.3 Tech Transfer: Moving to Industry

Some two years after constructing the rooftop model, Sandia built a second, larger wind turbine- this one on the ground. With a blade span of 17 m, the turbine's main purpose was to show that it could compete in cost with the more traditional horizontal-axis machines. An economic study from 1976 supported the research: vertical and horizontal axis wind turbines, or HAWTs, should indeed be comparable in performance and price if some improvements were made to the VAWT's design.

The 1976 study suggested these improvements. First, two blades would be better — the earliest design had three. Next, slimming down the shape of the turbine would improve its design, and the turbine's efficiency could be improved with better airfoil shapes. Finally, the study also found that a blade span of at least 17 m was best. During its first year of operation, 1976, this experimental machine was the largest VAWT in existence, and its performance compared favorably with that of a horizontal-axis machine.

The first VAWT blades were expensive because they were made of aluminum, fiberglass, and a man-made honeycomb-like material, all of which had to be carefully fitted together.

Alcoa Industries was interested in reducing manufacturing costs of VAWT blades and in the mid-1970s developed an extrusion process in which partially molten bars of aluminum are forced into a die cut into the shape of airfoil. The aluminum is under such pressure that it melts and flows through the die, where it cools and resolidifies. The result is a uniformly manufactured airfoil in the required shape. The process dramatically reduces the cost to manufacture VAWT blades, and it continues to be used today.

Alcoa won a DOE contract a few years later, in 1979, to construct four low cost VAWTs, each to have a 17-m blade span and to deliver 100 kW of electricity. Construction lasted from January 1980 until March 1981; however, because of DOE budget constraints, only three of the units were installed. Each of the sites was chosen for a specific application: Bushland, Texas, to demonstrate an agricultural application, Rocky Flats, Colorado, to confirm structural and performance tests, and Martha's Vineyard, Massachusetts, to demonstrate the VAWT's applicability to the utility grid.

Their successful operation — more than 10,000 hours for the Bushland machine — convinced two companies to commercialize this design. VAWTPOWER and FloWind each manufactured VAWTs for use in California, where weather conditions favor using the wind's power for electricity. The result was more than 500 VAWTs were operating in California and producing electricity by the mid-1980s (Fig. 2.5).

FIG. 2.5: FLOWIND COMMERCIAL 19-M VAWT COMMERCIALIZED IN COOPERATION WITH SANDIA

2.2.1.4 Using the Information for a New, Larger Machine

Because the 17-m VAWTs showed such success, the DOE Wind Program directed Sandia to develop an expanded research machine. System studies indicated 34 m was a good size for the blade diameter to test the new airfoils, and the size made economic sense. In cooperation with the Department of Agriculture, the culmination of planning began in 1984.

Called simply the 34-m Test Bed, this VAWT is a research tool for testing and developing advanced concepts. It can produce 500 kW of electricity, more than half of the local community's normal power needs, but its purpose is research, not power production. For this reason, instruments are strategically mounted on the VAWT to measure its parameters, especially stress on the blades. Weather conditions that affect the VAWT's performance are also recorded, including the wind direction and speed, ambient temperature, and barometric pressure.

A special feature of the Test Bed is that it can run over a continuously variable range of rotor speeds, from 25 to 40 rpm, whereas most wind turbines are designed to turn at a constant speed. The large, bowed aluminum blades are made of sections of specially designed airfoils that are bolted together; three different sizes and designs increase efficiency and regulate power through stall.

The work at Sandia and its Test Bed includes validating computer models, testing airfoil designs, and developing various control strategies. The work is part of improving the first-generation design, which has been commercialized in California, as well as developing next-generation VAWTs. Transferring technology from its national laboratories to the commercial sector is a major goal of the DOE, and Sandia's development of the VAWT and its subsequent adoption by industry is a good example of such a program.

2.2.1.5 The Future of VAWT Research

Within the DOE's Office of Energy Efficiency and Renewable Energy (EERE) is the Wind and Waterpower program, which oversees the current federal wind energy program, including wind research and development supported by the national laboratories. The DOE supported Sandia's efforts to develop VAWT technology, which serves as the basis for private industry to develop new generations of VAWTs with greater efficiency and longer life expectancy than any machine produced in the past. To this end, the Department supports its laboratories' forming cooperative research agreements with commercial firms to improve wind turbine designs.

The DOE's program for the vertical axis wind turbine came a long way since Sandia built its 30-cm-tall desktop version, and many of the elements which we see today on utility scale horizontal axis wind turbines (HAWT) were developed during this time.

In the mid-early 1990s, it was apparent that the industry had chosen a new path, and that it would convert primarily to the three-bladed HAWT. There are many reasons why that path was taken; in particular, the pursuit of higher wind resources at higher elevations, but it is difficult to quantify where VAWTs would be today if that decision would have been different.

2.3 SNL'S TRANSITION TO HAWT'S IN THE MID 1990S

Although VAWT technology had proven its feasibility to compete as a viable wind energy architecture, there was a fundamental shift in the early to mid-1990s that ended the investment of utility- scale VAWTs. Additionally, the industry in the U.S. had diminished given an expiration of the production tax credit, see Fig. 2.6. During this period, designers were seeking larger machines that could sweep a larger area and take advantage of the more benign higher velocity wind found at higher altitudes, see Fig. 2.7.

During this time, Sandia began to focus its research activities in HAWT technology and take the capabilities and corecompetencies of the laboratory and apply them synergistically to HAWTs. Although the industry in the U.S. had dwindled, Sandia transitioned and began applying their 20-year wind energy experience to wind rotors. Since that time and continuing today, Sandia has been engaged in developing next-generation blades that are designed to be innovative, low-cost, reliable, and maximize energy capture. Programs in aerodynamics, structural dynamics, materials and manufacturing, and testing and evaluation provided the foundation for the research program

2.3.1 Rotor Innovation

Wind turbine blades are designed to maximize energy capture and survive structurally the stochastic wind input for a 20-year design life. Although these structures appear quite simple from the exterior, there is immense innovation that has been applied over the history that have enabled blades to be efficient, reliable, and cost-effective. In order to maximize the efficiency of the rotor, designers focus on balancing structural requirements and aerodynamic efficiency to maximize the operational coefficient of performance, C_p.

Equation 2.1:

$$\text{Wind Power} = 1/2\, \rho A C_p V^3,$$

where ρ = air density, A = rotor swept area, C_p = coefficient of performance, and V = wind velocity. All utility-scale rotors today are comprised of three lift-based blades, which theoretically have a collective maximum efficiency of 59%, known as the Betz limit [1,2]. Advancements in computational fluid dynamic modeling coupled with airfoil evaluation and testing have enabled operational rotors today to have C_p in the high 40s to low 50s. That is quite a remarkable engineering accomplishment, given the random nature of the wind input and the fact that there is limited control authority in the system, variable speed and pitch.

Structurally, wind turbine blades are driven and designed to survive high fatigue cycles [4], and survive the environment conditions throughout the design life. Given these design constraints, composite materials lends themselves well for this application, and today, fiberglass dominates the market given its low cost and ease of manufacturing. Most wind turbine blades are designed and manufactured in three sections, a high and low pressure skin, and one or two shear webs as the main support member (Fig. 2.8). In order to save weight and prevent large unsuspended panel buckling, the panels are sandwich type structures with a core material, balsa wood or foam.

A large challenge for structural designers is the non-linear relationship as the weight of the blade scales to the third power of the length, see Fig. 2.9.

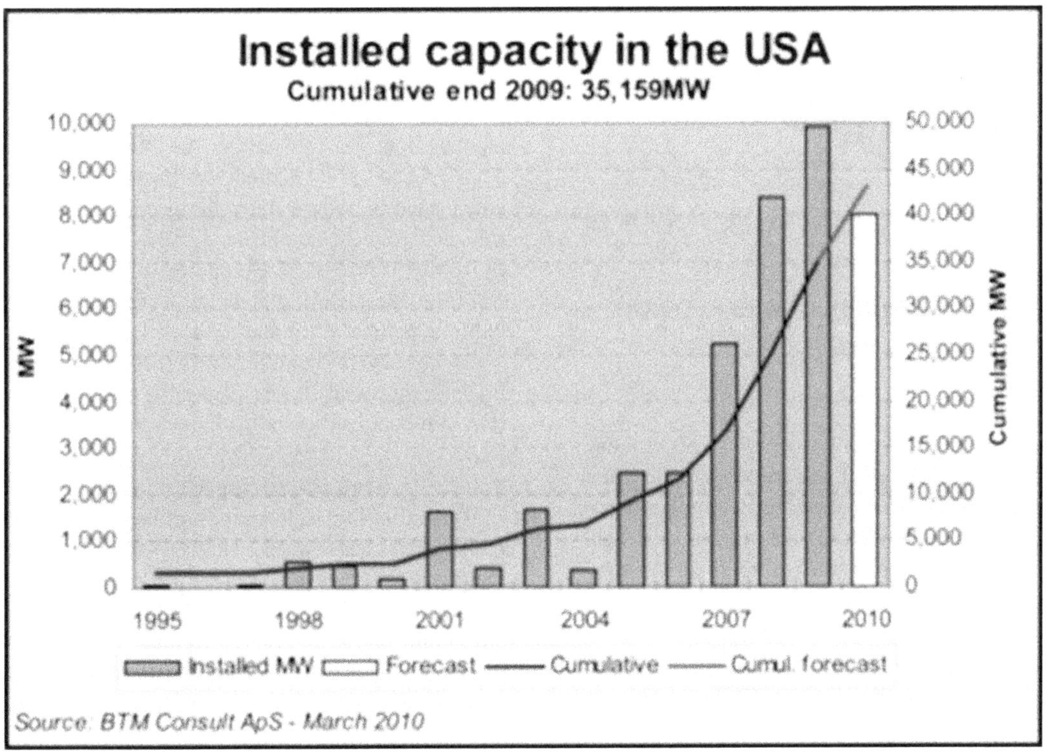

FIG. 2.6: BTM CONSULT U.S. ANNUAL INSTALLED CAPACITY

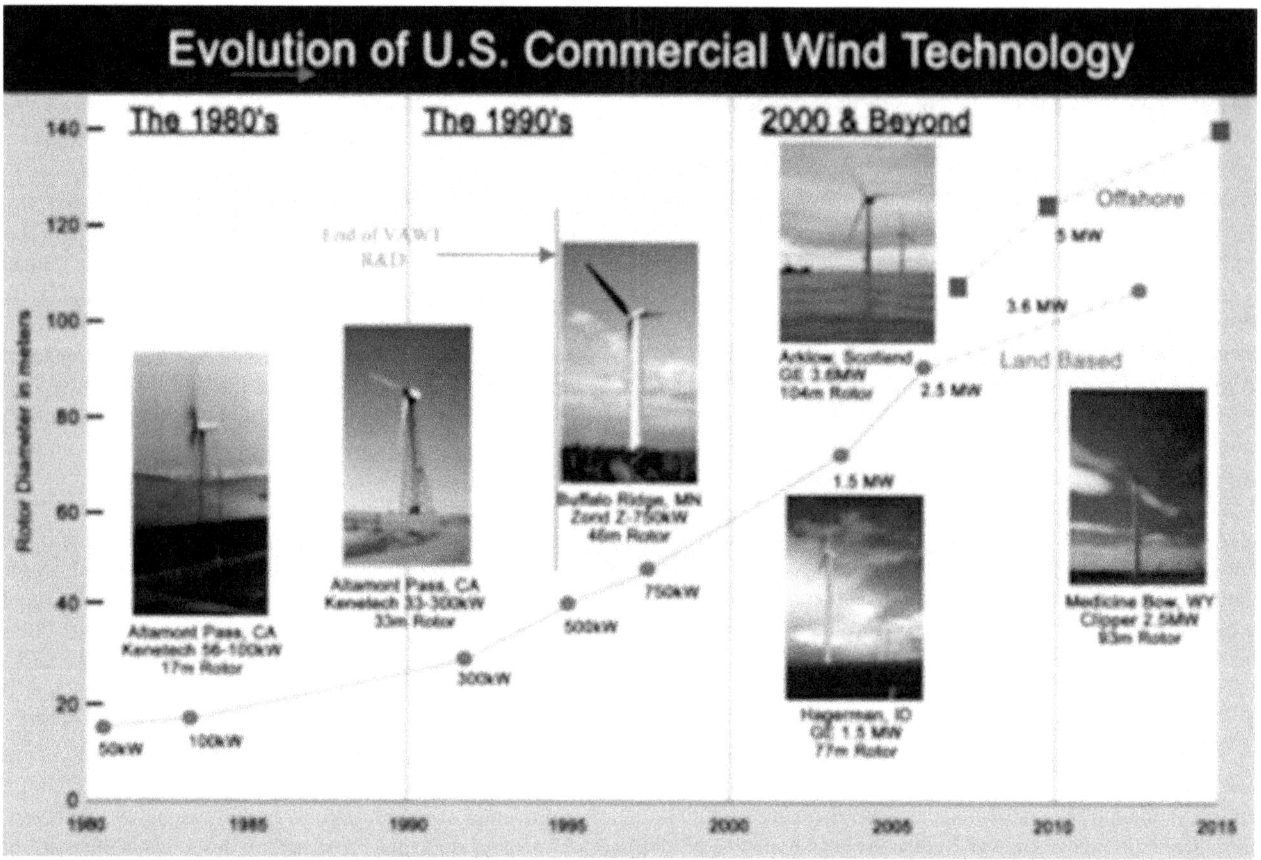

FIG. 2.7: WIND TURBINE EVOLUTION

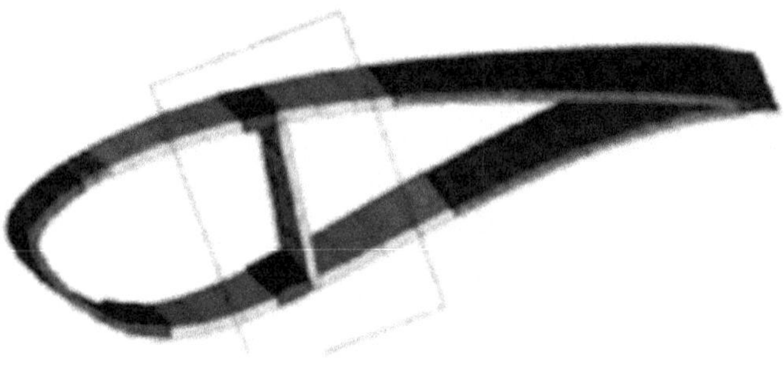

FIG. 2.8: ANSYS FEA WIND BLADE CROSS SECTION

As wind turbine blades have gotten larger (30 to 60 m today) innovative designs and utilization of advanced materials have enabled rotors to scale and remain competitive. A large portion of Sandia's research is targeted at evaluating the utilization of lighter-stronger materials, such as carbon fiber to optimize the structural integrity of the blade.

Over the past decade, Sandia's blade program has developed three-blade designs that have evaluated strategic methods for optimizing structural design, aerodynamics, and weight. All designs have taken into account economics, manufacturing, and performance to validate the next generation of blades for the industry (Fig. 2.10).

As an example, in 2002, Sandia developed a blade design utilizing "flatback" airfoils for the inboard section of the blade to achieve a lighter, stronger blade. Flatback airfoils are generated by opening up the trailing edge of an airfoil uniformly along the camber line, thus preserving the camber of the original airfoil. This process is in distinct contrast to the generation of truncated airfoils, where the trailing edge the airfoil is simply cut off, changing the camber and subsequently degrading the aerodynamic performance. Compared to a thick conventional, sharp trailing-edge airfoil, a flatback airfoil with the same thickness exhibits increased lift and reduced sensitivity to soiling [7].

Today, several manufacturers incorporate carbon fiber in their blade designs and are evaluating the utilization of inboard utilization of inboard structurally efficient flatback type airfoils.

2.3.2 Manufacturing Research

Typical utility-scaled wind turbine blades being manufactured today can range between 30 and 60+ m in length, but given that the majority of the installations are land-based, the range is between 30 and 45+ m (Fig. 2.11). Wind turbine blades pose manufacturing and supply chain challenges given their large size, large amount of raw materials, and significant labor content associated to the various accepted manufacturing processes. Additionally, in order to meet the demand and support large and emerging global markets, some utility scaled turbine manufacturers have their own blade manufacturing, while others have chosen to purchase them from

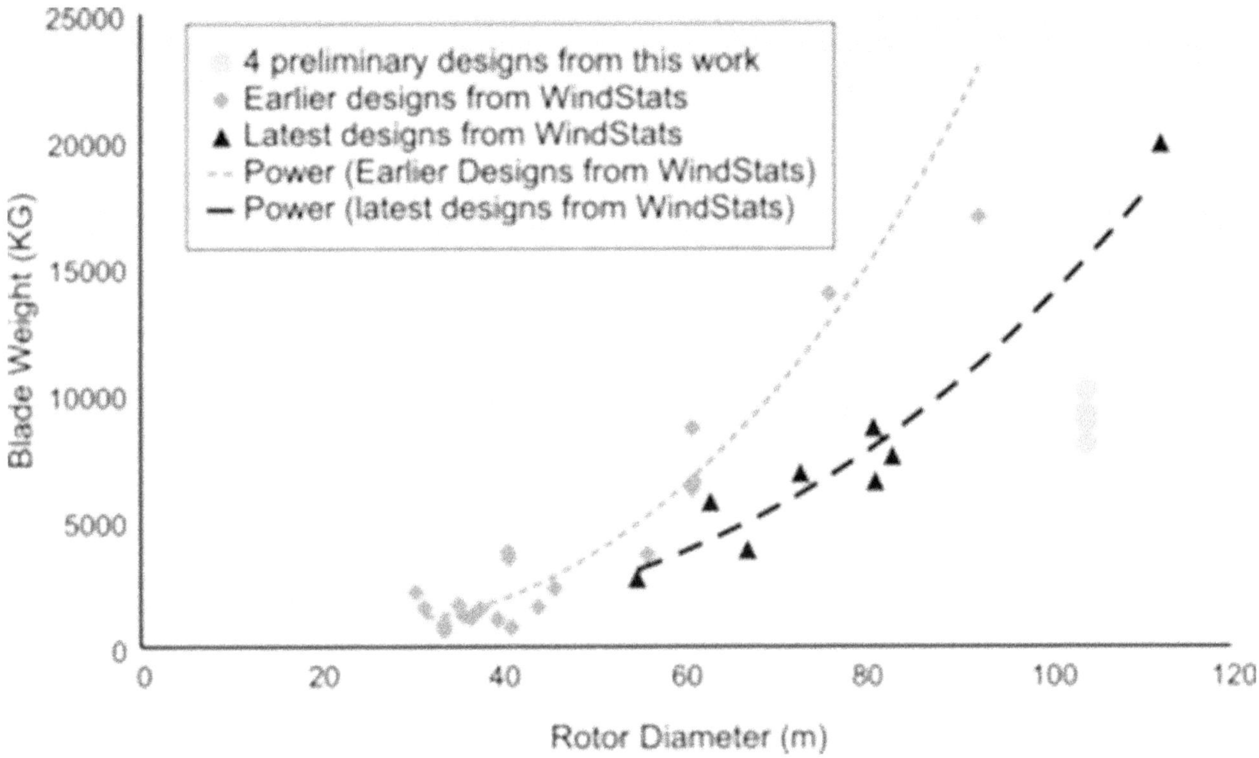

FIG. 2.9: WINDSTATS BLADE WEIGHT-VS-ROTOR DIAMETER

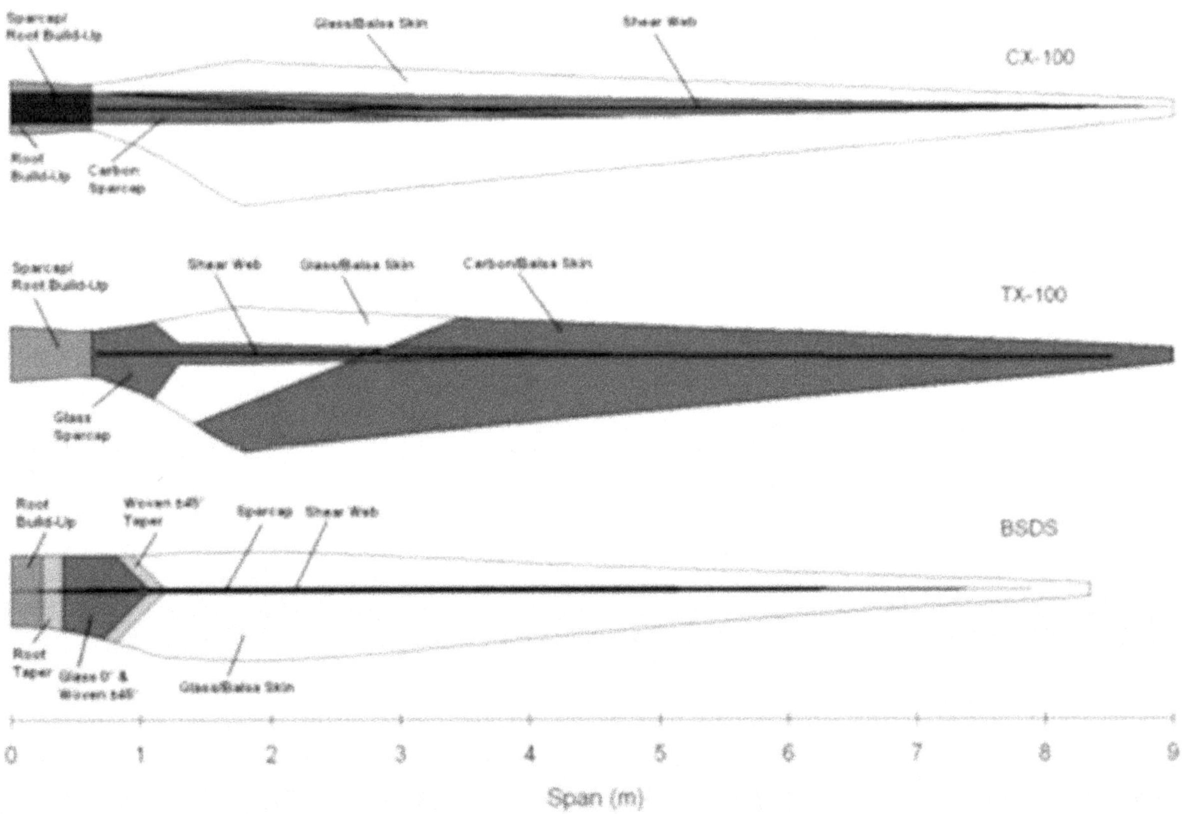

FIG. 2.10: SNL'S CARBON FIBER INNOVATIVE BLADE DESIGNS. TOP TO BOTTOM: CX-100 — CARBON SPAR BLADE, TX-100 — OFF-AXIS CARBON SKINS FOR AEROELASTIC TAILORING, AND BSDS — OPTIMIZED STRUCTURAL/AERO BLADE DESIGN

FIG. 2.11: PICTURE OF UTILITY SCALE BLADE MANUFAC-TURING (COURTESY OF TPI)

component suppliers to displace risk and large capital investment in manufacturing infrastructure.

As an example, focusing on a record year, 2009, where approximately 10,000 MW were installed across the U.S. and assuming an average machines being 1.5 MW in size (~40-m blades), ~20,000 blades were manufactured just to meet the U.S. installations. A typical 40-m blade weighs approximately 12,500 lbs and is composed of fiberglass, some OEMs have carbon fiber on spar cap, core material (balsa wood or foam), and a resin system (epoxy, polyester, or vinylester) and is primarily manufactured through an infusion process. Out of the total weight of a blade, the dry fiberglass can represent 70% of the total weight, the resin 25%, and the rest is the coring material. The raw material supply and delivered quality is crucial to manufacturing a high-quality product that can not only meet the certified requirements, but can survive the industry average design life of 20 years.

In manufacturing, Sandia, through the support from DOE, has embarked on a manufacturing program to address the challenges and opportunities of manufacturing high-quality cost-effective wind blades. The program is multi-disciplinary in nature, where quality, reliability, and cost-effectiveness are the primary metrics for success.

As blade length increases, the associated increase in blade weight places additional loads on both the rotor and the supporting structure. This increase in blade length has also resulted in scaling issues for structural aspects like bond lines, root attachments, and thick laminate infusion. In addition to gravitational loads, wind turbines also experience tens of millions of fatigue cycles during their operational lifetime due to turbulence in the wind, making fatigue resistant materials necessary for design. Wind turbines also often operate in difficult and harsh environments, which necessitate the use of coatings for protection. Finally, since wind must compete with other generation resources, there is a cost constraint on the blades of around \$5 to \$7/lb. These three factors create a uniquely challenging design problem for wind engineers.

To address and ensure quality, the program targets improvement opportunities in robust and lean manufacturing techniques to minimize human errors, given the labor intensiveness in manufacturing, and nondestructive inspection techniques (NDT) to indentify and address issues in the finish product prior to shipment and de- livery. Nondestructive techniques typically used for wind blades, ultrasonic and thermography, provide mixed results and vary in applicability given the complex geometry and internal architecture. Through experience and design, knowledgeable manufacturers inspect critical regions and developed guidelines for acceptable flaws. SNLs manufacturing program evaluates all available applicable NDT techniques to develop a portfolio of options that will minimize false-positive inspection results, which can lead to field problems where cost of repair grows exponentially.

Given expert projections or the results of industry studies, such as the DOE 20% by 2030 scenario where the analysis documents the viability and improvements needed to achieve 20% wind energy by the year 2030, it is clear that a robust, reliable, and high-quality wind blade supply chain is needed for the industry.

2.3.3 Materials Research

Sandia National Laboratories has performed research in the area of wind turbine materials for over 20 years. A primary effort of that work has been a partnership with Montana State University to produce the DOE/SNL/MSU Composite Material Fatigue Database. The database features the results of over 10,000 mechanical tests of wind turbine blade materials and is the largest publically available data set of its kind in the world [3]. The focus of much of this research has been in the area of high-cycle composite fatigue. This research has been broad in focus, with investigations of resins, fibers, resin–fiber interfaces, fabrics, adhesives, and design/manufacturing implementations. Through this research, the turbine OEMs have been able to discover material solutions to challenging design problems, and material suppliers have been able to evaluate their products and fast-track them into the industry.

2.4 MOVING FORWARD: STATE OF THE INDUSTRY

Although the U.S. experienced a large influx of installations during the 1980s, it is not until recent years that wind energy in the U.S. has achieved large market installations and continued market acceptance. Over the past few years, the Federal government has continued to provide a production tax credit (1.8 ¢/kW/h), and it is the combination of large amount available land with adequate resource, renewable portfolio standards, tax credit, renewed market pull for clean energy, and technology viability that has spurred this growth.

As can be seen in Fig. 2.6, the U.S. industry has experienced an exponential growth over the last five years, and many states have chosen to have a significant percentage of wind in their system (Fig. 2.12). Today, the U.S. has the largest installed capacity, but there is significant competition from emerging countries, such as China, which installed over 13 GW in 2009.

Through the third quarter of 2010, the U.S. has approximately 37,000 MW of installed capacity, which approximately represents 2% of our energy consumption [5]. Additionally, the growth of the industry has enabled wind energy to emerge as the leader of the new-generation clean energy portfolio (Fig. 2.13).

The growth and acceptance of the industry has disrupted the installation trends of the more traditional forms of generation, such as natural gas. Over the past three years, wind energy has represented on average 35% of all new-generation installations (Fig. 2.14).

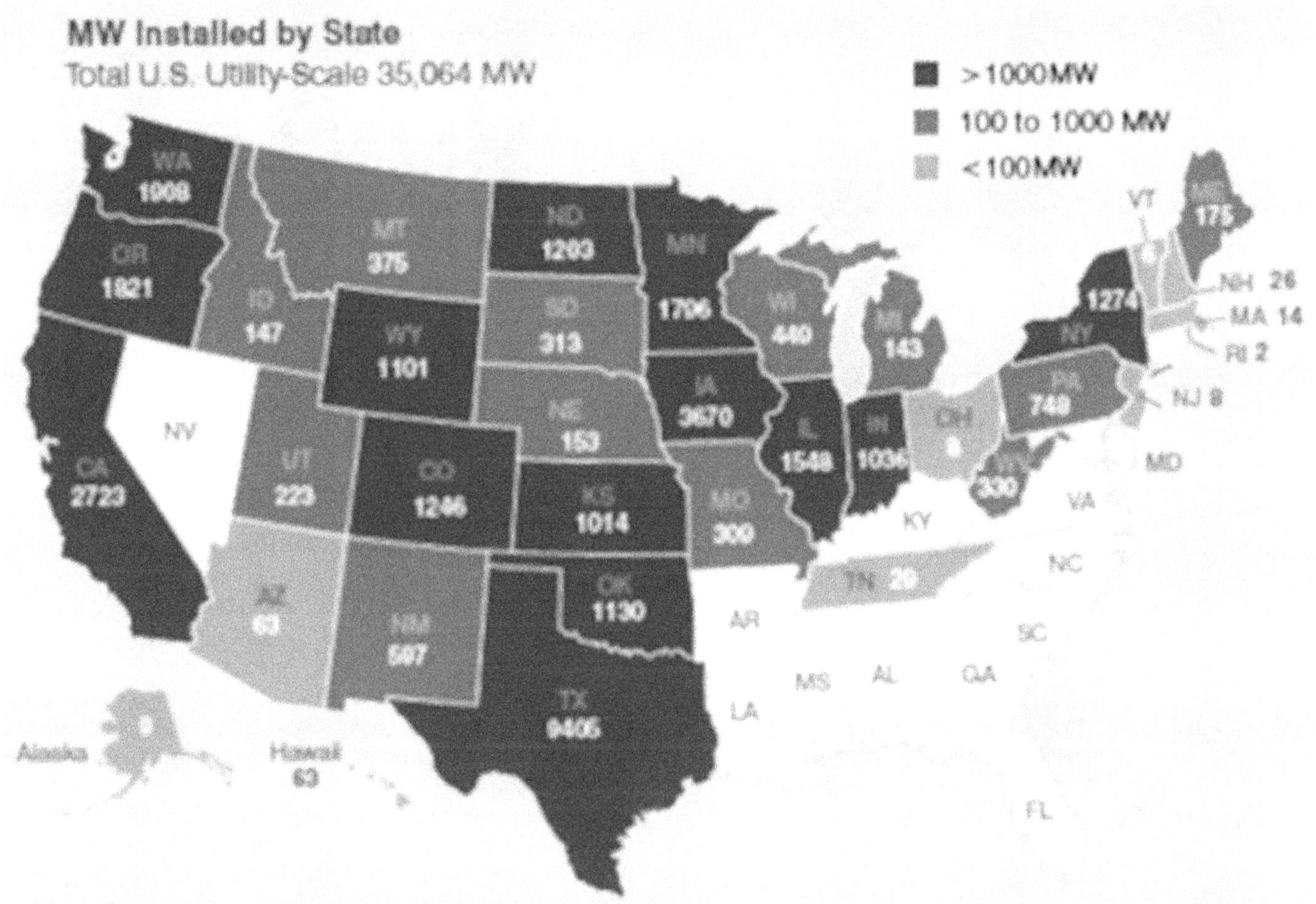

FIG. 2.12: WIND ENERGY INSTALLED BY STATE — END OF 2009

In 2010, overall energy demand is down and combined with low natural gas prices, it is expected to be a low year for wind.

2.4.1 DOEs 20% by 2030 Scenario

In May of 2008, the Department of Energy (DOE) sponsored a National Study to analyze the feasibility of 20% of the U.S. energy by 2030 delivered from wind energy [6]. The 20% by 2030 scenario report outlines the required advancements in technology, as well as improvements in the siting process, the manufacturing scale-up, and the integration needs in order to achieve the goal.

Since being published, the report has provided the wind industry with a target that has become a unifying objective that is both recognized and acknowledged as an achievable goal. The analysis results captured a scenario that predicts the need for 305 GW to be online by 2030 (Fig. 2.15). Although mostly land-based technology, the capacity target does include 54 GW of offshore wind as well. In order to achieve the 305 GW, the analysis projects a scale-up to a steady state ~16 GW annual installation by 2016. As an example of feasibility, in the record year in 2009 ~9.9 GW were installed; showing the ability of the industry to scale to meet the demand.

Unfortunately, 2010 is not continuing the growth of the last few years, and as a matter of fact, it is expected that 2010 will be a difficult year for the industry (Fig. 2.16). This change is not solely due to the deep economic crisis being experienced, it is also a function of lower energy demand, lower-cost energy sources, and available transmission.

The 20% by 2030 report also has serves as key document to drive DOE's activities and investment in research and development. Through a balance program portfolio, Sandia, in coordination with other laboratories, academia, and industry, supports the industry and develops the next generation of technology targeted at improving the efficiency and reliability and promote a larger market acceptance.

2.4.2 Aeroacoustics

As wind turbines continue to be deployed across the nation, the likelihood of wind farms being sited near inhabited areas increases. An important constraint on wind turbine placement arises due to the consideration of wind turbine noise. As a key design metric, the noise generated by a turbine can determine its required setback distance from residences or buildings and depends on local community noise regulations. Noise is typically measured on a logarithmic or decibel scale [9, 10]. As an example, a six-decibel

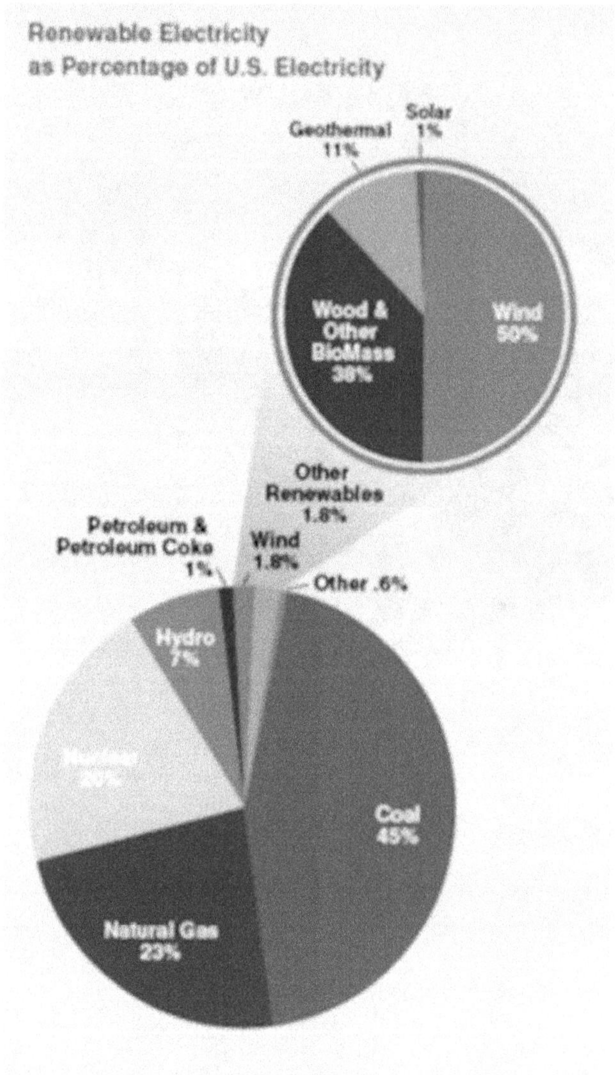

Data Source: Energy Information Administration

FIG. 2.13: EIA U.S. INSTALLED CAPACITY BY GENERATION SOURCE

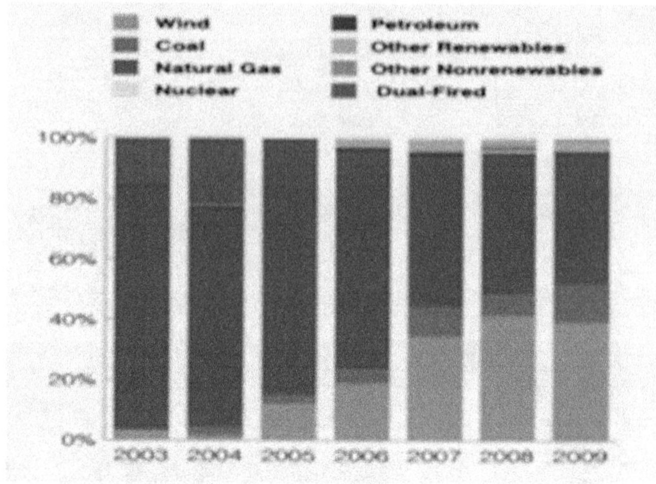

FIG. 2.14: EIA ANNUAL INSTALLED CAPACITY BY GENERATION SOURCE

increase in the noise of a turbine would double the required turbine setback distance; likewise, a six-decibel decrease in noise may allow the turbine to be half as far away. Wind developers seek to place turbines in locations with the optimal wind resource, but as installations encroach populated areas, the noise constraint can prevent the optimal placement and adversely impact the economics of a wind farm.

Noise involves several distinct elements, including the source, the propagation through the atmosphere, and the perception, all of which are relevant to wind turbine acoustics and design. It is important to recognize that not all noise is the same, and that not all noise is perceived in the same way. Tones, or noise at discrete frequencies, tend to be perceived as more bothersome to humans than broadband noise, which is spread over a continuous range of frequencies. Low-frequency noise propagates through the atmosphere more efficiently than high-frequency noise; hence, it can travel over large distances.

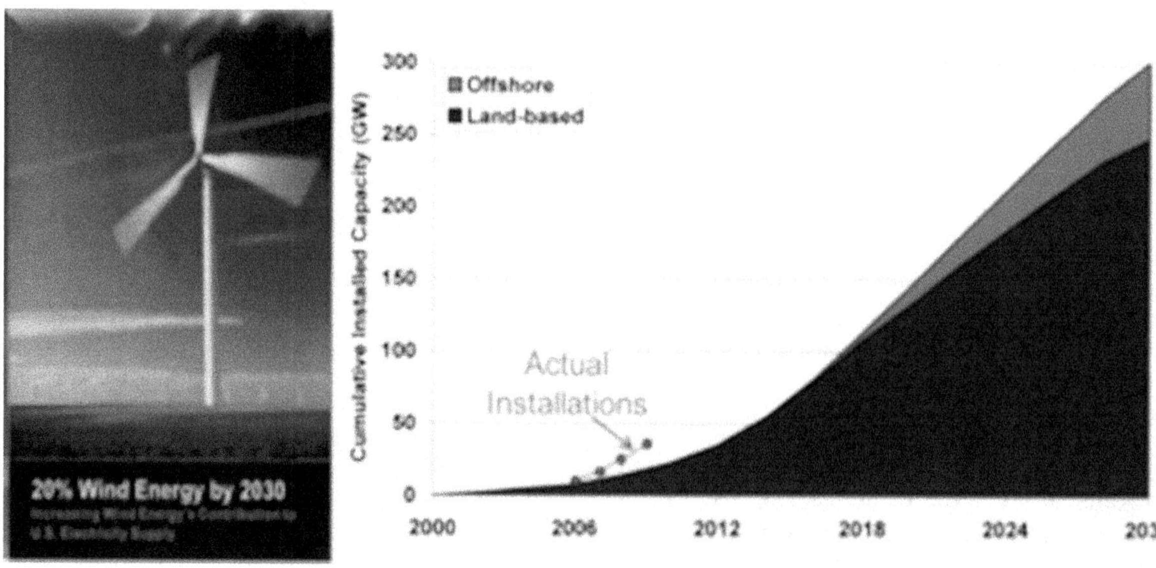

FIG. 2.15: DOE 20% BY 2030 CUMULATIVE TREND

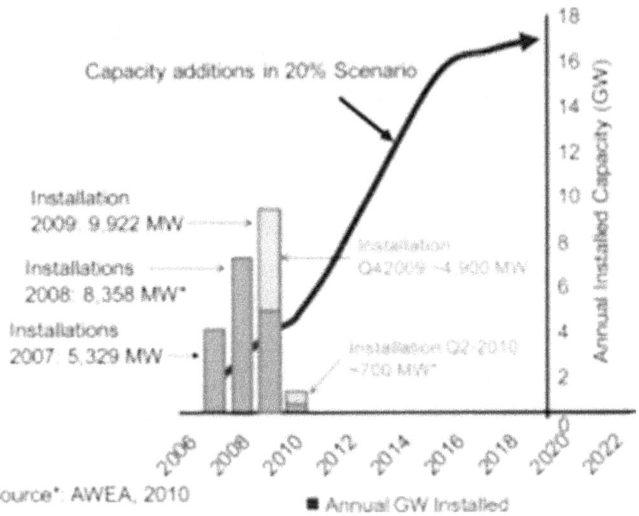

FIG. 2.16: DOE 20% BY 2030 ANNUAL CAPACITY ADDITION AND U.S. ANNUAL WIND INSTALLATIONS

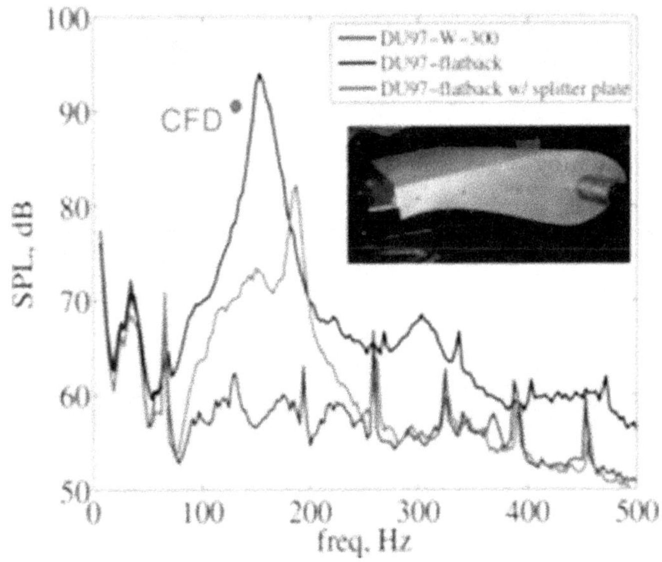

FIG. 2.17: MEASURED SOUND PRESSURE LEVELS (DU97 AIRFOIL — SHARP TE, FLATBACK AND FLATBACK WITH SPLITTER PLATE)

There are two primary sources of noise generated by wind turbines: mechanical noise, and aero-acoustic noise. Mechanical noise involves machinery-generated noise from the gearbox, bearings and generator. This noise can directly radiate from the machinery components and cause vibration in the surrounding structures, such as the nacelle and tower (called "structure-borne" noise). Mechanical noise often occurs at well-defined tones associated with the rotational frequencies of the machinery components, such as gears and individual gear teeth. Unlike aero-acoustic noise, mechanical noise sources are often easier to isolate since the source and location is well known and can lend themselves to effective mitigation through the use of insulating material in the nacelle and vibration isolation to prevent structure-borne noise.

Aero-acoustic noise is the noise created due to the motion of the rotating turbine blades relative to the surrounding air. Aero-acoustic noise is the result of several complex fluid dynamical phenomena that occur on a wind turbine blade and is usually broadband in nature, meaning that the noise signal is spread over a continuous range of frequencies. A particularly important aero-acoustic noise source is trailing edge noise, which results from the flow of air past the aft or trailing edge of a blade. For an observer on the ground near a turbine, this noise is modulated by the passage of the rotating blades, resulting in a characteristic "swoosh, swoosh" sound. Trailing edge noise imposes a rather strict design constraint on the tip speed of wind turbine rotors, limiting how fast the turbine rotor can rotate (Fig 2.17).

A key scientific challenge involves the fact that the precise relationship between the shape of a blade design and its aero-acoustic noise signature is not well understood, which makes blade designers apprehensive to large changes that could result in a higher acoustic signature. This constraint tends to limit innovation in blade design.

Key acoustic research being conducted at national labs, universities, and industry is targeted at developing the underpinning technology and analytical tools to better understand the phenomena (Fig. 2.18). Once successful, we can expect that not only will wind turbines be able to be sited closer to populated areas but also the overall efficiency of wind systems will increase.

2.4.3 Aerodynamics

The discipline of aerodynamics plays a critical role in wind turbine design for two main reasons: first, aerodynamic blade lift forces are responsible for creating the torque on the rotor necessary to drive the generator; and second, aerodynamic forces create the primary loads that drive the structural design of the turbine. The aerodynamic forces are the result of complex fluid-dynamical processes occurring in the wind, over the blades themselves, and in the wake of the turbine. The fluid dynamical system surrounding a turbine is multi-scale in nature, creating a significant modeling challenge for aerodynamicists (Fig. 2.19). The scales range from rotor-scale fluctuations in the atmospheric boundary layer down to micron-scale turbulent fluctuations in the boundary layer surrounding the rotor blades.

The turbine inflow, or oncoming wind seen by the rotor, is not uniform; in fact, it varies both in time and space in a stochastic fashion that is typical of turbulent flows. This unsteady, stochastic operating environment is especially important when considering both fatigue and extreme loads on the turbine rotor, tower, and drive-train. The aerodynamic response to the wind inflow is determined by the blade aerodynamic characteristics.

Modern HAWT blades are comprised of a continuous sequence of two-dimensional airfoil sections defined along the span of the blade. Blades are typically designed with non-uniform twist, taper, and chord distributions along the span in order to maximize energy capture. Power-generating torque is generated by lift forces, while drag acts to decrease torque and power. Thus, airfoil sections with high lift-to-drag ratios are critical for good turbine performance. However, other considerations also govern the selection and design of airfoils for blades, such as structural constraints on the thickness of the airfoils and performance of the airfoils under soiled conditions.

Under design conditions and steady, uniform inflow, air flows smoothly over the wind turbine blades, leading to predictable performance and loads and optimal energy capture. However, various phenomena, such as sudden wind gusts or insect and dirt

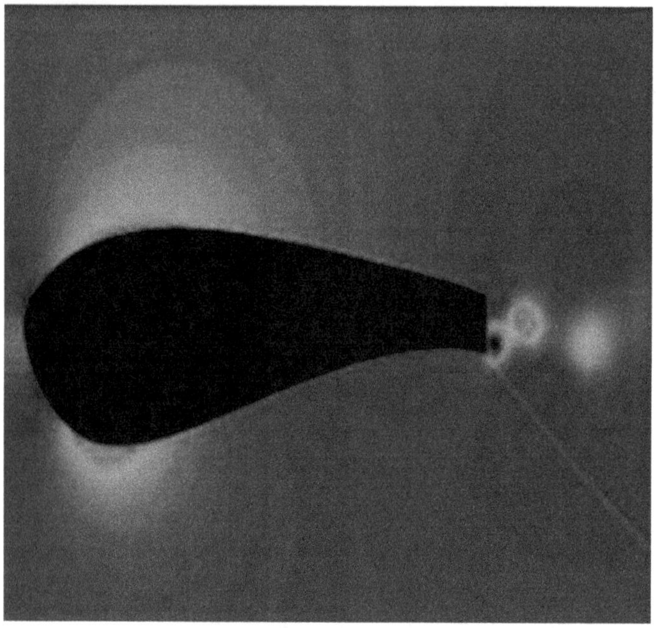

FIG. 2.18: FLATBACK CFD ANALYSIS: NOTE THE ASYM-METRIX VORTEX SHED OF THE TRAILING EDGE WHICH RESULTS IN HIGH ACOUSTIC EMISSION

build-up on the blade leading edge, can lead to dramatic changes in the flow over the blades that have important consequences for both loads and energy capture. Parts of the blade may enter stall, or an un- steady form of airfoil stall known as dynamic stall. In either case, the flow detaches, or "separates," from part of the blade, leading to large changes in lift and drag forces. Separated flow regions may become three-dimensional, such that near the blade surface air moves radially along the blade instead of flowing from the leading to trailing edge. These complex phenomena are difficult to measure experimentally at full scale, as well as difficult to model (Fig. 2.20). However, modeling approaches based on Computational Fluid Dynamics (CFD) offer the ability to model these off-design aerodynamic phenomena important to wind turbine design.

Wind turbine rotors slow down the wind as they extract kinetic energy from it, resulting in a region of low momentum air downwind of the turbine known as the wake. The flow in the wake is also quite complex and three-dimensional. Flow perturbations caused by fluid motion in the wake are felt by the blades, affecting rotor performance and loads. Thus, any engineering model for a wind turbine rotor must consider both the blade forces as well as the behavior of the wake. Wakes tend to persist for large distances downwind from turbines (Fig. 2.21). They interact with and modify the inflow seen by turbines placed downwind of turbines in a wind farm array. This can have important implications for energy capture and reliability of large wind farms.

2.4.4 Sensors and Condition Health Monitoring

Although most machines today share the same architecture as the older machines, advancements in sensors, controls, and power electronics have provided opportunities for designers to develop algorithms and operational strategies that continually attempt to maximize energy capture, load management, and reliability.

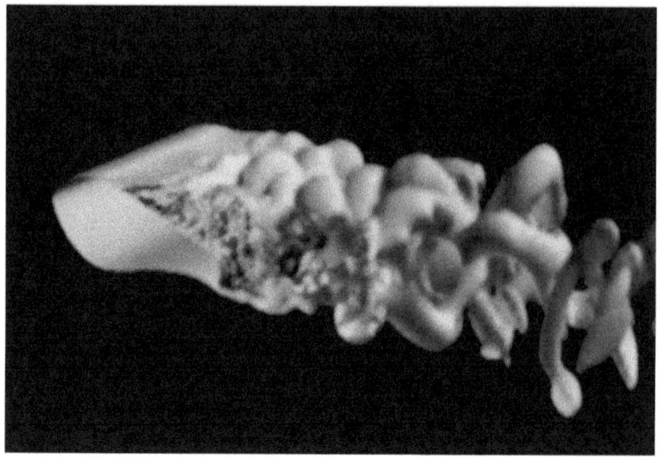

FIG. 2.19: CFD-BASED RESULTS ON FLATBACK AIRFOIL

A typical turbine today relies on hundreds of sensors for their effective operation and survivability. The role of those sensors vary from control observers (i.e., wind speed, high-speed shaft RPM, pitch position, etc.), fault detections (generator over temp, cable twist, etc.), to conditional health monitoring (gearbox lubricant quality, vibration levels, etc.). The effective operation of these sensor systems or networks is crucial for the safe operation of the machine and must operate reliably throughout the design life, which is typically 20 years. This strategy is increasingly important for offshore deployed systems, as machines are more complicated and have limitations in access when compared to land-based systems.

In the future sensor systems may play an even larger role on wind turbines. Currently, Sandia as well as other European laboratories are all engaged in the development and application of sensor and operational measurement methods. Some of the key objectives include: determination of inflow loads and damage state (Sandia National Laboratories), advanced condition monitoring of gearboxes (National Renewable Energy Laboratory), and monitoring of localized aerodynamic flow conditions (Ris∅ DTU National Laboratory) [7a]. These technologies are all targeted at building a smarter wind turbine that can itself, identify the loads being applied by the wind, the damage created by these loads, and deploy control strategies to mitigate the loads while maintaining optimal power productions.

In order for newer, higher fidelity sensors to be adopted, there are several challenges/observations that must be addressed: sensor arrays and interrogator must have minimal cost, simple installation, and an operational life on the order of years and tens of years. Over these long durations of application, the sensor must also maintain calibration and sensitivity, otherwise Type 1 and Type 3 errors (false positive and false negative) will reduce the reliability and usefulness of the technology. Sandia's sensor program is focused on identifying sensor technologies that can potentially fulfill these design requirements. Currently, Fiber Bragg strain sensors interrogated over fiber optic lines, ruggedized accelerometers, hot-film aerodynamic sensors, and aerodynamic surface pressure taps are all simultaneously being investigated (Fig. 2.22). Each sensor technology is evaluated to determine the relative cost which is dictated by the number of sensors required to accurately monitor the rotor blade, the cost of

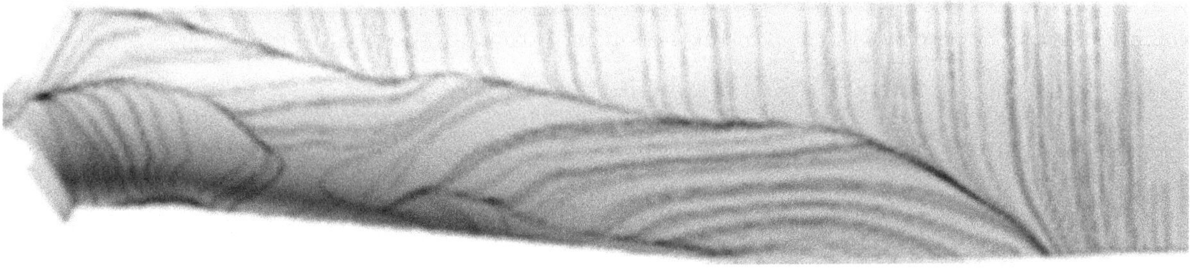

FIG. 2.20: CFD SOLUTION SHOWING NEAR-SURFACE FLOW STREAMLINES OVER THE INBOARD REGION OF A UTILITY-SCALE WIND TURBINE BLADE, SHOWING A RE- GION OF THREE-DIMENSIONAL SEPARATED FLOW

the interrogator used to measure the sensor signal, and the optimal/reliable method for integrating and protecting the sensor to maximize survivability.

Examples of adoption of new sensor technologies can be seen in several commercial machines today [8].

As an example, several wind manufacturers rely on fiber optic networks on the blades to enhance operation and control strategy. These sensors offer the flexibility in that many sensors can be placed in a single fiber line and can be incorporated in the manufacturing process.

As we foresee future designs, it is important to acknowledge that innovation will continue to play a key role in making wind systems more reliable and cost-effective. Sensor technologies are just one of those key elements that will continue to contribute to turbine optimization. It is conceivable that sensors will not only contribute to single turbine improvements in the future, and can be utilized for wind plant operations, as machines could have the ability to adapt to address real-time conditions.

2.4.5 Advanced Control Strategies

In order to continue to reduce wind turbine costs, future large multi-megawatt turbines must be designed with lighter-weight rotors, potentially implementing active controls strategies to mitigate fatigue loads while maximizing energy capture and adding active damping to maintain stability for these dynamically active structures operating in a complex inflow environment. Development, evaluation, and testing of advanced controls to mitigate fatigue loads caused by complex turbulent inflow are crucial for future designs.

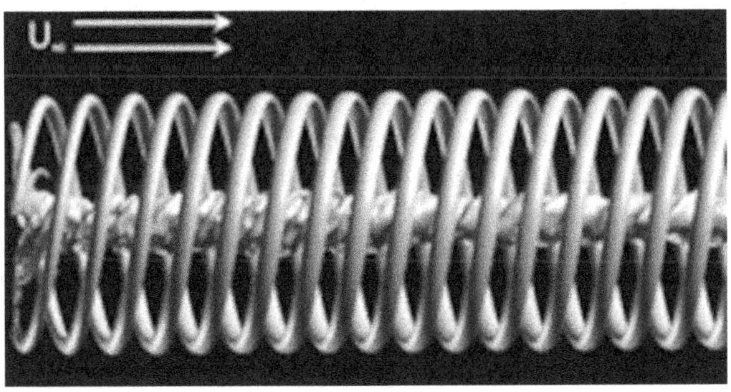

FIG. 2.21: CFD SOLUTION VISUALIZING THE HELICAL STRUCTURE OF THE WAKE OF A WIND TURBINE OPERATING IN A UNIFORM WIND, DEMONSTRATING THE PERSISTENCE OF THE WAKE WELL DOWN-WIND FROM THE ROTOR, WHICH IS LOCATED AT THE LEFT END OF THE FIGURE

The wind turbine is a highly non-linear dynamic machine that operates over a large turbulent wind regime. Current conventional designs are limited to linearized models about nominal wind speed operating points that require gain scheduling to transition between each nominal wind speed operating point [7]. Today, commercial machines rely on either classical single-input–single-output (SISO) controllers or state-space multiple-input–multiple-output (MIMO) controllers based on linearized models. While adequate for controlling the "stiff" machines of the past, these methods are not ideal for stabilizing future large multi-megawatt turbines that will experience greater dynamic coupling due to greater flexibility and lower rotor speeds. To meet these future challenges, Sandia focuses on advanced control methods and paradigms that are designed to meet multiple control objectives with a single unified control loop, where multiple control actuators and multiple sensors can be used to greatest advantage to reduce fatigue loading, stabilize the complex structure, and maximize power.

Moving forward, the possibility of designing full non-linear dynamical system in a non-linear/adaptive control design may allow for the potential to capture more energy in below rated-power conditions, efficient transition between below and rated-power conditions, and for above rated-power conditions to mitigate and reduce fatigue loads on turbine components and blades. This results in longer operational life for the wind turbine components (gearboxes, blades, etc.)

2.4.6 Advanced Architecture

Recent technology innovation in rotor technology — including individual blade pitch control, passive bend-twist and sweep-twist coupling (aero-elastic tailoring), and fast-acting active aerodynamic load control — offer the potential for further enhancing turbine energy capture and decreasing turbine cost of energy (COE). There is a significant amount of research domestically and globally that showcase the value of these innovation, and ongoing research in both controls and sensing will provide the operational architecture to make them a reality.

Advanced control architectures that fully take advantage of these innovations can provide the technology pathway to continue to refine these large machines and ensure that safety, efficiency, economics, and reliability metrics are fully realized.

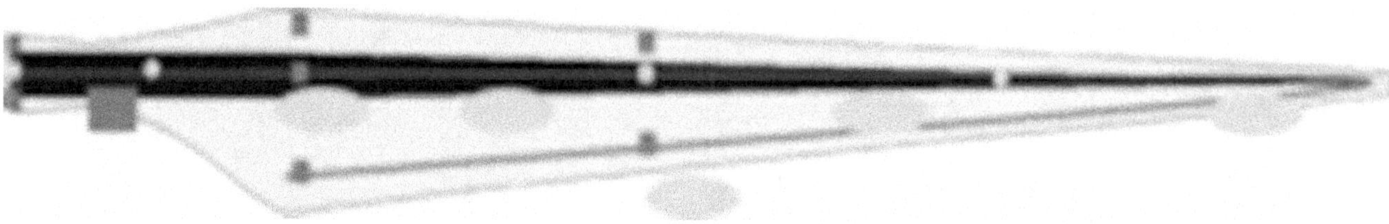

FIG. 2.22: SANDIA'S SENSOR BLADE — INCLUDES STRAIN GAUGES, FIBER OPTICS FBG'S, THERMOCOUPLES, AND ACCELEROMETERS

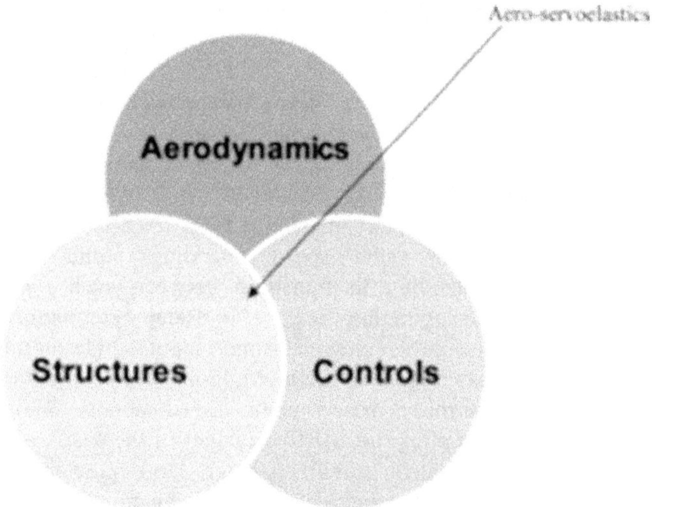

FIG. 2.23: SNL INTEGRATED AEROSERVOELASTIC STRATEGY

With today's computational capabilities and the researchers' ability to model the integrated system, multidisciplinary approaches are a key to improving the technology and ensuring that the maximum efficiencies are attained (Fig. 2.23).

2.4.7 Revitalizing U.S. Clean Manufacturing

In order to continue to support the growth of the wind industry, local and cost-competitive manufacturing is an important aspect to provide sustainability and the green economy. The wind industry provides a series of jobs that will require a trained a robust workforce. Like any other energy industry, the opportunities span from engineering, service and operation, and manufacturing. A big challenge for the U.S. in the manufacturing area is the labor cost when compared to other parts of the world. While the industry in the U.S. was ramping up, a large percentage of the components were being imported. Today given the persistence of the U.S. market, many companies have established U.S. manufacturing, but in order to continue to capitalize on this opportunity, a cost-effective manufacturing strategy must be developed to displace the higher labor costs. Unlike the semiconductor industry, which has predominantly left the U.S., wind turbine components have the opportunity to stay given their large size, the countries manufacturing capabilities, and transportation and logistics cost (Fig. 2.24).

Sandia's Advanced Manufacturing Initiative (AMI) is targeted at evaluating the manufacturing process, optimizing the process flow, identifying opportunities for automation, and improving quality and plant output. The initial program is targeting wind blade manufacturing due to its high labor content, the size of the components, and need for improve quality and reliability.

FIG. 2.24: 2MW WIND TURBINE BLADE BEING TRANSPORTED

2.4.8 Testing and Evaluation

As wind turbines have grown in size and capacity over the last three decades, the importance of reliability and technology innovation have been quite apparent. Even though, these "Gentle Giants" look much like their predecessors of the 1980s — three-bladed, upwind configuration — technology improvements have been vital for the success of this vibrant industry. Every component and subcomponent of the turbine including airfoils, materials, structures, and sensor and control systems have to be tested and evaluated prior to being deployed and accepted.

All engineered components and systems must go through a testing phase in order to validate the engineering assumptions made in the design, analysis and manufacturing processes. The key question for testing is why, when, and is some cases how? As the wind industry has gone through its maturity phase, the trial-and-error days of going straight to the field and patching flaws real-time are hopefully long gone. This has been driven and enabled by both the fidelity of today's engineering tools and computing capacity, as well as the requirement to certify components and systems to standards and the shear cost of building and testing structures of this magnitude (Fig. 2.25).

Wind energy components pose many challenges when it comes to testing and evaluation. Not only does wind have a series of unique components, the size of the components and the measurement requirements can make testing quite costly and challenging. Take for example a typical utility-scale wind turbine blade, which is 30 to 60 m in length, weighs 10 to 20 tons, and is quite complex in shape. As previously explained,

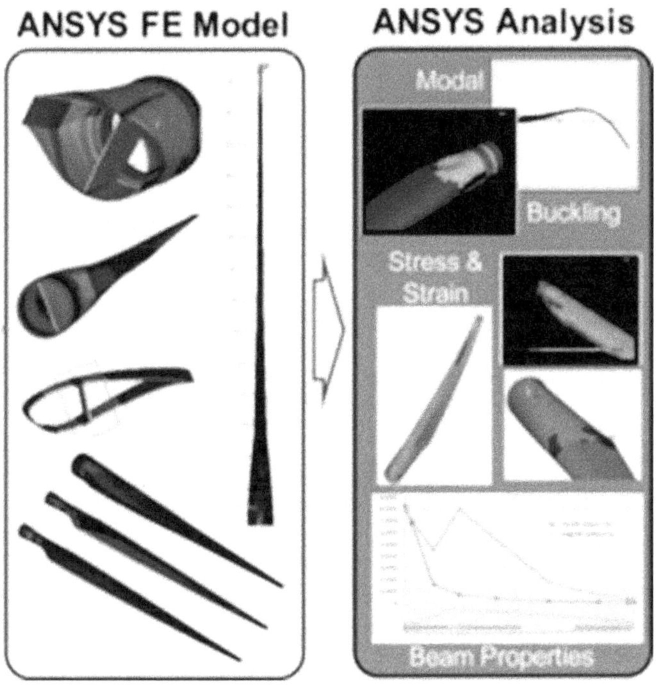

FIG. 2.25: ANSYS FEA BLADE MODEL AND CALCULATED ANALYSIS

modern blades are predominately made of a combination of fiberglass and carbon fibers, resin, and balsa or foam. Each one these materials has to be certified and tested by their respective manufacturer and must be brought together to design a blade which itself has to be certified. Airfoils are now optimized for aerodynamic and acoustic performance and are tested rigorously in wind tunnels to ensure this performance under a variety of simulated field conditions. Moreover, the shape must lend itself to not only localized aerodynamic performance, but must also be coupled to the structural and manufacturing design, where the internal structure is designed to take advantage of the materials available and how they will be organized or stacked to develop a structurally efficient blade. Also, as part of the design phase, the manufacturability must be continually evaluated to ensure that the blade can be manufactured to the specifications without the introduction of defects.

Although, there will be several sub-scale testing phases throughout the engineering process, the entire completed blade must also be tested for structural strength and aerodynamic performance to validate the computer models and receive certification. These tests can be both complex and costly due to the size of the structure and the complex testing and loading requirements, which must replicate such things as, in the case of fatigue loading, the number of and magnitude of loading cycles that a blade will see over its 20- year life (Fig. 2.26). It is key at this point that the tested article is close to the final design, as it can cause significant cost increases and project delays to redesign, rebuild, and retest.

Coincidently, these complex testing processes must also be completed as applicable for other components of a wind turbine, including gearboxes, generators, controllers, etc. In order to completely be sure that the system will operate as predicted, all components should be evaluated together to validate that the systems will work reliably. Sandia's technology program relies on this approach and evaluates the technology at the sub-scale test site in Bushland, TX (Fig. 2.27).

FIG. 2.26: UTILITY SCALE STATIC TESTING AT NREL'S NWTC

The tools used to develop and evaluate these designs are only as good as the data used to validate and improve the fidelity of the code. Additionally, the tools are only able to model the article to a certain degree of fidelity within a certain operational envelope, and many practical elements can diverge from the model in the as-manufactured, as-installed final product. The lessons learned and data gathered from lab and field testing, at both full-scale and sub-scale enables engineers to continually improve their designs, the result of which can be clearly seen in the viability and resilience of the wind industry.

2.4.9 Design Tools

Wind turbines are designed and optimized to capture as much energy as possible in a given wind resource. Research into innovative wind turbine design improvements is being performed on all components of the wind turbine in order to improve the overall reliability and efficiency of the machines in the field. Example research topic areas include towers, generators, drive trains, and blades. With the component size of today, the ability to build and test these components is cost prohibited and would delay the product cycle significantly.

FIG. 2.27: SANDIA'S SUB-SCALE TEST SITE IN BUSHLAND, TX

Additionally, changes to individual component technology affect system behavior and result in both costs and benefits for any given innovative idea. It is vital to understand not only the benefits resulting from an innovation but also to understand the nature and magnitude of resulting costs that may be present elsewhere in the system. Common system costs may include the following: increased forces and moments elsewhere in the system, increased complexity or decreased energy capture. Use of system dynamic models enables researchers to assess overall cost and benefit of new ideas even to the point that effects on final cost of electricity (COE) may be determined.

The system dynamics model of a wind turbine includes physics representing three major areas as seen in Fig. 2.23. It includes elements to describe aerodynamic and structural aspects as well as wind turbine controls interactions. The combination of these elements is known as an *aeroelastic*, or *aeroservoelastic*, problem due to the coupled interactions of the aerodynamics, structural deflections, and controls that are involved.

The system model is used to assess the wind turbine design with respect to typical design requirements (Fig. 2.28). Wind turbines are designed for a 20-year life and are subjected to stochastic turbulent wind input, extreme loads, and operation through component faults. Potential failures are assessed in terms of ultimate loads, fatigues loads, and functional requirements, such as blade-tower clearance. Finally, the system model can be used to simulate the efficiency of energy capture during normal operations.

2.5 FUTURE TRENDS

Although the wind energy industry has experienced a large growth over the last decade, both nationally and globally, technology improvements continue to be required in order for the industry to be competitive and continue to be a key part of the energy mix of today and the future. Technology innovations that balance efficiency improvements and reliability are especially attractive, and national labs, industry, and academia are engaged to develop and implement these technologies.

Additionally, although the U.S. has an immense resource in particular in the Great Plains region (Fig. 2.29), technology that is viable in lower resource sites and offshore is needed.

2.5.1 Smart Rotor

Since the global acceptance of the utility-tied, three-bladed up-wind configuration over 30 years ago, engineers and scientist across the globe have developed innovative techniques and improvements to increase the efficiency, availability, and reliability of wind tur- bines. Efficiency improvement options are always attractive given the strong coupling to cost-of-energy (COE) and because of the ease of calculating the return on the investment. Revolutionary examples of innovation over the last 30 years include the use of laminar airfoils, the transition to variable speed and pitch from stall-regulated designs, and many more. These innovations have enabled the wind industry to become globally cost-competitive and to install products that are designed for a 20-year lifespan.

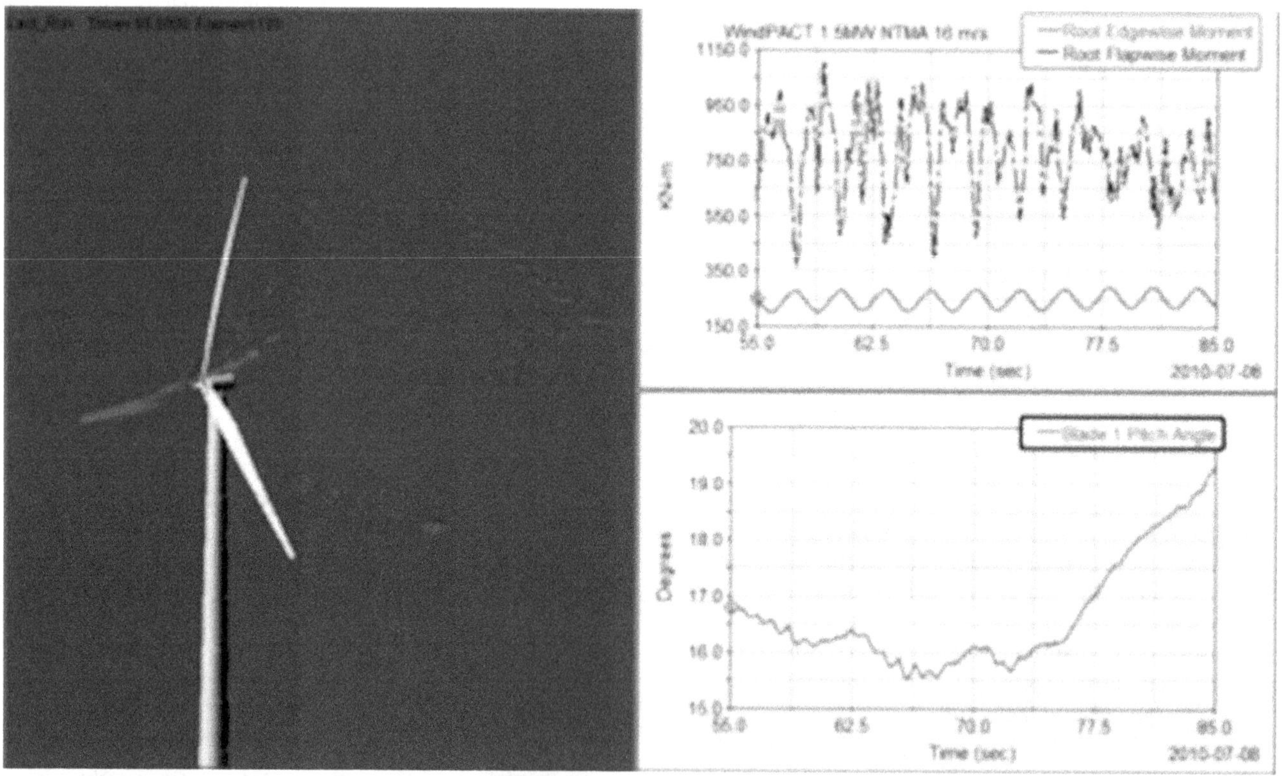

FIG. 2.28: GRAPHICAL REPRESENTATION OF A SYSTEM DYNAMICS SIMULATION AND RESPONSES

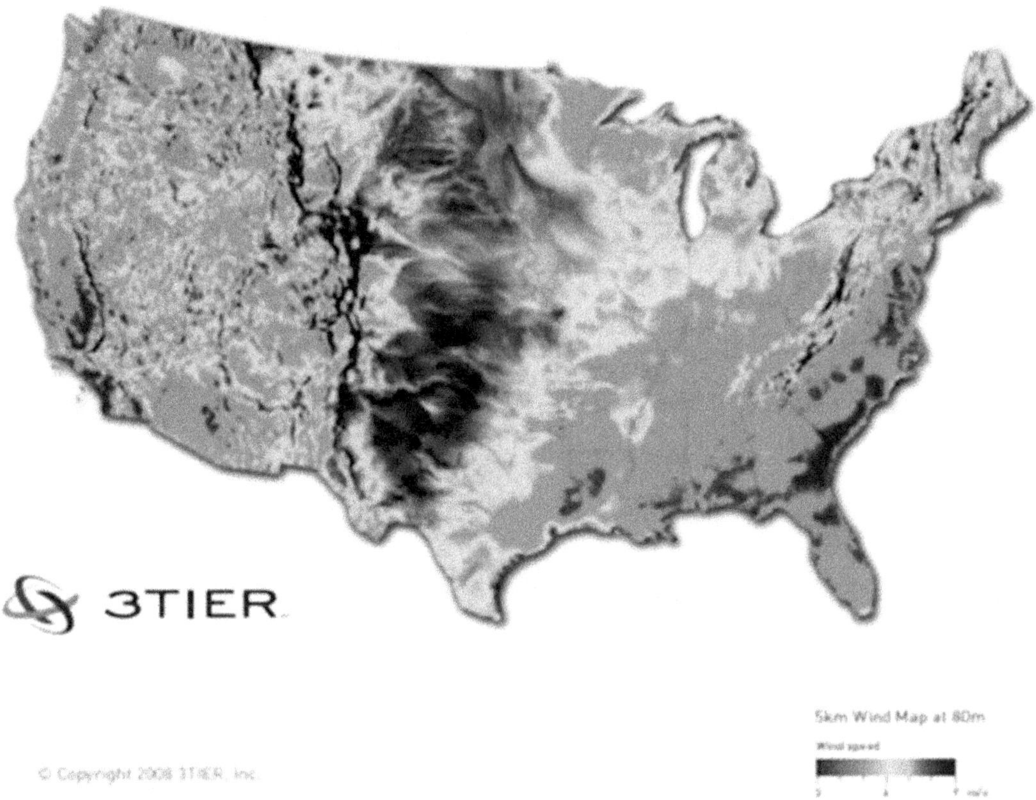

FIG. 2.29: U.S. 80- M WIND MAP

As we look to the future, the large "low-hanging fruit" of efficiency improvement areas for land-based deployment are no longer there, and designers, engineers, and scientist at national labs, universities, and manufacturers are evaluating, designing, and implementing concepts that are focused on refining and improving the technology.

Ongoing research — taking place both domestically and internationally — focused on next-generation concepts has identified the viability and feasibility of both passive and active aerodynamic surfaces on wind turbine blades.

There are two ways to implement passive aeroelastic coupling on a wind turbine blade: off-axis materials and geometric sweep (Fig.2.30).

A key advantage of the passive methods is the simplicity versus authority that the design enables. By passively modifying the blade, the rotor is able to change the incident angle resulting in an overall load reduction over the entire power curve spectrum. The result is the ability of designing a larger rotor that captures more energy and manages the system loads. The disadvantage is that the wind is not constant and is quite random in nature. Since the system is passive, designers must identify the primary design point and try to maximize performance throughout the operational envelope. Initial field evaluation studies of both methodologies have shown energy capture improvements ranging from 5% to 10% [11].

Sandia's active aerodynamics program focuses on designing and implementing low energy-consuming, fast-acting, and simple aero surfaces that can modify the localized flow in order to affect the high-frequency content in the wind (Fig. 2.31). This capability will provide designers with a new set of actuators that can be managed to fine-tune the performance of machines and will be able to adapt to local atmospheric phenomena that are difficult to resolve with current actuation.

Initial results from Sandia's SMART rotor program have shown the ability of these methodologies to significantly reduce system loads and enable designers to increase the rotor size for a given architecture, yielding a net annual energy increase (Fig. 2.32).

In order to fully realize the benefits of active aerodynamics, the localized conditions must be understood. Cost-effective sensor technologies that provide the necessary information (load, pressure, etc.) and have the appropriate resolution must be collocated near the actuators to be able to control the surface effectively and efficiently. Although the initial results are quite promising, it is important to keep cost, reliability, and maintainability in mind in order to ensure the implementation and acceptance by the industry.

Innovation, such as the SMART program, will always be a part of technology development. Although cost-competitive, the wind industry must continue to identify improvement areas to increase viability and ensure that wind can compete in a diverse energy sector of the future.

2.5.2 Offshore Wind

Although developers will continue to pursue economically feasible land-based sites to install projects, concerns or limitations in our transmission system, NIMBY, and limited land on most coastal

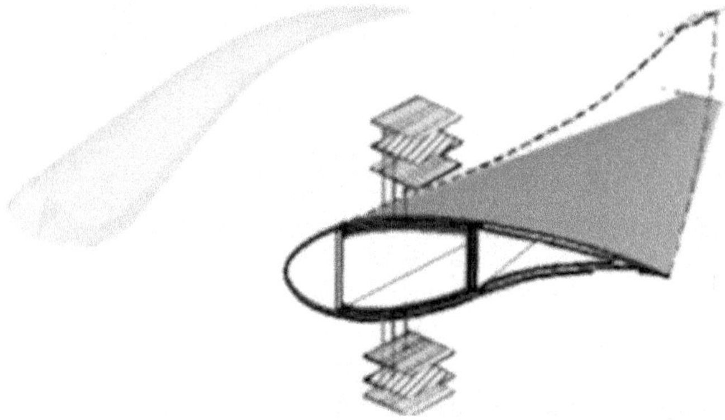

FIG. 2.30: LEFT: SWEPT K&C STAR DESIGN AND RIGHT: SNL OFF AXIS DESIGN

FIG. 2.31: MICROTAB ACTIVE AERO CONCEPT AND CFD CALCULATION

states are all reasons why a strong interest has been spurred in exploiting our coastlines for offshore wind installations. There are many attractive reasons why offshore wind installations should be pursued on the Gulf coast, the Atlantic or Pacific coast, and/or the Great Lakes. The U.S. has an excellent offshore wind resource, and approximately 78% of our total energy consumption is consumed by the 28 states with a coastline (Fig. 2.33). Currently, there are no installed offshore projects in the U.S., but there are 13 projects being proposed, totaling to 2.4 GW.

Several European countries have already leveraged their coastlines for offshore projects; there are currently 39 installed offshore projects totaling over 2 GW (Fig. 2.34). Although the modern wind industry has many decades of experience, the offshore wind industry is quite young, and several of the initial projects have experienced several technical premature reliability challenges since their installation.

Fundamentally, offshore machines are quite similar to the land- based systems that we have become accustomed to, but both are driven by economic and environmental differences. Offshore machines must be equipped with systems and an operational architecture that provide accessibility and enable them to survive the challenges of the offshore environment (Fig. 2.35).

From a resource perspective, it is well understood that wind over the water is often more consistent and less turbulent than on land. This typically translates to higher capacity factors and morepredictable electrical output. Coincidently, the design of the machine from the foundation to the rotor must take these differences into account, as well as the impact of the hydrodynamic loading induced by the ocean in order to design a machine that is efficient, reliable, and cost-effective. All current offshore installations in Europe today have been installed in fairly shallow waters (<30 m), which have provided the opportunity to leverage well-known foundation designs (primarily monopole) from other industries. Unfortunately, if the U.S. wants to capitalize on its vast offshore resource, research and development must be performed for deeper water depths where jacketed or floating structures will be needed (Fig. 2.36).

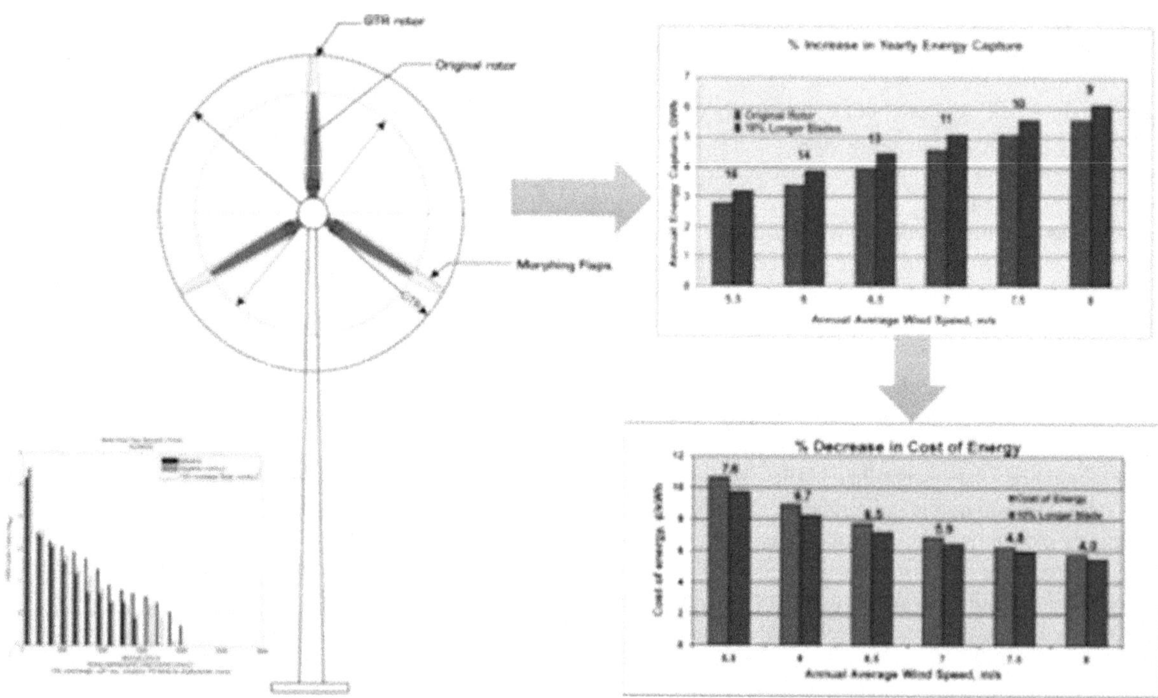

FIG. 2.32: ACTIVE AERODYNAMIC STRATEGY — IMPROVEMENTS IN ENERGY CAPTURE AND COST OF ENERGY

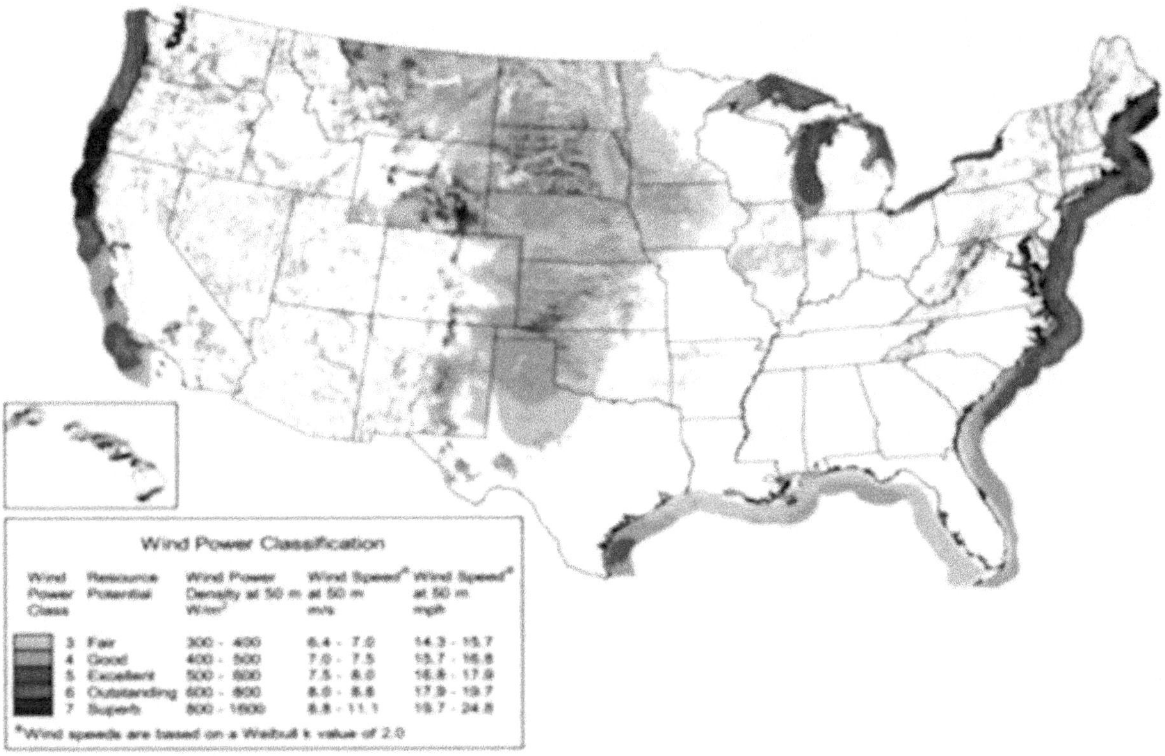

FIG. 2.33: LAND-BASED AND OFFSHORE WIND RESOURCE MAP

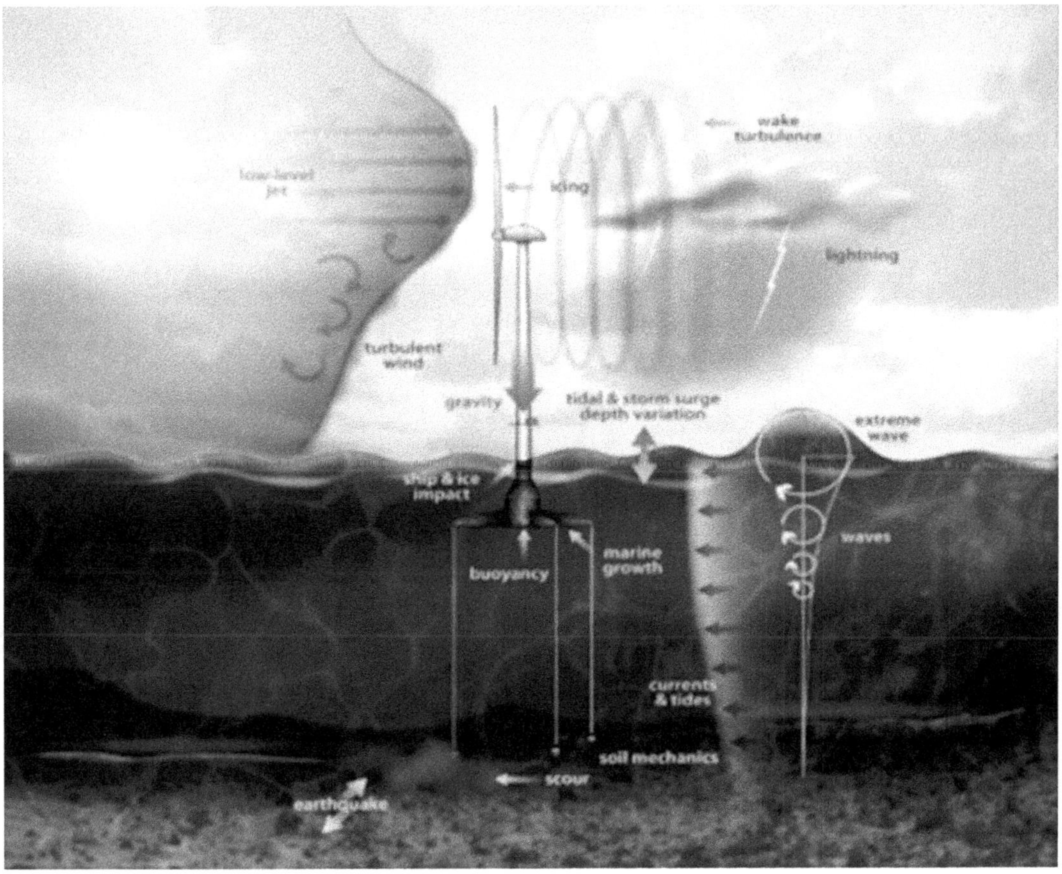

FIG. 2.34: OFFSHORE WIND TURBINE WITH COMPLEX DESIGN CONDITION SHOWN (COURTESY OF NREL)

FIG. 2.35: SIEMENS OFFSHORE MACHINES — COPENHAGEN HARBOR (SOURCE: WIKIMEDIA COMMONS)

There are many advantages as to why offshore projects should be pursued in the U.S. Outside of a key advantage of proximity to large load centers, offshore machines can be significantly larger than the typical land-based machines being installed today (1 to 2.5MW) since the limitations in both infrastructure and transportation can be mitigated by having coastal manufacturing and barging the components to the installation sites. Typical offshore machines today range from 2 to 5 MW, but larger turbines are being designed and tested. There are several challenges associated with offshore wind as well. In comparison to the land-based machines, the cost balance of an offshore project is not the same, as the turbine rep- resents a much smaller percentage of the total cost (~25%). Cost associated with the support structure, the electrical infrastructure, and operations and maintenance (O&M) are significantly higher for offshore projects; hence which is why current offshore projects come in at a higher cost (Fig. 2.37).

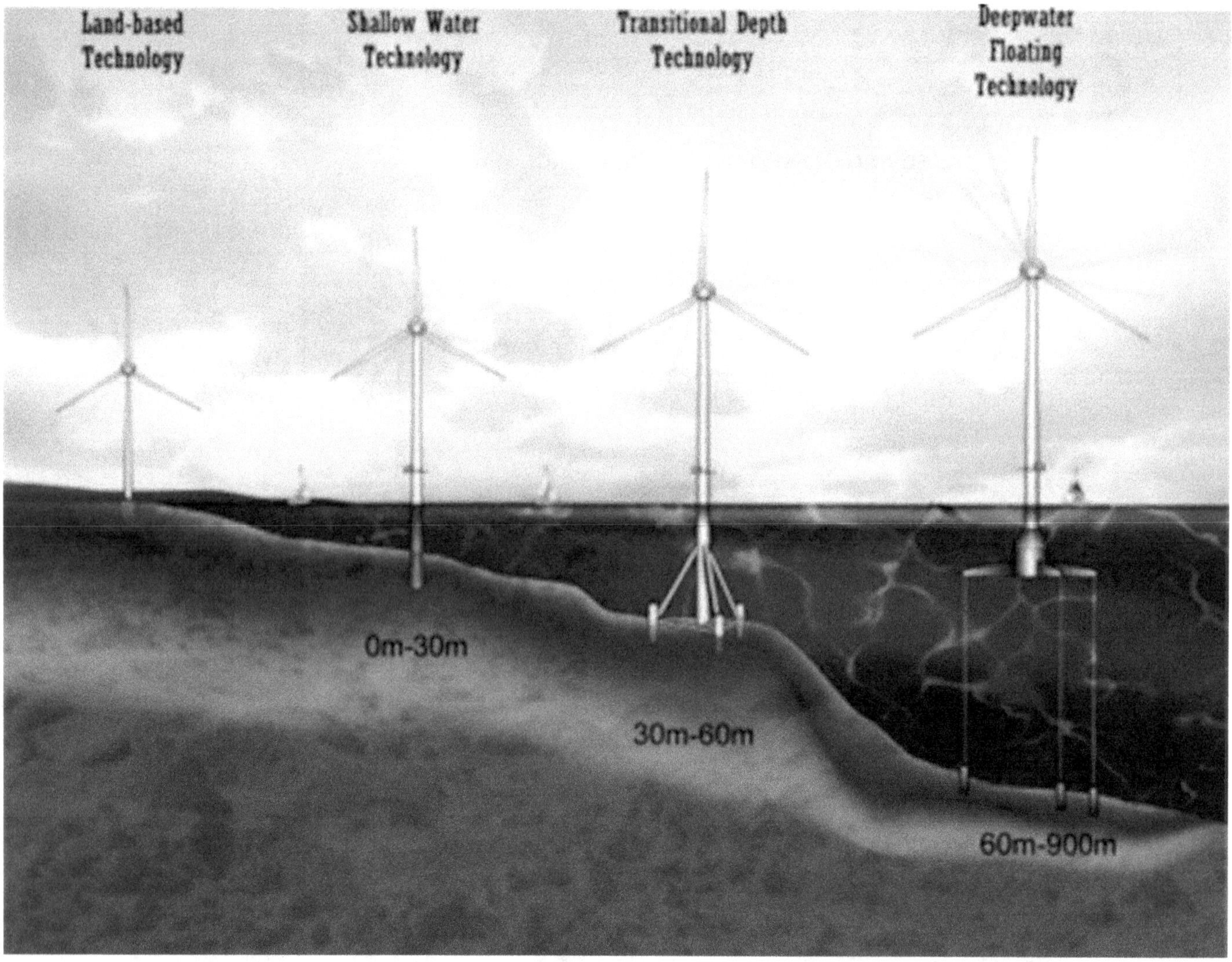

FIG. 2.36: OFFSHORE WIND FOUNDATION/PLATFORM DESIGNS (COURTESY OF NREL)

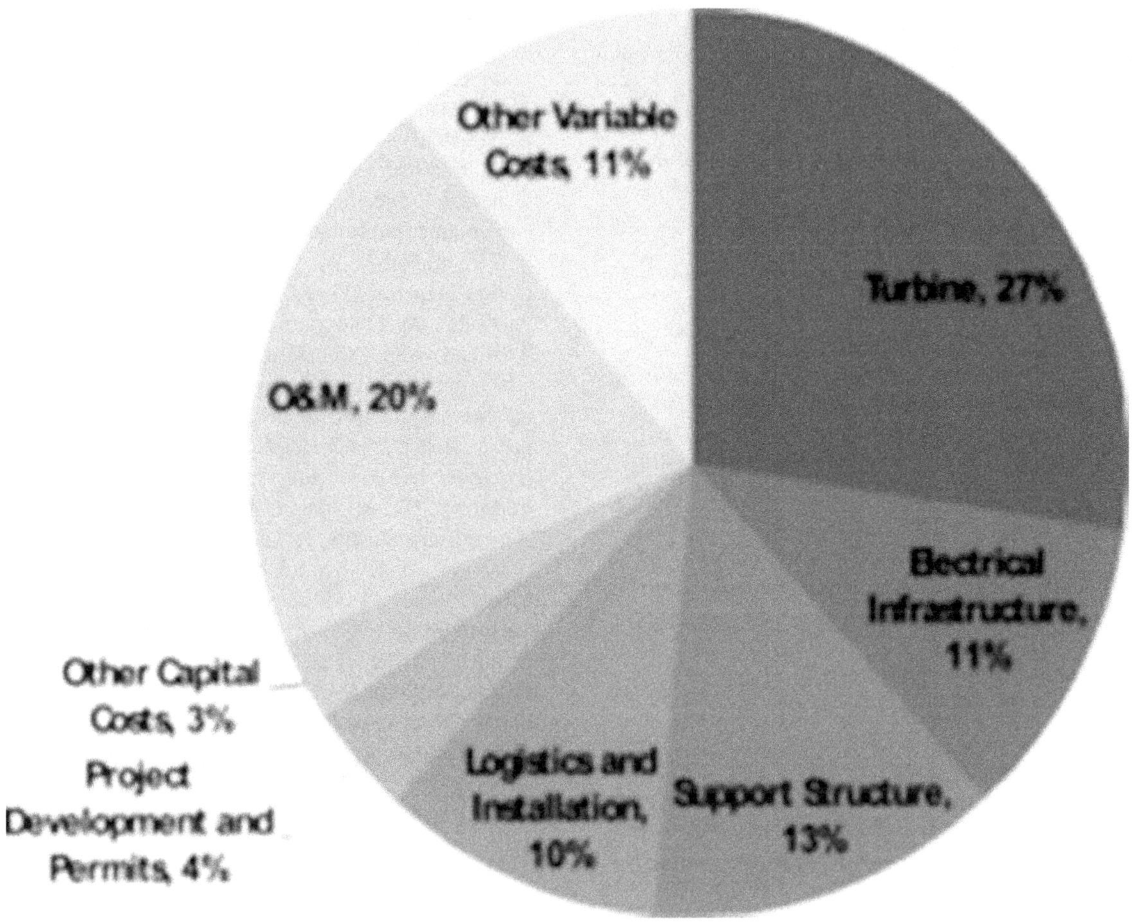

FIG. 2.37: OFFSHORE LIFE-CYCLE COST BREAKDOWN

To outweigh these challenges, future designs must be smarter and able to operate and report upcoming failures and service requirements prior to a catastrophic system failure. As an example, new machines could incorporate a sensor network that increases the fidelity in operation and condition health monitoring. Given that the turbine cost does not dominate the total cost of the installation as the land-based system does, innovation in this area is critical and a cost-effective way to enable reliable offshore turbines.

2.6 CONCLUSION

The last 30 years of investment in the wind industry have transformed this old technology into the fastest-growing clean energy source in the world. Through record setting years, wind energy deployment both worldwide and in the U.S. continues to show the effectiveness of policies and incentives, coupled with a clean, affordable, and reliable energy supply.

There is still an immense set of technology options that can potentially improve wind systems, such as condition health monitoring systems, distributed sensor networks, advance materials options, such as nanoparticles to strengthen local areas, etc., but evaluating and balancing these technologies on an economic basis is the key. Sandia National Laboratories in conjunction with other labs, academia, and the industry will continue to explore these options in order to continue to innovate.

Like the leadership and direction that DOE's 20% by 2030 report outlined, there are several new ongoing studies targeted at evaluating and calculating the feasibility of large penetrations of renewable in the future energy mix, with all studies showing that wind energy will represent a significant percentage of the clean energy portfolio. Additionally, with current administration goals, which aggressively suggest up to 10% renewable energy by 2012 and 25% by 2025, it is important to understand that not only clean technology viability and feasibility is needed, but also a robust supply chain and manufacturing sector is imperative to meet these goals.

As we foresee, a future for the wind industry where there will be offshore and land-based machines available and installed glo- bally, it is important to acknowledge that technology innovation will need to play a key role in making wind systems more reliable and cost-effective. As the leader of the clean energy portfolio, the wind industry must continue to find ways to improve the technology and pave the road for other upcoming technologies. It is hard to predict what the future energy picture will look like, but there is a high probability that if this industry continues to innovate, grow, and lead, it will have a key role in our energy future.

2.7 ACRONYMS

AMI: Advanced Manufacturing Initiative
CFD: Computational Fluid Dynamics
COE: Cost of Energy
DOE: Department of Energy
FBG: Fiber Bragg Grating
FEA: Finite Element Analysis
GW: gigaWatt
HAWT: Horizontal Axis Wind Turbine
MIMO: Multiple-Input Multiple-Output
MW: megawatt
NDT: Non-Destructive inspection Technology
NREL: National Renewable Energy Laboratory
O&M: Operations and Maintenance
SISO: Single-Input Single-Output
SMART: Structural Mechanical Adaptive Rotor Technology
SNL or Sandia: Sandia National Laboratories
USDA: United States Department of Agriculture
VAWT: Vertical Axis Wind Turbine

2.8 REFERENCES

1) http://en.wikipedia.org/wiki/Betz%27_law
2) Gijs A.M. van Kuik, *The Lanchester–Betz–JoukowskyLimit*, Wind Energy Journal. 2007; 10:289–291
3) John F. Mandell and Daniel D. Samborsky, *DOE/MSU Composite Material Fatigue Database:Test Methods, Materials, and Analysis*, SAND97-3002 UC-1210, Sandia National Laboratories
4) H. J. Sutherland and John F. Mandell, *Application of the U.S. High Cycle Fatigue DataBase to Wind Turbine Blade Lifetime Predictions*, Energy Week 1996, Book VIII: Wind Energy, ASME, January-February, 1996, pp. 85-92.
5) http://www.awea.org
6) http://www1.eere.energy.gov/windandhydro/wind_2030.html 7a. http://www.risoe.dtu.dk/?sc_lang=en
7) Rush D. Robinett, III and David G. Wilson, *Maximizing the Performance of Wind Turbines with Nonlinear Aeroservoelastic Power Flow Control*, AWEA 2009
8) Dale E. Berg and Jose R. Zayas, *Aerodynamic and Aeroacoustic Properties of Flatback Airfoils*, AIAA Symposium 2008
9) Mark A. Rumsey and Joshua A. Paquette, *Structural Health Monitoring of Wind Turbine Blades*, SPIE Conference 2008
10) M. Barone, D. Berg, W. Devenport, and R. Burdisso, *Aerodynamic and Aeroacoustic Tests of a Flatback Version of the DU97-W-300 Airfoil.* SAND2009-4185, Sandia National Laboratories, August 2009.
11) S. Wagner, R. BareiB, G. Guidati, *Wind Turbine Noise*. Springer- Verlag, Berlin, 1996.
12) D. Berry, T. Ashwill, *Design of 9-Meter Carbon-Fiberglass Prototype Blades: CX-100 and TX-100,* SAND07-0201, Sandia National Laboratories, 2007

2.9 SCOPE OF WIND ENERGY GENERATION TECHNOLOGIES SINCE 2011

2.9.1 Introduction

This section 2.9 covers the Scope of Wind Energy Generation Technologies since 2011, the earlier period was covered in "Scope of Wind Energy Generation Technologies" by Jose Zayas in "Energy and Power Generation Handbook- Established and Emerging Technologies" Edited by K. R. Rao published by ASME, 2011 [1]. Since Jose Zayas was unavailable for this publication Editor-in-Chief of ASME Renewable Energy Series K. R. Rao contributed this update. K. R. Rao had a comprehensive classic "Wind Energy for Power Generation - Meeting the Challenge of Practical Implementation" in two volumes published by Springer in 2019 [2].

This section 2.8 has four parts: 2.9.2 dealing with the Next Generation Issues of Wind Energy"; 2.9.3 covers "Next Generation – Analysis"; 2.9.4 addresses "Wind Vision and Wind Map" and lastly 2.8.5 ends with a Conclusion. In 2.9.2 discussion starts with topics which impacted Wind Energy Generation since 2010 to-date, which are issues with impact on next generation issues. The topics are "Bigger the Better", "Hub Height"; "Rotor Diameter"; "Inverters"; "Life Extension and Recycling and Circular Economy"; "Name Plate Capacity"; "Transportation Challenges"; "Artificial Intelligence" and "Hybrid and Storage Technologies". In 2.9.3 coverage starts with "Market Analysis - On Land"; "Market Analysis - Offshore"; "Distributed Wind"; "Grid Integration"; "Environmental Aspects of Wind Turbine Siting"; "Cyber Security"; "Digital Related Issues"; "High Fidelity – HPC & HPM, Compressible and Incompressible flow models and Nonlinear Structural Flows". Section 2.9.4 picks up information in "Wind Vision" contributed by Jose Zayas as Director Wind & Water Technology Office. Finally in 2.9.5 a conclusion is provided elaborating the points discussed in section 2.9, summarizing what the future of wind energy as contemplated by US DOE and NREL.

2.9.2 Next Generation Issues

Key activities are in Offshore Wind, Distributed Wind, Atmospheric Electrons, Wind Resource Assessment & Characterization, Next-Generation Wind Technology, Research in Testing & Certification, Wind Manufacturing & Supply Chain, Renewable Systems Integration, Environmental Impacts & Siting, Workforce Development & Education.[2]

The Wind Energy Technologies Office (WETO) works with industry partners to increase the performance and reliability of next-generation wind technologies while lowering the cost of wind energy. Wind energy costs have been reduced from over 55 cents (current dollars) per kilowatt-hour (kWh) in 1980 to an average of under 3 cents per kWh in the United States today. To ensure future industry growth, wind industry technology must continue to evolve, building on earlier successes to further improve reliability, increase capacity factors, and reduce costs. Wind energy technologies available for licensing from U.S. Department of Energy laboratories and participating research institutions can be found on the Department of Energies (DOE's) Office of Energy Efficiency and Renewable Energy's Energy Innovation Portal and WETO Projects Map. During the past two decades, the office has worked with industry to develop a number of prototype technologies, one example is the GE Wind Energy 1.5- megawatt (MW) wind turbine.[3]

2.9.2.1 Component Development

WETO worked with industry partners to improve the performance and reliability of system components. Department of Energy's Sandia National Laboratories develop an innovative wind turbine blade that has led to an increase in energy capture

[2] https://www.energy.gov/eere/next-generation-wind-technology
[3] https://www.energy.gov/eere/wind/listings/articles-about-next-generation-technologies

FIG. 2.38: A PICTURE OF WIND ENERGY COMPONENT DEVELOPMENT (SOURCE:[4])

by 12%.The most distinctive characteristic of the Sweep Twist Adaptive Rotor (STAR) blade is a gently curved tip, which, unlike the vast majority of blades in use, is specially designed to take maximum advantage of all wind speeds, including slower speeds. More recently, development of more reliable gearboxes, has incorporated in the design innovative drivetrain concepts. DOE funding has fostered one of the world's most advanced wind energy testing at Clemson University to speed the deployment of next generation energy technology in addition to additive manufacturing, commonly known as 3D printing for the production of wind turbine blade molds. The traditional method of blade design requires the creation of a plug, or a full size representation of the final blade, which is then used to make the mold. Creating the plug is one of the most time-intensive and labor intensive processes in wind blade construction, so 3D printing saves these critical resources.[5]

Carbon Fiber Composites for Blades by DOE's Sandia National Laboratories, Oak Ridge National Laboratory, and Montana State University demonstrated the commercial viability of a novel cost-competitive carbon fiber composite material for use in wind turbine blades. The analysis found commercial viability and system-level benefits for using carbon fiber composites to reduce the overall cost of wind energy and manufacture long, slender wind turbine blades. The project revealed a 25% blade mass reduction when using carbon fiber spar caps compared to fiberglass. The textile-based carbon fiber material used for spar caps in this study cost 40% less than commercial carbon fiber to reduce system costs.[6]

Open-Source Performance Optimization Tool with funding from WETO, NREL released a new version of its FLOw Redirection and Induction in Steady State (FLORIS) model for wind plant performance optimization. FLORIS, was co-developed by NREL and the Delft University of Technology, is an open-source platform available for download and collaborative development to enhance

FIG. 2.39: PICTURE OF NEW TURBINE BLADES DEVELOPED BY NWTC (SOURCE:[7])

FLORIS's ability to accurately design and analyze wind plant control strategies for larger arrays of turbines. Since 2018, more than 2,000 users have accessed the tool to inform turbine operation as well as wind plant design. By optimizing flow control strategies like wake steering, FLORIS enables existing wind energy facilities to improve productivity and increase profits.[8]

2.9.2.2 Utility-Scale Research Turbines

The National Renewable Energy Laboratory's National Wind Technology Center (NWTC) has helped pioneer wind turbine blades, component, systems, and modeling methods that have driven industry acceleration, see figure 2.39. Research at the NWTC complements DOE's Atmosphere to Electrons (A2e) initiative, targets reductions in the cost of wind energy through an improved understanding of the complex physics governing wind flow into and through wind farms. Innovative wind energy research at the NWTC includes using computational fluid dynamics to develop the Simulator for Wind Farm Applications and other modeling and controls tools, to minimize the impact of turbine wake effects so that the overall wind power plant output could be increased by 4%–5%.

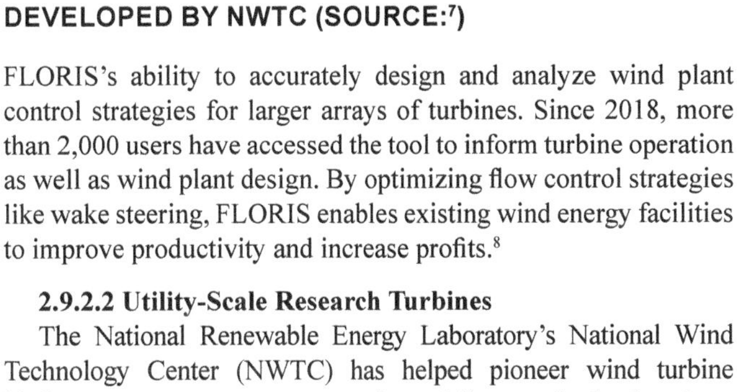

FIG. 2.40: VARIATION OF TURBINE SIZES SINCE 1985 THROUGH 2015 (SOURCE:[9])

[4] https://www.energy.gov/eere/wind/listings/articles-about-next-generation-technologies?page=1
[5] https://www.energy.gov/eere/wind/articles/no-time-waste-circular-economy-strategy-wind-energy
[6] https://cleantechnica.com/2021/01/05/2020s-top-wind-energy-rd-achievements/
[7] https://www.energy.gov/eere/wind/listings/articles-about-next-generation-technologies
[8] https://cleantechnica.com/2021/01/05/2020s-top-wind-energy-rd-achievements/
[9] https://www.lanl.gov/discover/publications/1663/2014-april/_assets/docs/1663_21_wind.pdf

The largest wind turbine in use today is 135 meters tall at the hub and 126 meters in rotor diameter.

FIG. 2.41: PICTURE OF A CURRENT LARGE TURBINE (SOURCE:[10])

FIG. 2.42: PICTURE SHOWS BIGGER TURBINES ABOUT WHICH "EXPERTS PREDICT 50% LOWER WIND COSTS THAN THEY DID IN 2015" (SOURCE:[14])

Utilizing the controllable grid interface test system, photovoltaic inverters, and energy storage systems react to disturbances on the electric power system in addition to analyzing offshore wind energy. As a member of the International Energy Agency (IEA) Wind Energy Executive Committee participates in 12 areas of wind energy research.[11] DOE announced wind-energy-related projects for national labs and their research partners using funding from DOE's Competitiveness Improvement Project (CIP). These filled critical gap for "Distributed Wind Industry" and Intergrid LLC which introduced the first inverter, developed specifically for small- and medium-scale wind turbines.[12]

2.9.2.3 Bigger The Better[13]

Modern wind turbines have a design life of 20 years and yet break down 2–3 times in the first 10 years. The bigger the wind turbine, the more wind power it generates. The largest in operation today stands one-and-a-half football fields tall at the hub and spins with a diameter as wide as two 747 jumbo jets, wingtip to wingtip. But with their great size comes equally great, unwanted side effects: great downstream turbulence, great sensitivity to it, and great expense to repair. New Los Alamos research indicates how to design and operate wind farms to maximize output while minimizing the likelihood of damage.[15]

Since the early 2000s, wind turbines have grown in size in both height and blade lengths and generate more energy. DOE announces 2021 Technology Commercialization Fund Selections. DOE announced $1.95 million in wind-energy-related projects for national labs and their research partners as part of more than $30 million in awards for 68 projects supported by the Office of Technology Transitions Technology Commercialization Fund,[16] (see figures 2.40, 2.41).

---

[10] https://www.lanl.gov/discover/publications/1663/2014-april/_assets/docs/1663_21_wind.pdf
[11] .https://www.energy.gov/eere/wind/listings/articles-about-next-generation-technologies
[12] https://www.energy.gov/eere/next-generation-wind-technology, April 15, 2021
[13] https://www.energy.gov/eere/articles/wind-turbines-bigger-better
[14] Ibid.
[15] https://www.lanl.gov/discover/publications/1663/2014-april/_assets/docs/1663_21_wind.pdf
[16] https://www.energy.gov/eere/wind/listings/articles-about-next-generation-technologies?page=1, June 25, 2021

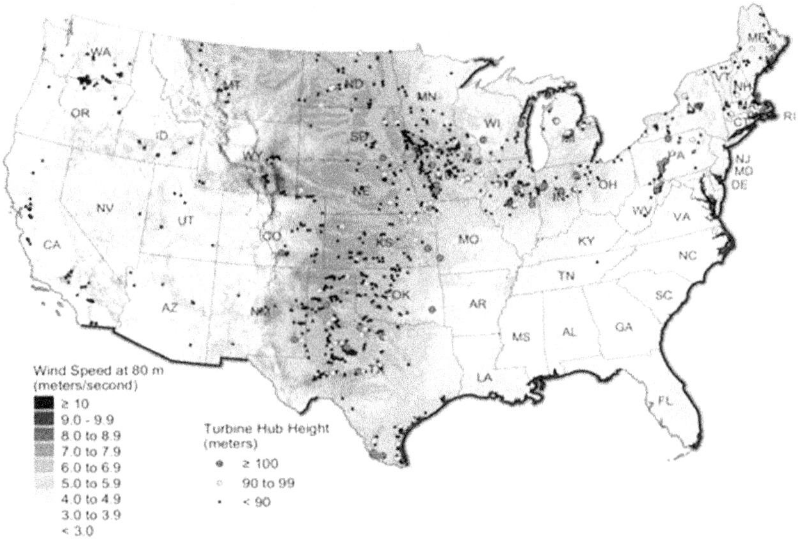

FIG. 2.43: LOCATION OF TALL-TOWER TURBINE INSTALLATIONS (SOURCE: LAND-BASED WIND MARKET REPORT: 2021 EDITION[17])

2.9.2.4 Hub Height[18]

A wind turbine's hub height is the distance from the ground to the middle of the turbine's rotor. The hub height for utility- scale land-based wind turbines has increased 59% since 1998–1999, to about 90 meters (295 feet) in 2020. That's about as tall as the Statue of Liberty! The average hub height for offshore turbines in the United States is projected to grow even taller, from 100 meters (330 feet) in 2016 to about 150 meters (500 feet), or about the height of the Washington Monument, in 2035, see fig. 2.42.

Turbine towers are becoming taller to capture more energy, since winds generally increase as altitudes increase. The change in wind speed with altitude is called wind shear. At higher heights above the ground, wind can flow more freely, with less friction from obstacles on the earth's surface such as trees and other vegetation, buildings, and mountains. Most wind turbine towers taller than 100 meters tend to be concentrated in the Midwest and Northeast, two regions with higher-than-average wind shear.

Based on an advanced turbine concept and assuming hub heights of 110 meters (m) (which are already in wide commercial deployment in Germany and other European countries), the technical potential for wind deployment is estimated to grow to 4.3 million square kilometers, a 54% increase compared to current technology with 80-m hub heights. By pursuing hub heights of 140 m, the technical potential for wind deployment is estimated to grow to 4.6 million square kilometers, a 67% increase compared to current technology with 80-m hub heights (see fig. 2.43 and 2.44).[19]

Based on first-order cost estimates informed by current technology, the most wind-rich regions of the country generally show an economic preference for the lowest considered tower height; higher hub heights (e.g., 110 m and 140 m) are often preferred in more moderate wind speed regions. This result is consistent with industry experience to date. Mesoscale resource data indicate an increase in annual average wind speed of 0.5 to 1.0 meters per second (m/s) when moving from 80 to 110 meters (m) and 1.0 to 1.5 m/s when moving from 80 to 160 m. Wind speed differences translate to sizable capacity factor improvements. Although the observed variance is broad, median capacity factor gains with higher hub heights are estimated at approximately 2 to 4 percentage points when going from 80 to 110 m and an additional 2 to 4 percentage points when going from 110 to 140 m. Between 140 and 160 m, median capacity factor gains are approximately 1 percentage point. Relatively larger gains occur east of the Rocky Mountains, with the greatest gains sprinkled throughout the Heartland, the Midwest, and the Northeast.[21]

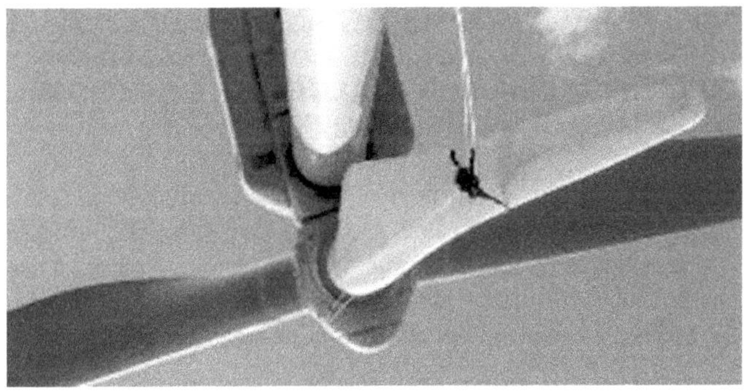

FIG. 2.44: WITH INCREASING TOWER HEIGHTS OPPORTUNITIES AS WELL AS CHALLENGES ARE GREAT! (SOURCE:[20])

Pursuing higher hub heights can truly enable wind to be deployed across all 50 states as shown in figure 2.45[22]

[17] https://www.energy.gov/eere/articles/wind-turbines-bigger-better
[18] https://www.energy.gov/eere/articles/wind-turbines-bigger-better
[19] https://www.energy.gov/sites/prod/files/2015/05/f22/Enabling-Wind-Power-Nationwide_18MAY2015_FINAL.pdf
[20] https://www.energy.gov/eere/wind/downloads/increasing-wind-turbine-tower-heights-opportunities-and-challenges, 2019
[21] https://www.energy.gov/eere/wind/downloads/increasing-wind-turbine-tower-heights-opportunities-and-challenges
[22] https://www.energy.gov/sites/prod/files/2015/05/f22/Enabling-Wind-Power-Nationwide_18MAY2015_FINAL.pdf, 2015

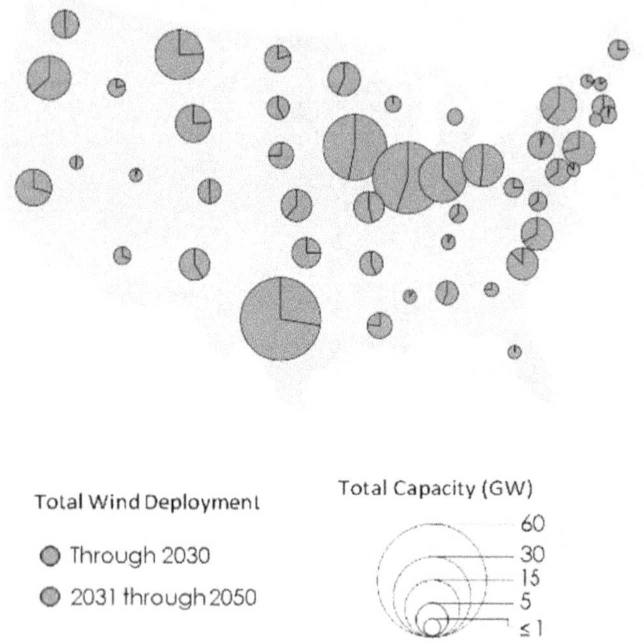

FIG. 2.45: LAND AREA ACHIEVING A MINIMUM 30% NET CAPACITY FACTOR BY GRID CELL, BASED ON CURRENT TECHNOLOGY, LARGER ROTOR DESIGNS AND A 140-M HUB HEIGHT (SOURCE:[23])

Note: Dark blue coloring identifies high quality wind resource areas that see no change in available land area meeting the 30% minimum net capacity factor threshold with technology advancement because the entire area is capable of achieving this threshold today. Light blue coloring identifies land area that meets the capacity factor threshold today but sees an increase in the proportion of the area able to achieve this threshold as a result of turbine and hub height improvements. Orange coloring identifies new land area able to achieve the minimum 30% net capacity factor level as a result of turbine and hub height improvements.

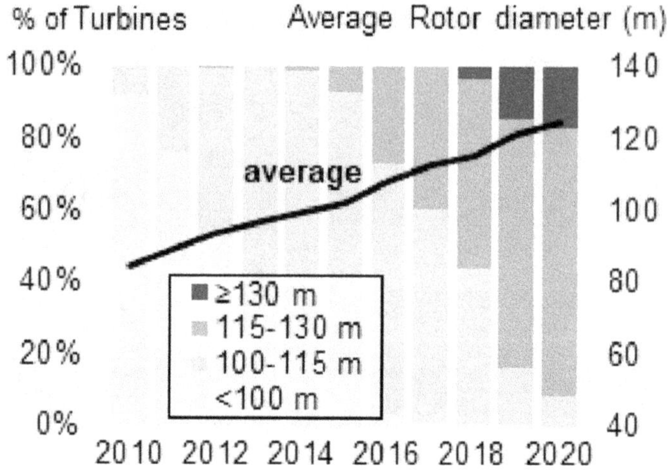

FIG. 2.46: VARIATION OF ROTOR DIAMETER SINCE 2010 (SOURCE:[24])

2.9.2.5 Rotor Diameter

Wiser and Bolinger (2014) illustrate in figure 2.47 that average rotor diameter has increased from approximately 80 m in 2008 to nearly 97 m in 2013. This represents an increase in rotor diameter of just over 20% and an increase in rotor swept area of nearly 50%.

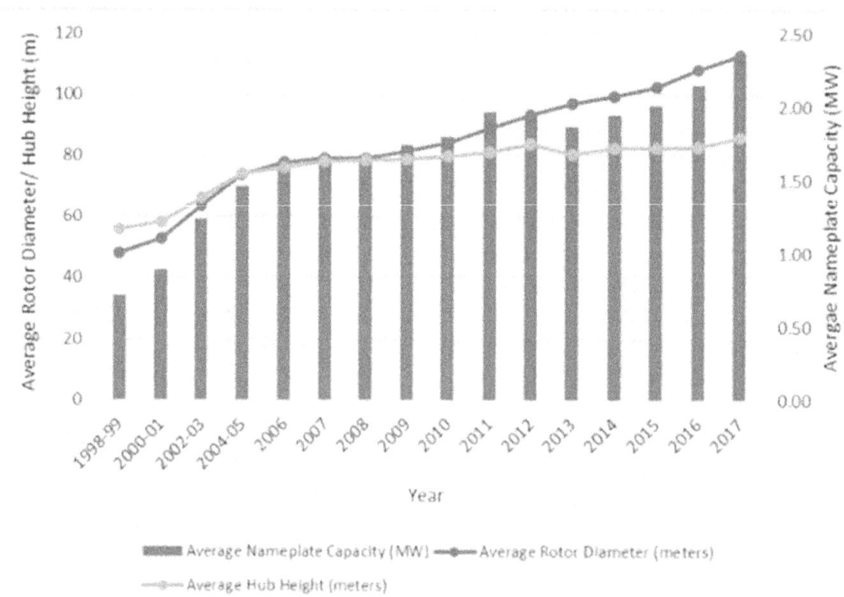

FIG. 2.47: HISTORICAL TRENDS IN U.S. WIND TURBINE ROTOR DIAMETER, HUB HEIGHT, AND NAMEPLATE CAPACITY SOURCE: WISER AND BOLINGER 2014[25]

[23] https://www.energy.gov/sites/prod/files/2015/05/f22/Enabling-Wind-Power-Nationwide_18MAY2015_FINAL.pdf, 2015

[24] https://www.energy.gov/eere/articles/wind-turbines-bigger-better

[25] https://www.energy.gov/sites/prod/files/2015/05/f22/Enabling-Wind-Power-Nationwide_18MAY2015_FINAL.pdf.

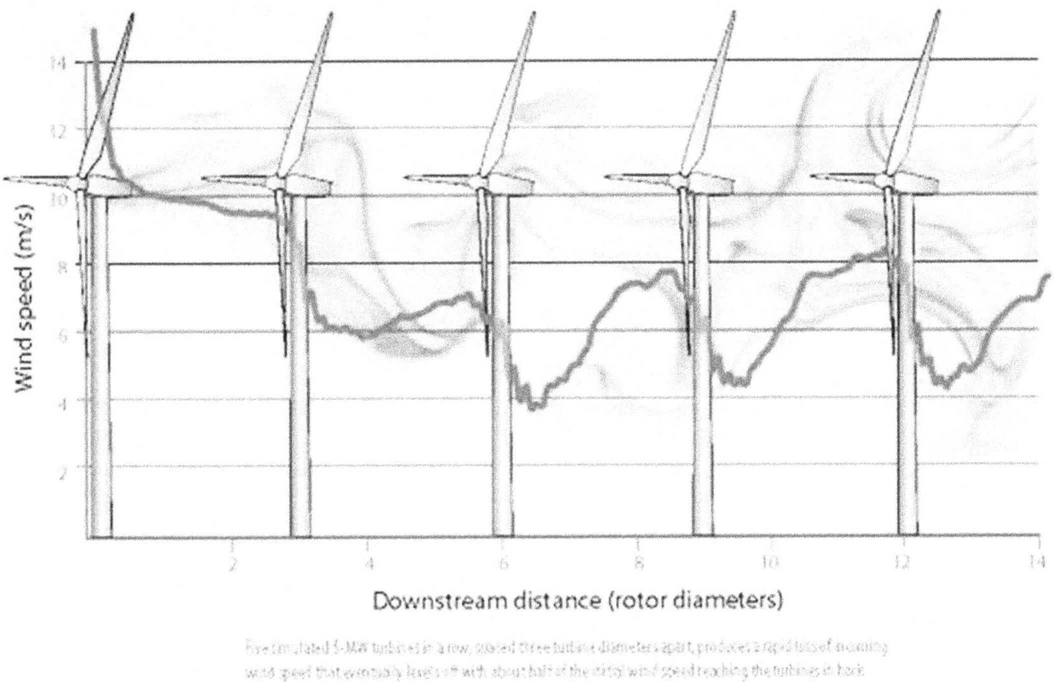

FIG. 2.48: WIND SPEEDS DOWNSTREAM DISTANCE FROM TURBINES (SOURCE:[26])

A turbine's rotor diameter, or the width of the circle swept by the rotating blades (the dotted circles in Figure 2.46), has also grown over the years. Back in 2010, no turbines in the United States employed rotors that were 115 meters (380 feet) in diameter or larger. In 2020, 91% of newly installed turbines featured such rotors. The average rotor diameter in 2020 was about 125 meters (410 feet), longer than a football field.[27]

Larger rotor diameters allow wind turbines to sweep more area, capture more wind, and produce more electricity. A turbine with longer blades will be able to capture more of the available wind than shorter blades, even in areas with relatively less wind. Being able to harvest more wind at lower wind speeds can increase the number of areas available for wind development nationwide. Due to this trend, rotor swept areas have grown 570% since 1998 – 1999.[28]

Most of the downtime-causing damage suffered by wind turbines in the field afflicts the gear boxes (60%), generators (14%), and rotor blades (7%). A long blade of a turbine makes it easy for turbulent wind to monkey with the gearbox and generator which is partly why it is so common to see that they are not spinning. Arrangement of turbines and the path of the wind can combine to produce temporary dead spots in the air flow, see figure 2.48. But when a turbulent stream twists the rotor out of plane of rotation, the leverage can turn into damage. It costs more than a quarter $1,000,000 just to get a crane large enough to reach the hub of the damage turbine out to the site and the cost to repair or replace the broken component comes on top of that. It is more economical to wait until all of them are broken before getting a crane to fix them all at once.

2.9.2.6 Inverters

Inverters change the voltage and frequency produced by a wind turbine's generator into power that can be used by homes and businesses or supplied to the electric grid. Until 2016, manufacturers could obtain ready-made inverters for their small wind turbines. But growth of the solar industry around that time led to an unintended side effect: inverter manufacturers abandoned the distributed wind market to focus on inverters for distributed solar. That's where Intergrid, and the CIP came in. Intergrid received a CIP component innovation award in 2014 to develop a modular power inverter for wind turbines under 25 kW. The inverter meets the latest grid interconnection standards while incorporating the ability to operate independently from the grid when combined with storage, providing resilient backup power. Under the Competitiveness Improvement Project (CIP), Intergrid's new 25-kW inverter

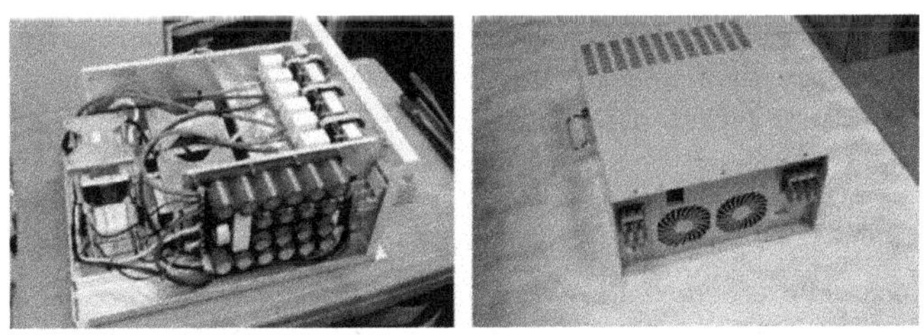

FIG. 2.49: THE MODULAR POWER INVERTER FROM INTERGRID, FOR THE DISTRIBUTED WIND INDUSTRY. (SOURCE: *PHOTOS BY ROB WILLS, INTERGRID*[29])

[26] https://www.lanl.gov/discover/publications/1663/2014-april/_assets/docs/1663_21_wind.pdf
[27] https://www.energy.gov/eere/articles/wind-turbines-bigger-better
[28] https://www.energy.gov/eere/wind/listings/articles-about-next-generation-technologies?page=1, June 2, 2021
[29] https://www.energy.gov/eere/wind/articles/competitiveness-improvement-project-awardee-fills-critical-gap-distributed-wind

completed certification to grid interconnection standards. As the first inverter developed specifically for small- and medium-scale wind systems, it's truly a game changer for the distributed wind industry. Managed by NREL on behalf of the Wind Energy Technologies Office, the CIP supports innovation to advance wind energy as a low-cost distributed energy resource option. Distributed wind, or on-site wind power, holds promise for numerous applications, from homeowners seeking low-cost renewable energy for their homes and electric-vehicle charging to farmers and other small businesses wanting more resilient on-site power. Intergrid's 25-kW inverter will go into mass production to start in 2021 with initial capacity to produce more than 1,000 units per year. The modular design of Intergrid's inverter also enables it to be scaled up for use with medium-scale turbines with capacities of hundreds of kilowatts. Intergrid wind power inverter platform develop an inverter and control system for wind generators in the 50-kW to 300-kW size class. This will support a broader range of wind turbine sizes for the distributed wind market and enable wind technology to become a more versatile distributed energy resource option.,[30] see figure 2.49.

Inverters Fill the Critical Gap for "Distributed Wind Industry". Competitiveness Improvement Project (CIP) Awardee Intergrid fills critical gap for "Distributed Wind Industry" using funding from DOE's program. Intergrid LLC has introduced the first inverter developed specifically for small- and medium-scale wind turbines.[31]

2.9.2.7 Life Extension, Recycling & Circular Economy

Emerging technologies promise to increase opportunities for reuse and recycling. Wind energy is one of the fastest-growing sources of electricity generation growth that is vital for reducing carbon emissions from electric power. However, building wind turbines requires large quantities of materials and components. These components will eventually reach the end of their design lifetimes and could go to landfills unless innovation provides an answer. Using, reusing, recycling, and remanufacturing wind turbine materials will reduce waste and create a "circular economy." A circular economy for energy materials also means that technology should be engineered from the start to require fewer materials, resources, and energy while lasting longer and having components that can easily be broken down for use in subsequent applications. "The earliest point in a product lifecycle that circular economy strategies can be applied is in the design of products from the ground up literally," said Tyler Christoffel, Oak Ridge Institute for Science and Technology Fellow at the Department of Energy (DOE). "Wind turbines must be designed sustainably from the concrete base, through the tower and nacelle, all the way to the blade tips."[32]

According to a recent NREL publication, as renewable energy generation and storage become central components of the clean energy economy, new materials and manufacturing strategies will be necessary to ensure that such technologies are not contributing to end-of-life challenges. Shifts in recycling technologies, new materials, extended useful life, and innovative reuse are needed to move toward a circular economy for wind turbine blades. The Wind Energy Technologies Office (WETO) is working with researchers across industry, academia, and DOE national laboratories to focus on wind blade recycling and enable a circular economy for the entire wind turbine,[34] see figure 2.50.

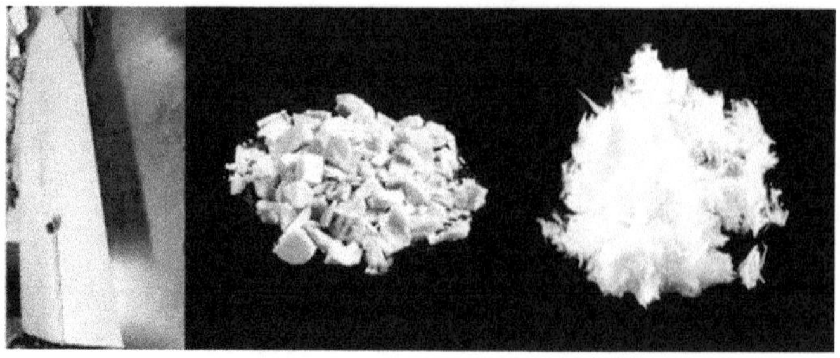

FIG. 2.50: THE CARBON RIVERS AND UNIVERSITY OF TENNESSEE, KNOXVILLE, BLADE RECYCLING PROCESS TAKES DECOMMISSIONED WIND TURBINE BLADES (LEFT), MECHANICALLY REDUCES THEM INTO SMALL PIECES (MIDDLE), AND THEN USES A NOVEL PYROLYSIS TECHNIQUE TO RECOVER HIGH-QUALITY GLASS FIBER (RIGHT). (*IMAGES COURTESY OF CARBON RIVERS,* SOURCE[33])

Recycling and upcycling wind turbine blade materials, such as fiberglass or carbon fiber, are difficult to separate from other components and their value is often degraded even when successfully extracted. With WETO funding, researchers from the University of Tennessee and Carbon Rivers are developing a novel pyrolysis, or thermal decomposition, method of reclaiming fiberglass from retired wind turbine blades. Extending life cycles for wind turbine components like blades can fail and require replacement earlier than expected, indicating a need for manufacturing methods and materials that optimize longer life cycles. National Renewable Energy Laboratory (NREL) researchers recently assessed the current end-of-life treatment for wind turbine blades and identified methods for achieving a circular economy for wind energy, including design for circularity, lifetime extension, mechanical recycling/upcycling, and uses in cement production. Advanced drone and robotic maintenance and repair strategies for older blades, as well as diagnostic and inspection techniques that support certifying wind turbines to operate for an extended lifetime, could help support sustainability. With WETO funding, Sandia National Laboratories is developing advanced blade inspection techniques and improving repair capabilities as part of its Blade Reliability Initiative. Incorporating new and fewer innovative materials such as carbon fiber, are making wind turbine blades lighter, more durable, and easier to recycle. As part of this effort, DOE is supporting the following:[35]

- Arctura's development of a blade coating that can mitigate lightning strike damage.
- Resodyn Corporation's development of a blade coating that resists erosion.
- Joint research by Oak Ridge National Laboratory and NREL into advanced manufacturing processes to make blades and generators with fewer materials.

---

[30] https://www.energy.gov/eere/wind/articles/competitiveness-improvement-project-awardee-fills-critical-gap-distributed-wind
[31] https://www.energy.gov/eere/wind/articles/competitiveness-improvement-project-awardee-fills-critical-gap-distributed-wind
[32] https://www.energy.gov/eere/wind/articles/no-time-waste-circular-economy-strategy-wind-energy
[33] https://www.energy.gov/eere/wind/articles/no-time-waste-circular-economy-strategy-wind-energy
[34] https://www.energy.gov/eere/wind/articles/no-time-waste-circular-economy-strategy-wind-energy
[35] https://www.energy.gov/eere/wind/articles/no-time-waste-circular-economy-strategy-wind-energy

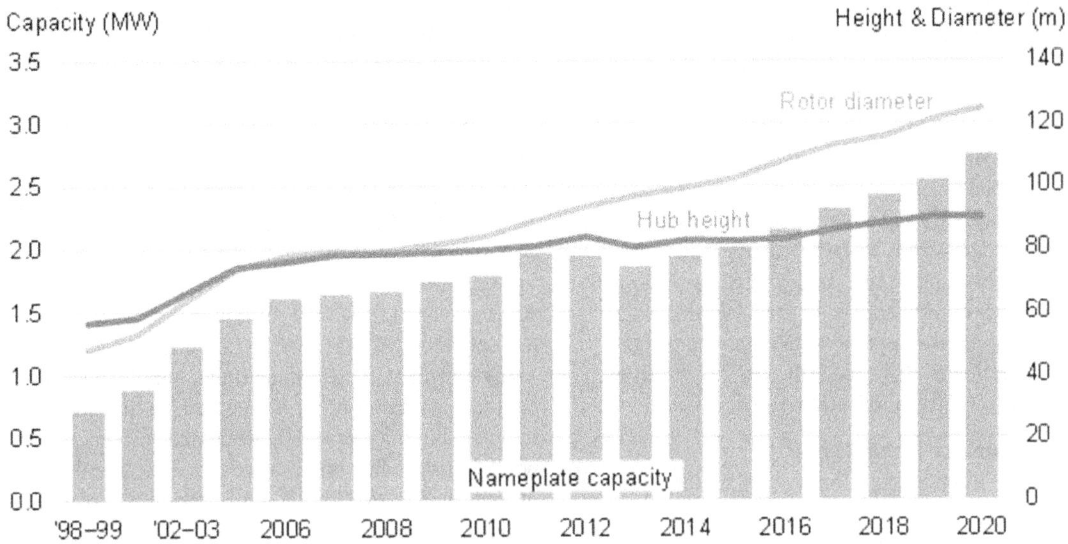

FIG. 2.51: AVERAGE TURBINE HUB HEIGHT, ROTOR DIAMETER, AND NAMEPLATE CAPACITY FOR LAND-BASED WIND PROJECTS (SOURCE[36])

- "WETO efforts will continue to reduce material and energy usage and cost, supporting expansion of cost-effective, sustainable wind energy deployment," Christoffel said.[37]

2.9.2.8 Name Plate Capacity

In addition to getting taller and bigger, wind turbines have also increased in maximum power rating, or capacity, since the early 2000s. The average capacity of newly installed U.S. wind turbines in 2020 was 2.75 megawatts (MW), up 8% since 2019 and 284% since 1998 - 1999. In 2020, there was a sharp increase for turbines installed in the 2.75 –3.5 MW range. More wind energy per turbine means that fewer turbines are needed to generate a desired capacity across a wind plant, ultimately leading to lower costs,[38] see figure 2.51.

These high-energy-capture turbines have lower specific-power ratings, meaning they have larger rotors relative to the size of their generator, which helps them capture more wind and perform better at lower wind speeds, thereby increasing the turbine capacity factor. Building a better (highly flexible) blade has capacity for greater swept area. A greater swept area helps capture more wind energy and access higher wind speeds at elevated heights. Increased capacity factors with increasing the rotor size while maintaining the nameplate generation size as intended gives sustained energy output of a turbine at peak performance can increase the amount of time the turbine operates at rated power. In some circumstances, higher capacity factors may increase the value of wind energy to the electricity system. Expanded wind energy development and more efficient production may be possible with turbines with low-specific-power ratings and expand wind energy development into new areas like the southeastern United States by allowing plant operators to tap into lower-speed wind areas at a competitive price.

2.9.2.9 Transportation Challenges

If bigger is better, why aren't even larger turbines used currently? Although turbine heights and rotor diameters are increasing, there are a few limitations. Transporting and installing large turbine blades for land-based wind is not easy, since they cannot be folded or bent once constructed. This limits the routes trucks can take and maneuver radius of their turns. Turbine tower diameters are also difficult to manage since they may not fit under bridges or highway overpasses. DOE is addressing these challenges through its research projects. For instance, DOE is designing turbines with more slender and flexible blades that can navigate through curves in roads and rail lines that conventional blades cannot. DOE is also supporting efforts to develop tall turbine towers that can be produced on site, thus eliminating tower transportation issues.[39]

High energy capture turbines with larger rotors have many advantages, but the length and width of their blades creates transportation challenges that must be overcome to make true on the adage, bigger is better.[40] Slender, Flexible Blades Enable high-energy-capture wind turbines by resolving transportation challenges, the big adaptive rotor (bar) project works to design and enable the next generation of land-based wind turbines with rotors that stretch 206 meters (m) in diameter, which will increase capacity factors by 10% or more over a typical land-based turbine. There are fewer transportation constraints with slender, flexible blades which can navigate curves in rail lines that conventional blades cannot at elevated heights. The ability to deliver a single-piece, 100-m blade to site by rail likely represents the lowest cost and most reliable option for blades of that size.[41]

---

[36] https://www.energy.gov/eere/articles/wind-turbines-bigger-better
[37].https://www.energy.gov/eere/wind/articles/no-time-waste-circular-economy-strategy-wind-energy
[38].https://www.energy.gov/eere/articles/wind-turbines-bigger-better
[39] https://www.energy.gov/eere/articles/wind-turbines-bigger-better
[40] https://www.energy.gov/eere/wind/downloads/flexing-limits-land-based-wind-turbine-rotor-growth
[41] https:// www energy gov/ sites/default/files 2021/01/f82/bar- fact- sheet- 2021.pdf

2.9.2.10 Artificial Intelligence

Open Operational Assessment (OpenOA) is an open-source data toolkit for analyzing the performance of operating wind power plants. The OpenOA toolkit uses artificial intelligence and big data solutions to help wind plant operators characterize, predict, and optimize wind plant operations, in turn helping to maximize revenue.[42] Thus, applications of artificial intelligence exist across wind energy science. Artificial intelligence (AI) and machine learning (ML) has transformed the process of data-driven science, leading to new fundamental insights and improved predictive models that can address many outstanding challenges in wind energy. In this emerging paradigm, AI/ML techniques provide a new pathway to obtaining computationally efficient surrogate models that encode the key insights from high fidelity models to apply AI/ML across wind energy science.[43]

2.9.2.11 Hybrid and Storage Technologies[44]

Research under DOE's Technology Commercialization Fund support promising energy technologies, including those that support hybrid power systems. Field Testing and Validation of Hybrid Optimization and Performance Platform (HOPP) with GE Global Research is a HOPP is a software tool that enables detailed design, analysis, and optimization of hybrid power plants down to the component level. It has the capability to assess and optimize projects containing combinations of wind (land-based and offshore), solar, storage, geothermal, and hydro.[45]

NREL Researchers Study Synergistic Value Streams in Hybrid Power Plants.[46] Ultimately, advancing hybrid systems research at NREL and other national laboratories will require more coordination with industry. "Bringing the key players together will help us to recognize their motivations and close the gap in how to represent them in our models."[47] Hybrid renewable energy systems combine multiple renewable energy and or energy storage technologies into a single plant, and they represent an important subset of the broader hybrid systems universe. These integrated power systems are key to unlocking maximum efficiency and cost savings in future decarbonized grids, but NREL analysis indicates there are challenges in the benefits of hybrids with the tools used to help plan those future grids. In comparing hybrids to standalone alternatives, it is important to tackle: Is it always beneficial to combine renewable and storage technologies, instead of siting each technology where

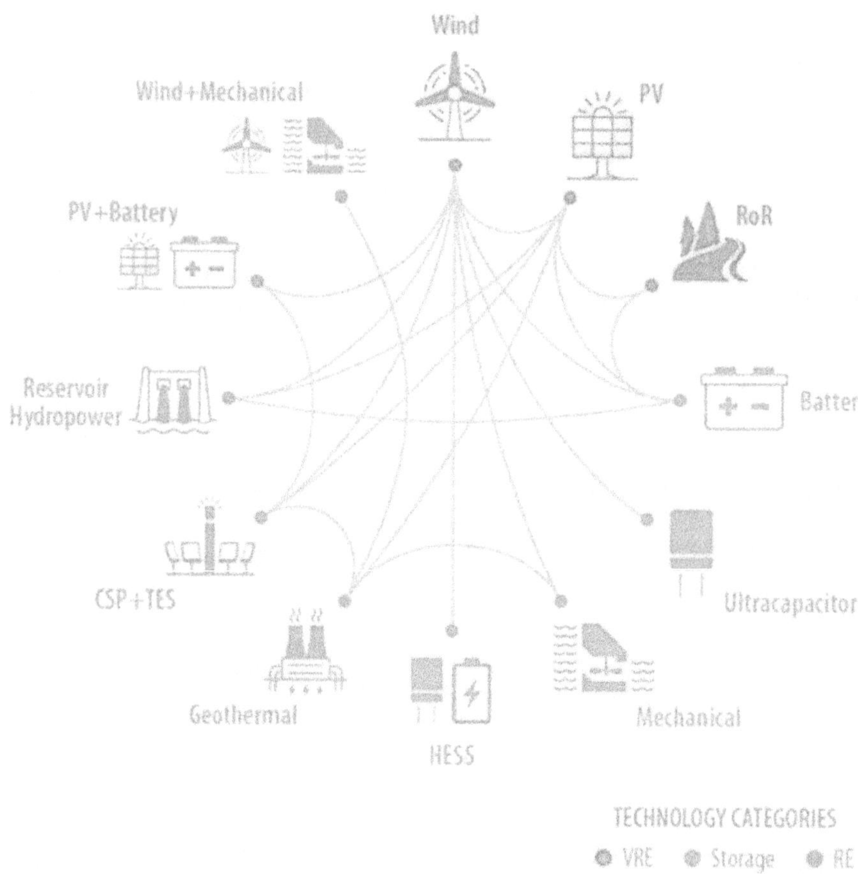

FIG. 2.52: DOE HYBRID POWER SYSTEMS SUPPORTING WIND ENERGY (SOURCE: IMAGE: DENNIS SCHROEDER, NREL[48])

[42] https://www.energy.gov/eere/wind/articles/doe-announces-2021-technology-commercialization-fund-selections
[43] https://www.youtube.com/watch?v=dAI_3bn1Wk
[44] https://www.nrel.gov/news/features/2021/are-hybrid-systems-truly-the-future-of-the-grid.html
[45] https://www.energy.gov/eere/wind/articles/doe-announces-2021-technology-commercialization-fund-selections
[46] https://www.nrel.gov/news/features/2021/are-hybrid-systems-truly-the-future-of-the-grid.html
[47] *Citing Devonie McCamey from energy analysis and grid modernization report* https://www.nrel.gov/news/features/2021/are-hybrid-systems-truly-the-future-of-the-grid.html
[48] https://www.energy.gov/eere/wind/articles/doe-announces-2021-technology-commercialization-fund-selections

their individual contributions to the grid can be maximized? Or are only certain hybrid designs beneficial? Does the energy research community consistently represent the characteristics of hybrids in power system models? The results of this body of work highlight some gaps between what different models show and what many in the energy community have proclaimed when it comes to the value of hybrid systems to the future grid. "Hybridization creates opportunities and challenges for the design, operation, and regulation of energy markets and policies and current data, methods, and analysis tools are insufficient for fully representing the costs, value, and system impacts of hybrid energy systems," said Paul Denholm, NREL principal energy analyst. First, What Do We Mean When We Talk About Hybrid Systems? NREL Proposes a Taxonomy To Delineate What Makes a System a True Hybrid which predicts, and optimizes wind plant operations, in turn helping to maximize revenue.[49]

In figure 2.52 blue nodes represent variable renewable energy (VRE) technologies, green nodes represent energy storage technology types, and orange nodes represent less-variable renewable energy (RE) technologies or systems; arcs indicate technology pairs that have been proposed in the literature. PV: photovoltaic; RoR: run-of-river; HESS: hybrid energy storage system; CSP + TES: concentrating solar power with thermal energy storage; the Mechanical storage icon encompasses compressed air energy storage and flywheels, both of which ultimately convert the stored energy to electricity.[50]

"Ability to quantify hybrids' potential impacts could be hindered by inconsistent treatment of these systems and incomplete understanding of which aspects of hybridization will have the greatest influence," Murphy said. After a thorough literature review, the team developed a new organization scheme for utility-scale systems that combine renewable and energy storage technologies, only a subset of which can truly be called "hybrids." They came up with three categories based on whether the systems involve locational or operational linkages, or both. "We found that technology combinations do not represent a meaningful delineation between hybrids and non-hybrids, only the nature of the linkages are more important distinctions," Murphy said. The resulting categories can help inform policy considerations, as they define system characteristics that could challenge existing siting, interconnection process, and policy implementations. The taxonomy is also helpful in informing model development efforts, as the categories identify the unique characteristics to represent hybrid systems in a model, including the effects of the linkages on both a project's costs and the values it can deliver to the grid. NREL analysts homed in on a set of technology combinations and linkages that are consistent with a true hybrid system, co-optimizing the design and self-scheduling of linked technologies to maximize net economic benefits. NREL modeled hybrid systems using three different tools that underpin many of the laboratory's forward-looking power system studies.[51]

"By creating opportunities to directly solicit insights from industry, utility planners, and other stakeholders, move toward a deeper understanding of what sources of value are driving industry interest in hybrids," Murphy said. "Is there inherent value that can only be unlocked through hybridization, or is some of the value embedded in the familiar? By adding storage to variable resources, we can make them look and participate more like the controllable resources we are used to having on the power system.[52]

2.9.3 Next Generation Scope - Analysis

The topics addressed in this section 2.7.3 follow US DOE's efforts and are sequentially Market Analysis – On Land; Market Analysis – Offshore; Distributed Wind; Grid Integration; Environmental - Wind Turbine Siting; Cyber Security; Digital; High Fidelity - HPC & HPM; Compressible & Incompressible Flow Models and Nonlinear Structural Flows.

"Next-Generation Wind Technologies" Publications addressed[53]

- "Flexing the Limits of Land-Based Wind Turbine Rotor Growth" which included "The Big Adaptive Rotor" project to design and enable the next generation of land-based wind turbines.
- 2020 Wind Energy Research and Development Highlights 2020's most notable wind energy research and development accomplishments.
- The 2019 report presented opportunities, challenges, and potential associated with increasing wind turbine tower heights
- 2018 Offshore Wind Technologies Market Report.
- "Enabling the SMART Wind Power Plant of the Future Through Science-Based Innovation" explains how new energy science and technological breakthroughs could cut the cost of wind energy in half by 2030. and finally
- "Enabling Wind Power Nationwide" Report shows how the United States can unlock the vast potential for wind energy deployment in all 50 states.
- "FASTFarm", a new modeling tool from NREL,[54] will help users predict the power performance and structural loads of wind turbines within a wind farm, enabling them to maximize the farm's power production and reliability, see figure 2.53.

FIG. 2.53: PICTURE SHOWS HOW DOE ANNOUNCED TEAMS COMPETED IN THE 2022 COLLEGIATE WIND COMPETITION (SOURCE:[55])

[49] https://www.energy.gov/eere/wind/articles/doe-announces-2021-technology-commercialization-fund-selections
[50] Source: "A Taxonomy of Systems that Combine Utility-Scale Renewable Energy and Energy Storage Technologies", https://www.nrel.gov/news/features/2021/are-hybrid-systems-truly-the-future-of-the-grid.html
[51] https://www.nrel.gov/news/features/2021/are-hybrid-systems-truly-the-future-of-the-grid.html
[52] https://www.nrel.gov/news/features/2021/are-hybrid-systems-truly-the-future-of-the-grid.html
[53] https://www.energy.gov/eere/next-generation-wind-technology, April 15, 2021
[54] https://www.energy.gov/eere/wind/articles/new-modeling-tool-helps-maximize-wind-farm-power-production-and-reliability
[55] https://www.energy.gov/eere/wind/listings/articles-about-next-generation-technologies?page=1, February 18, 2021.

FIG. 2.54: PICTURE FROM LAND-BASED WIND MARKET REPORT: 2021 EDITION RELEASE (SOURCE:[56])

FIG. 2.55: LAND-BASED WIND TURBINES FOR UTILITY-SCALE LAND-BASED WIND DEVELOPMENT WHICH ARE RAPIDLY EXPANDING IN THE UNITED STATES. PHOTO BY DENNIS SCHROEDER, NREL (SOURCE:[57])

[56] https://www.energy.gov/eere/wind/articles/no-time-waste-circular-economy-strategy-wind-energy

[57] https://cleantechnica.com/2021/01/05/2020s-top-wind-energy-rd-achievements/

DOE encouraged wind turbine tests, at the collegiate level with 11 teams selected to compete in the 2022 Collegiate wind competition, where students design, build, and test a model wind turbine, see figure 2.53.

2.9.3.1 Market Analysis – Turbines On Land

The Big Adaptive Rotor (BAR) project works to design and enable the next generation of land based wind turbines with rotors that stretch 206 meters (meters) in diameter, which will increase capacity factors by 10% or more over a typical land based turbine. These high energy capture turbines have lower specific power ratings, meaning they have large rotors related to the size of the generator, which helps them capture more wind and perform better at lower wind speeds, thereby increasing the turbine capacity factor.

Land-Based Wind and Cross-Cutting Wind Technology developed by High-Efficiency, Lightweight Wind Turbine Generators of WETO supported three projects to develop next-generation wind turbine drivetrain technologies that will facilitate the continued growth of wind turbines. Lightweight generators are important because the size and weight of the generator impact the weight and cost of the wind turbine tower and foundation, as well as the specialized equipment needed to transport and install the large components. General Electric (GE) Research designed a prototype high-efficiency ultra-light low temperature superconducting generator that leverages investments from the magnetic resonance imaging (MRI) industry, eliminates the need for foreign- sourced rare earth materials, and reduces generator mass as compared to current technologies. WEG Energy Corporation developed a high-efficiency permanent magnet direct drive lightweight generator to integrate into its existing platform of American Superconductor Corporation developed a high-efficiency lightweight wind turbine generator that incorporates high- temperature superconducting materials to replace permanent magnets in the generator rotor.[58]

2.9.3.2 Market Analysis – Offshore

Significant WETO-supported offshore wind demonstration projects are:

- New England Aqua Ventus Demonstration Project in which Diamond Offshore Wind is a subsidiary of the Mitsubishi Corporation, and RWE Renewables bring global expertise in offshore wind project with project's floating concrete foundation.
- Lidar Buoy Deployment in September 2020, WETO leveraged funding from the Bureau of Ocean Energy Management (BOEM) to deploy two offshore wind research buoys off the coast of California in water that is 2,000–3,000 feet deep.[59]
- Webinar Series Improves Understanding of Offshore Wind Turbine-Radar Interference. Offshore wind turbines can interfere with radar systems used for air traffic control, weather forecasting, and homeland security. To understand potential impacts, the Wind Turbine Radar-Interference Mitigation Working Group hosted a series of webinars.[60]
- The Offshore Wind Market Report 2021 Edition provides detailed information on the domestic offshore wind industry to provide a U.S. context and help navigate technical and market barriers and opportunities.[61]
- The National Offshore Wind R&D Consortium was established by DOE and the New York State Energy Research and Development Authority with the State agencies in Maryland, Virginia, and Massachusetts as Consortium for industry's prioritized offshore wind R&D topics.[62]

FIG. 2.56: AN ILLUSTRATION OF OFFSHORE WIND TURBINES IN THE OCEAN, ILLUSTRATION BY JOSHUA BUAER, NREL (SOURCE:[63])

[58] https://cleantechnica.com/2021/01/05/2020s-top-wind-energy-rd-achievement
[59] https://cleantechnica.com/2021/01/05/2020s-top-wind-energy-rd-achievements/_
[60] https://cleantechnica.com/2021/01/05/2020s-top-wind-energy-rd-achievements/_
[61] https://cleantechnica.com/2021/01/05/2020s-top-wind-energy-rd-achievements/
[62] https://cleantechnica.com/2021/01/05/2020s-top-wind-energy-rd-achievements/_
[63] https://cleantechnica.com/2021/01/05/2020s-top-wind-energy-rd-achievements/_

- In early 2020, DOE's National Renewable Energy Laboratory (NREL), funded by WETO, developed a 15-megawatt (MW) offshore wind reference turbine, an open-access design of a complete wind turbine system. The reference turbine accommodates multiple software tools and features separate modules for fixed-bottom wind turbines in shallow water and deep-water, floating wind turbines.[64]

2.9.3.3 Distributed Wind

Technology innovation pathways for distributed wind balance-of-system is meant for cost reduction of Wind Energy Program. Figure 2.58 is a photo of CBF's Brock Environmental Center against a blue sky with its solar panels and residential wind turbines, CBF's Brock Environmental Center produces nearly twice as much energy as it using distributed energy technology.

FIG. 2.57: PICTURE SHOWS TOP TRENDS IN OFFSHORE WIND MARKET WITH A DECADE OF UNPRECEDENTED GROWTH. (SOURCE:[65])

Competitiveness Improvement Project, NREL, with funding from WETO, works with dozens of small businesses across the United States to advance wind technology as a distributed energy resource through the Competitiveness Improvement Project (CIP). Launched in 2013, the CIP supports manufacturers of distributed wind technology, typically small and medium scale wind turbines through competitively-awarded, cost-shared projects aiming to: optimize designs for increased energy production and grid support; 2) test turbines and components to national standards to verify performance and safety; and 3) develop advanced manufacturing processes to reduce hardware costs. Between 2013 and 2020, NREL was leveraging with private-sector involvement. In August 2020, DOE selected eight projects to receive DOE funding and leveraging for industry support.[66]

The Microgrids, Infrastructure Resilience, and Advanced Controls Launchpad (MIRACL) project is part of WETO's portfolio of investments aimed at enabling wind technology to be an affordable, accessible, and compatible distributed energy resource DER). In 2020, this collaborative research project which includes DOE's National Renewable Energy Laboratory, Pacific Northwest National Laboratory (PNNL), Sandia National Laboratories, and Idaho National Laboratory established a high-speed data link called the MIRACL Hub to leverage nearly a billion dollars' worth of research capabilities and DER infrastructure across the four laboratories. This enables research, development, and industry validation of DER design and planning tools, components, and hybrid energy systems across multiple scales and configurations in geographically diverse operating conditions. A 2020 PNNL report found that many user facing distribution modeling and simulation tools do not accurately calculate the performance or reflect the locational, and resilience value of distributed wind identifying a future area of work for MIRACL R&D.[67]

FIG. 2.58: THIS IS A PHOTO OF CBF'S BROCK ENVIRONMENTAL CENTER, PHOTO BY PRAKASH PATEL, PNNL. DISTRIBUTED WIND MARKET REPORT: 2021 EDITION RELEASED (SOURCE:[69])

---

[64] https://cleantechnica.com/2021/01/05/2020s-top-wind-energy-rd-achievements/_

[65] https://www.energy.gov/eere/wind/articles/no-time-waste-circular-economy-strategy-wind-energy

[66] https://cleantechnica.com/2021/01/05/2020s-top-wind-energy-rd-achievements/

[67] https://cleantechnica.com/2021/01/05/2020s-top-wind-energy-rd-achievements/

[68] https://windurance.com/2020/01/23/6-advantages-of-distributed-wind-generation/

[69] https://cleantechnica.com/2021/01/05/2020s-top-wind-energy-rd-achievements/

How is Distributed Energy related to Small-Scale Wind Power System?[68]

Distributed wind is "behind the meter," meaning the power generation is on the customer's side of the electric meter. Power created by a distributed wind generator is consumed by the customer, but it's not lost if that customer's demand is low. The rest can be sold back to the provider through a process called net metering. Distributed wind is often called *small wind* because it produces limited amounts of energy compared to the large utility-scale wind farms. Distributed wind is ideal for small-scale operations and residences. The total potential of available distributed wind sites in the United States (8,000 GW, including larger systems)

Advantages of Distributed Wind Generation[70]

Schools, farms, businesses, and other institutions have the perfect conditions for distributed wind. According to the Distributed Wind Energy Association (DWEA), the issues of small-wind solutions are:

- Capture untapped potential
- Profit from buy-backs
- Avoid paying for shipping
- Increase the electric grid's resilience
- Take advantage of shrinking pitch control technology
- Help renew American manufacturing!

Benefits of Distributed Wind Generation[71]

Benefit #1: Captures Untapped Wind Energy

Not every location has room for large-scale utility developments. Small-scale wind power generation taps sources of wind energy that might otherwise go unused. For example, a small community may not allow installation of a massive, 80-meter wind tower to serve a school. However, more attractive small-scale wind turbines may be approved that are half that height. Distributed wind breaks down the barriers that prevent communities from adopting renewable energy.

Benefit #2: You Can Sell It Back

Your average electric utility runs power-generating facilities to distribute and sell electricity to consumers. In a distributed wind setting, the consumer owns the asset, uses that asset to generate power for his own consumption, then sells the excess back to the utility. The percentage of refund you get back from the utility varies by state. In some states, such as New York, you can sell the energy directly to another nearby consumer through remote net metering. In this scenario, a farm could generate wind power, then sell his extra energy to the farm 30 miles down the road through a cooperative generation sharing program

Benefit #3: You Don't Pay to Ship the Energy

One of the challenges of utility-scale wind energy is that the resource is usually very far away from the end user. High-voltage transmission lines are necessary to transmit power from any power plant to the point of use. The greater the distance, the greater the transmission losses and customer costs. Consumers who choose to use distributed wind energy generated nearby are often able to actively lower their power bill by avoiding traditionally inefficient and more expensive power generation options.

Benefit #4: Increase Grid Resilience

Small-wind energy producers do their part to relieve the pressure on the electrical grid. If more consumers rely on independent sources of energy, big fluctuations in energy usage will be less likely to cause grid instability or even blackouts. The future of power generation will include advanced, resilient microgrid systems that allow power production and distribution even when outages or disaster may strike. A diverse energy portfolio will be important in future decades as the world's population battles climate change.

Benefit #5: Distributed Energy is a rising star in American clean-energy solutions.

And it's not offshore wind but Distributed Wind Generation which produces green energy, contributes to the health of the power grid, and creates low-cost alternatives for people and businesses. Distributed wind generation could make a dent in climate change, and over 90% of wind turbines installed in America last year for use in distributed wind setups were built right in the U.S., whereas over 90% of the solar modules installed last year were imported from China.

Benefit #6: Advances in Wind Turbine Technology

Today's distributed-scale wind turbines are a far cry from the old, metal Aermotor windmills that still peppered across America from a 100 years ago. Thanks to advances in wind turbine technology, small-scale systems can enjoy most of the same features as the utilities. Whether the application is a stand-alone or a grid-connected system it is worth considering the many advantages of distributed generation through small-scale wind energy.

Benefit #7: Helping the Renaissance of American Manufacturing

A recent NREL study estimated that 49 million sites were suitable for small-scale wind energy. Distributed wind energy is an American commodity. Small scale wind power is helping to elevate American manufacturing and with sufficient policy support from the government, distributed wind could be a big part of the next clean tech boom. Distributed wind opens a whole new market to match the demand of smaller entities like ranches, communities, and universities. It's efficient, cost effective, and environmentally conscious.

2.9.3.4 Grid Integration[72]

Grid Services Field Demonstration[73]

As energy providers explore new ways to reliably integrate renewable energy sources into the electric grid, they supply more than just power from wind. With support from WETO NREL, Avangrid Renewables, the California Independent System Operator, and General Electric conducted a series of tests to provide a full suite of grid services and respond to the dynamic needs of the electric grid. These tests performed in an active grid-operating environment over several days were promising. They show active power control capabilities for inverter-connected plants that combine wind, photovoltaic solar, and energy storage technologies could further accelerate the movement toward the electric grid of the future.

---

[70] https://windurance.com/2020/01/23/6-advantages-of-distributed-wind-generation/
[71] https://windurance.com/2020/01/23/6-advantages-of-distributed-wind-generation/
[72] https://cleantechnica.com/2021/01/05/2020s-top-wind-energy-rd-achievements/
[73] https://cleantechnica.com/2021/01/05/2020s-top-wind-energy-rd-achievements/

FIG. 2.59: NREL'S NEW 19.9 MW SUBSTATION, USED FOR ARIES UTILITY-SCALE GRID INTEGRATION RESEARCH. PHOTO BY DENNIS SCHROEDER, NREL (SOURCE:[74])

Continental Transmission Study[75]

Through the Interconnections Seam Study, NREL joined national lab, university, and industry partners to examine the value of increasing transfers of electricity across the U.S. electric grid using high-voltage direct-current transmission. The study, sponsored by EERE and the DOE Office of Electricity, quantifies the costs and benefits of strengthening the connections (or seams) between the Eastern and Western Interconnections, which could create a more integrated power system and increase efficient development and use of the nation's abundant energy resources, including solar, wind, and natural gas.

Tools to Optimize Transmission Utilization[76]

DOE's Idaho National Laboratory (INL) licensed its General Line Ampacity State Solver (GLASS) software to WindSim Americas. GLASS is a tool to dynamically rate overhead transmission lines, optimizing the use of electricity infrastructure for the integration of wind energy

2.9.3.5 Environmental Wind Turbine Siting and Atmospheric Applications[77]

Deploying wind in ways that increase its value for the nation includes the use of domestically manufactured materials, surrounding communities, the environment, wildlife and the general public. DOE quantified the economic benefits and impacts of wind deployment in terms of greenhouse gas (GHG) reductions, pollution reductions (sulfur dioxide [SO2], nitrogen oxide [NOx], and fine particulates), water use and withdrawal reductions, price stability economic impacts, land use, local land lease, local tax generation, U.S. jobs and U.S. manufacturing impacts.[78] Energy market pricing recognizes the value of low carbon, low emissions power sources which are environment related. As part of the Environmental Research, Siting, and Workforce Development, *wind energy R&D accomplishments over the past several decades are cited in https://www.energy.gov/eere/wind/downloads/wind-energy-technologies-office-lasting-impressions.*[79]

Defense Deployable Disaster Wind Turbine

The WETO-funded Defense Deployable Disaster Wind Turbine project evaluates the technical and market potential for rapidly deployable wind energy systems to meet the energy needs of defense and disaster response activities. A 2020 report found there is a significant opportunity for wind turbines to provide on-site power and overcome the risks and limitations of existing diesel-powered generators used to support foreign defense missions or humanitarian activities, such as disaster response. The report's analysis suggests that low wind speed optimized and rapidly deployable wind turbines, when integrated with battery storage, could be used to offset 80% of diesel generator use at certain forward operating bases.[80]

---

74 https://cleantechnica.com/2021/01/05/2020s-top-wind-energy-rd-achievements/

75 https://cleantechnica.com/2021/01/05/2020s-top-wind-energy-rd-achievements/

76 https://cleantechnica.com/2021/01/05/2020s-top-wind-energy-rd-achievements/

77 https://www.nrel.gov/wind/atmosphericscience.html#:~:text=Atmospheric%20Science%20NREL%27s%20technical%20experts%20are%20improving%20weather, water%2C%20have%20been%20studied%20for%20thousands%20of%20years. _8-29-21

78 https://www.energy.gov/sites/prod/files/2015/05/f22/Enabling-Wind-Power-Nationwide_18MAY2015_FINAL.pdf

79 https://www.nrel.gov/news/video/grand-challenges-in-the-science-of-wind-energy-text.html

80 https://cleantechnica.com/2021/01/05/2020s-top-wind-energy-rd-achievements/

FIG. 2.60: PHOTO OF CBF'S BROCK ENVIRONMENTAL CENTER WHICH PRODUCES NEARLY TWICE AS MUCH ENERGY AS IT USING DISTRIBUTED ENERGY TECHNOLOGY. PHOTO BY PRAKASH (SOURCE:[81])

Addressing Environmental Concerns associated with wind energy development 'The Wind Vision Report" indicates that under a scenario in which wind energy grows to serve 10% of the nation's electricity demand by 2020 and 20% by 2030, wind power is estimated to provide annual benefits of $9 billion in 2020 and $30 billion in 2030 (in terms of 2013$) from air pollution reductions greenhouse gas emissions reductions. These environmental benefits are bolstered by consumer savings for natural gas purchases outside the electric sector, wind investment derived jobs totaling 330,000-426,000 by 2030, and direct land-lease and property tax payments totaling hundreds of millions of dollars annually by 2020 and billions of dollars annually by 2030.

Bat Deterrent System Commercialized

A WETO-supported bat deterrent system developed by NRG Systems, Inc. of Vermont has been successfully commercialized. NRG announced in July 2020 that Siemens Gamesa Renewable Energy (SGRE) was providing installation support of the bat deterrent system, which uses nacelle-mounted ultrasound-generating devices and a controller integrated with the turbine's SCADA system to deter bats from wind turbines. SGRE has installed the system on its wind turbines in Ontario, Canada, and Maui, Hawaii.[82]

3D Thermal Tracking of Birds and Bats

Commissioned by WETO, DOE's Pacific Northwest National Laboratory developed ThermalTracker-3D, a stereo-vision solution for evaluating flight tracks and other data on birds and bats around offshore wind turbines. The technology uses thermal video that records movement day or night, even in limited visibility conditions. ThermalTracker-3D continuously transmits satellite data showing if and when birds and bats are active near proposed wind farms. New tool protects bats while increasing energy production. Natural Power has developed the first-ever smart curtailment technology that simultaneously reduces bat fatalities and energy losses at wind farms. Also, developing the correct innovative technology to combat right whales mitigating the offshore wind energy development for this endangered species is a challenge.[83]

Eagle Research to Mitigate Impacts (see figure 2.61).

In 2020, Purdue University and the University of Minnesota with WETO-funded research focused on exploring eagle physiology to improve the effectiveness of deterrents used around wind energy facilities. Purdue evaluated eagle hearing and vision ranges,

FIG. 2.61: PHOTO OF A WIND FARM WITH A MAN STANDING IN THE FOREGROUND AND A HAWK PERCHED ON THIS HAND, TEST BIRD FLIGHTPATHS AROUND WIND TURBINES AT FLATIRONS CAMPUS. PHOTO BY JASON ROADMAN, NREL (SOURCE:[84])

[81] https://cleantechnica.com/2021/01/05/2020s-top-wind-energy-rd-achievements/
[82] https://cleantechnica.com/2021/01/05/2020s-top-wind-energy-rd-achievements/
[83] https://cleantechnica.com/2021/01/05/2020s-top-wind-energy-rd-achievements/
[84] https://cleantechnica.com/2021/01/05/2020s-top-wind-energy-rd-achievements

while the University of Minnesota assessed eagle hearing. Purdue found that eagles have a blind spot near the tops of their heads that hinders their ability to see ahead, when looking down and hunting. This finding suggests the need for a deterrent that is sufficiently alarming to an eagle to cause it to look up when hunting. Purdue also found that blue/indigo and orange/red would be most visible to eagles against various backgrounds. The study concluded that golden eagles are more likely to respond to visual signals, whereas bald eagles are more likely to respond to a combination of sight and sound. The University of Minnesota explored hearing range and response to various auditory stimuli, including eagle vocalizations, finding that eagle deterrent developers should consider varying frequency and volume patterns to achieve the strongest and least-habituated responses.[85]

As wind deployment reaches new heights and new regions, additional environmental and human use factors will need to be considered and addressed. This includes the potential for new or additional interactions with wildlife such as birds and bats, as well as effects on their habitats. Additionally, impacts related to human use concerns, such as civilian and military radars, must be also evaluated will support efforts to enable -wind power nationwide. For instance, the Federal Aviation Administration (FAA) is creating lighting guidelines for wind turbines with total heights above 500 feet.[86]

Environmental Concerns Associated with Wind Energy Development

The deployment and operation of wind turbines present unintended environmental effects. For land-based wind energy, the focus is primarily on collision risk, during the operational phase, to certain species of bats and raptors. Offshore wind energy development may pose related risks for bats and birds but also impact marine mammals and deep-water flora and fauna during construction. To achieve renewable energy production goals, it is necessary to fully understand and, in certain situations, mitigate these effects. Therefore, the National Renewable Energy Laboratory is using engagement and research to address environmental concerns, showcased in https://www.nrel.gov/news/progra2021/beyond-technical-potential-nrel-explores-the-challenges-of-siting-wind-in-a-low-carbon-future.html. Beyond Technical Potential, NREL explores the challenges of siting wind in a low- carbon future. Detailed geospatial and power-sector modeling points to need for more integrated planning to balance local deployment decisions with decarbonization goals.[87] Achieving decarbonization goals, including the Biden administration's climate and energy plan will require rapid and sustained deployment of wind and solar power. However, the interaction between siting considerations and clean energy development at this scale is not well understood. What is more, this interaction is poorly reflected in data and tools

FIG. 2.62: WIND DEVELOPERS FACE UNIQUE SITING CONSTRAINTS THAT COULD HAVE A MEASURABLE IMPACT ON OVERALL U.S. WIND POTENTIAL, ACCORDING TO RECENT NREL ANALYSIS. (SOURCE:[88])

[85] https://cleantechnica.com/2021/01/05/2020s-top-wind-energy-rd-achievements/

[86] https://www.energy.gov/sites/prod/files/2015/05/f22/Enabling-Wind-Power-Nationwide_18MAY2015_FINAL.pdf, March 10, 2021

[87] https://federallabs.org/events/addressing-environmental-concerns-associated-with-wind-energy-development_

[88] https://www.nrel.gov/news/program/2021/beyond-technical-potential-nrel-explores-the-challenges-of-siting-wind-in-a-lowcarbon-future.html

used to inform these goals.In two complementary research efforts recently published in the journal *"Energy"*, National Renewable Energy Laboratory (NREL) analysts combined detailed geospatial and power-sector modeling to shed light on this critical, yet underappreciated, topic. "Integrated analysis reveals the impact that wind siting constraints could have on U.S. wind development and broader power-system planning," said Anthony Lopez, NREL senior geospatial data scientist. "The research points to a need for more integrated land use, energy, and technology R&D planning to balance local clean-energy deployment decisions with mitigating global climate change."[89]

How Do Land Use and Turbine Technology Influence U.S. Wind Potential?

To reflect the complex dynamics of wind resource potential, NREL analysts used uniquely detailed geospatial modeling and high-resolution datasets to examine onshore wind resource potential for the continental United States. Findings are published in an *"Energy"* journal article. 2040–2050 electricity prices averaged 4% higher and there was an additional 240 metric tons (10% of 2005 emissions) of carbon dioxide emissions in BAU scenarios with limited access compared to reference access. At the same time, carbon dioxide abatement costs increased by about $23 per metric ton in the carbon-constrained cases. "Not only do the scenarios point to how siting can lead to substantial differences in emissions and the cost to decarbonize the grid, but they also show how wind siting can interact with future transmission expansion and plant design in complex ways, which requires collaboration among energy modelers, wind technology and plant engineers, land-use experts, ecologists, social scientists, and others." NREL's understanding of the marine environment helps improve offshore wind energy forecast models and offers expertise in marine and atmospheric measurements, wind resource assessments offshore, and extrapolation of low-level winds to hub height.[90]

2.9.3.6 Cyber Security Roadmap

More than 50,000 wind turbines with a combined capacity of over 100 gigawatts operate in the United States, creating a serious need to protect wind infrastructure from cyberattacks. With support from its national laboratories, WETO published the Roadmap for Wind Cybersecurity, which outlines the increasing challenges of cyber threats to the wind industry, its technologies and control systems, and present activities and best practices that the wind industry can use the improve cybersecurity. More than 50,000 wind turbines with a combined capacity of over 100 gigawatts operate in the United States, creating a serious need to protect wind infrastructure from cyberattacks. With support from its national laboratories, WETO published the Roadmap for Wind Cybersecurity, which outlines the increasing challenges of cyber threats to the wind industry, its technologies and control systems, and present activities and best practices that the wind industry can use the improve cybersecurity.[91]

2.9.3.7 Digital

Offshore Wind Turbine Digital Twin is used for the Prediction of Component Failures with Stiesdal Offshore Technologies. The offshore wind industry is challenged by the high cost of operations and maintenance. Using a cloud-based model known as a "digital twin" can help mitigate those costs by allowing engineers to assess different turbine configurations and better predict component failures, *Sandia National Laboratories.*[92]

The U.S. Department of Energy Collegiate Wind Competition prepares students from multiple disciplines to enter the wind energy contest by providing real-world technology experience. DOE announced the winners of the 2020 Collegiate Wind Competition. The winner of the Turbine Digital Design contest was California State University Maritime Academy, and the winner of the Project Development contest was James Madison University. The 2021 Collegiate Wind Competition focus on "outreach and adaptability".

2.9.3.8 High Fidelity - HPC & HPM, High Fidelity - HPC & HPM; Compressible & Incompressible Flow Models and Nonlinear Structural Flows[93]

Grand Challenges in the Science of Wind Energy are HPC and HFM which are the foundation for next-generation engineering models.[94] "The future of high-performance computing for wind energy"[95] also depict Compressible & Incompressible Flow Models and Nonlinear Structural Flows.

Motivation for next generation high fidelity models at low cost is worthwhile for wind energy generation. High penetration of wind energy requires large wind farms composed of megawatt scale turbines for both land based and offshore wind energy generation. Wind farm flow dynamics and coupled turbine-structural dynamics are extremely complex and models are lacking for relevant dynamics which span many orders of magnitude. Only when we can model well the wind system can we optimize the system to maximize energy extraction and maximize turbine life minimizing the downtime.

HFM & HPC illuminate the path to reduce the cost of wind energy and answer the questions - Can we predict and understand Impact of wakes on downstream turbines? (ii) Evolution of the wakes? (iii) Formation of the Wakes? Indeed all of these in a highly complex, dynamic "metocean" environment.

As a Primer to understand what is "high-fidelity" modeling of HFM we have:

A Mathematical model describes a system using mathematical concepts i.e., an equation or a bunch of equations for HFM wind farm. Adhering to first principles to predict productivity via:

(a) Conservation of Mass and (b) Newton's Second Law of Motion. it did last Sean tickets want to live on the east

The accepted model for many fluid motions is embodied in the Navier-Stokes Equations which are a set of nonlinear, partial differential equations governing fluid velocity, pressure and temperature.

---

[89] https://www.nrel.gov/news/program/2021/beyond-technical-potential-nrel-explores-the-challenges-of-siting-wind-in-a-lowcarbon-future.html

[90] https://www.nrel.gov/wind/wind-energy-science-leadership-series.html

[91] https://cleantechnica.com/2021/01/05/2020s-top-wind-energy-rd-achievements/

[92] https://www.energy.gov/eere/wind/articles/doe-announces-2021-technology-commercialization-fund-selections

[93] https://www.nrel.gov/wind/assets/pdfs/future-of-hpc-webinar-2020-07-30.pdf

[94] https://www.nrel.gov/wind/assets/pdfs/future-of-hpc-webinar-2020-07-30.pdf

[95] https://www.nrel.gov/wind/assets/pdfs/ future-of-hpc-webinar-2020-07-30.pdf

Can solve Navier-Stokes Equations for wind energy systems?

Analytical solutions only exist for the simple laminar problems. Turbulence brings in orders of more complexity (many scales of which can be captured). "Big whirls have little whirls that feed on their velocity, and little whirls have lesser whirls and so on to viscosity". The wind energy system is NOT SIMPLE, and HIGHLY TURBULENT. The only way to solve Navier-Stokes equations for wind energy flow is: (a) Introduce some approximations; (b) "Discretize" the equations; (c) Solve using high- performance computing (HPC, eg., Supercomputers). This can be very difficult.

Development efforts are focused on open source "ExaWind" software stack.[96]

"ExaWind"[97] is an open source multi fidelity modeling and simulation software stack to run on laptops and next Gen supercomputers.

"Nalu Wind"[98] is an incompressible flow computational fluid dynamics (CFD) code; Unstructured grid finite volume discretization; Closely tied to Trillinos & hypre libraries; Blade resolved and actuator type simulations.

"AMR-Wind"[99] is incompressible flow CFD code; Structured grid finite volume atmospheric boundary layer "background solver"; Built on AMReX libraries; Coupled to 'Nalu-Wind" through overset meshes.

"OpenFAST"[100] is a whole turbine simulation Code. "ExaWind" modeling approach is defined in as follows:

It is a DOE strategic planning effort establishing the modeling and simulation requirements for predictive wind farm simulations:

- Compressible or incompressible flow model;
- Geometry resolving meshes;
- Fluid-structure interaction;
- Hybrid RANS/LES modeling;
- Nonlinear structural dynamics.

HPC performance portability is central to ExaWind development. Supercomputer architecture is evolving rapidly:

- USDOE supercomputers are increasingly relying on graphical processing units GPU's for computational speed at low power;
- The first ExaScale supercomputers will all have hybrid CPU-GPU architecture;
    - Coming online in 2021-2022.
    - Aiming for power requirements below 30 MW;
    - Exascales systems will be at least five times faster but require no more than three times the power.
- Hybrid CPU GPU architecture is expected to become more common among all clusters;
- CFD Codes will need to be able to utilize GPU;
- ExaWind Development is funded by two DOE offices:
- "High-Fidelity Modeling" project
- Period of performance: 2016 - 2023
- Partnership between NREL and Sandia National Laboratories (SNL) DOE NNSA/SC Exascale Computing project targets:
- "Exascale" Project
- Period of Performance 2016 - 2023
- Partnership between NREL, SNL, Oak Ridge National Laboratory, University of Texas at Austin and Parallel Geometric Algorithms, Inc.

Performance Measurement Examples are:

- Atmospheric boundary layer simulation
- Compute the complex flow in which turbines operate
- Simulation using millions of great points
- Computer resources (Cores) and measures of runtime

Conclusion and Goal:

"Exawind" framework is built on strong validation base with increasing levels of complexity

- Wind farm with flexible turbines in ABL are with complex terrain/offshore
    - Single turbine in ABL
    - Single turbine in uniform inflow
    - Fixed wing
    - Airfoil

2.9.4 Wind Vision and Wind Map[101]

2.9.4.1 Wind Vision

Recent literature demonstrated a need for higher-resolution wind resource and built environment data to better capture localized resource variability and land availability. Wind analysis advances technical potential assessments by using geospatial modeling and multidimensional datasets with unprecedented resolution. Analysts factored in siting constraints, regulations, landscape, and infrastructure, including 124 million buildings and every road, railway, transmission line, and radar tower in the United States. Three main scenarios were modeled, ranging in degrees of siting constraints from open access (least restrictive) to limited access (most restrictive). Analysts also considered three primary wind turbine designs to study the impact of technology innovation on wind potential.[102]

Simulations by NREL's Renewable Energy Potential Model[103] revealed a 7X difference in total U.S. wind technical potential across the three scenarios. The analysis shows that, under the reference case, wind has a technical potential of 7,800 gigawatts (GW) in the continental United States, but under other siting regimes, this potential could range from 2,280 GW to 15,175 GW. The limited access scenario reduced wind potential to 15% of the technical potential in the open access scenario and 10% of all land. Among factors considered, infrastructure and road setbacks, or the required boundary around infrastructure where wind turbines cannot be developed, have the biggest impact on technical potential. Because setbacks are influenced by turbine tip heights, the taller the turbine, the larger the setback and future technology innovation that results in taller turbines and larger rotor diameters might negatively impact technical

[96] https://iopscience.iop.org/article/10.1088/1742-6596/1452/1/012071/pdf
[97] https://www.exawind.org/
[98] https://github.com/exawind/nalu-wind
[99] https://github.com/exawind/amr-wind
[100] https://github.com/openfast
[101] https://www.energy.gov/sites/default/files/2015/03/f20/wv_full_report.pdf
[102] https://www.energy.gov/sites/default/files/2015/03/f20/wv_full_report.pdf
[103] https://www.nrel.gov/gis/renewable-energy-potential.html

potential in terms of capacity. However, because of the higher wind speeds at higher levels above ground and the greater energy capture potential with larger rotors, the change in estimated annual energy production is less than 1%. The analysis also demonstrates how siting considerations can dramatically impact wind plant sizes, particularly in more populated regions where conflicts with the built environment can challenge the development of large-scale wind power plants. These trade-offs in wind turbine and plant design and their interactions with other land uses should be considered in local energy and land planning.[104]

2.9.4.2 Four Principal Objectives Wind Vision:

1. Document the key technological accomplishments and societal benefits.
2. Explore the wind potential pathways for the future electricity needs of the nation, including reduced carbon emissions, improved air quality, and reduced water use.
3. Quantify costs, benefits, and impacts of the growth of U.S. wind power and
4. Identify future actions that could support growth of wind energy.

2.9.4.3 Key Findings of the Wind Vision Report:

- Wind energy can be a viable source of renewable electricity in all 50 states by 2050.
- Wind energy has the potential to support over 600,000 jobs in manufacturing, installation, maintenance, and supporting services by 2050.
- Wind energy is affordable with 20-year fixed pricing and the electric utility sector to be less sensitive to volatility in natural gas and coal fuel prices by reducing vulnerability to price spikes and supply disruptions and with long term pricing wind will to save consumers $280 billion by 2050.
- Wind energy reduces air pollution emissions. In 2013, wind energy avoided the emission of over 250,000 metric tons of air pollutants, which include sulfur dioxide, nitric oxide, nitrogen dioxide, and particulate matter. By 2050, wind energy could avoid the emission of 12.3 gigatons of greenhouse gases.
- Wind energy preserves water resources. By 2050, wind energy can save 260 billion gallons of water, the equivalent to roughly 400,000 Olympic-size swimming pools.
- Wind energy deployment increases community revenues. Local communities will be able to collect additional tax revenue from land lease payments and property taxes, reaching $3.2 billion annually by 2050.

2.9.4.4 Wind Map

The *Wind Vision* report[105] ends with a roadmap of technical, economic, and institutional activities to optimize wind's potential contribution to a cleaner, more reliable, domestic energy generation portfolio, as detailed in The *Wind Vision* Roadmap: "A Pathway Forward" and Appendix M (Detailed Roadmap Actions). Consistent with DOE's stated intention of updating the roadmap periodically, the National Renewable Energy Laboratory contracted Renewable Energy Consulting Services, Inc. to elicit feedback through a series of topical working sessions in 2016–2017, culminating in the *Wind Vision* Detailed Roadmap Actions: 2017 Update.[106]

2.9.4.5 Wind Road Map[107]

The roadmap is the beginning of an evolving, collaborative and a dynamic process. The *Wind Vision* depicts a future in which wind power has the potential to be a significant contributor to a cost-effective, reliable, low-carbon U.S. energy portfolio. The objective of the *Wind Vision* roadmap is to identify the challenges and actions necessary to increase the opportunities for U.S. wind deployment.[108]

Core Roadmap Actions are (i) To Reduce Wind Costs; and (ii) To Expand Wind Developable Areas and (iii) To Increase Economic Value for the Nation from Wind Energy.

High-level roadmap actions are summarized in Table 2.1. These are explained in detail in the *Wind Vision* report (Chapter 4 and Appendix M),[109] Also see "Wind Vision Detailed Roadmap Actions, 2017 Update"[110]

2.9.5 Conclusions

Jose Zayas was the original author of this chapter that appeared as Chapter 7 in the classic ASME publication "Energy and Power Generation Handbook - Established and Emerging Technologies" edited by K. R. Rao published in 2019. Jose Zayas was invited to update his original contribution but was unavailable for that purpose. The content of Jose Zayas's initial contribution is retained in its entirety since the textual content has relevance even to-day after two decades after the initial publication in 2011.

K. R. Rao contributed this update based on his 'on-line research' and findings from his Classic "Wind Energy for Power Generation - Meeting The Challenge of Practical Implementation" Published by Springer in two volumes in 2019. This update also picks up the essence of a report "WIND VISION - New Era for Wind Power in the United States", by Jose Zayas who was Director, Wind and Water Power Technologies Office U.S. Department of Energy. "WIND VISION - New Era for Wind Power in the United States" dated March 12, 2015. Wind Vison was a collaborative effort of "several hundred stake holders and individuals across the agency, industry, academia, and national labs for their strategic interest in a renewed vision for wind energy", which was a significant comprehensive contribution for Wind Energy Development. The essence of this report has been captured by K. R. Rao spread over in several sections of this current chapter update. Readers are urged to refer to the "Wind Vison" report to capture the message about the growth of the wind energy during the past decade and for the future up to 2050.

---

[104] https://www.energy.gov/sites/default/files/2015/03/f20/wv_full_report.pdf
[105] https://www.nrel.gov/wind/assets/pdfs/future-of-hpc-webinar-2020-07-30.pdf
[106] https://www.energy.gov/sites/prod/files/2018/05/f51/WindVision-Update-052118-web_RMB.pdf
[107] https://www.energy.gov/sites/prod/files/2018/05/f51/WindVision-Update-052118-web_RMB.pdf
[108] https://www.energy.gov/sites/default/files/2015/03/f20/wv_full_report.pdf
[109] https://www.energy.gov/sites/default/files/2015/03/f20/wv_full_report.pdf
[110] https://www.energy.gov/sites/prod/files/2018/05/f51/WindVision-Update-052118-web_RMB.pdf

TABLE 2.1: ROADMAP STRATEGIC APPROACH (SOURCE: TABLE ES 4.1)[111]

Core Challenge	Wind has the potential to be a significant and enduring contributor to a cost-effective, reliable, low carbon future for U.S. energy portfolio. Optimizing U.S. wind power's impact and value will require strategic planning and continued contributions across a wide range of participants.		
Key Themes	**Reduce Wind Costs** Collaboration to reduce wind costs through wind technology's capital and operating cost reductions, increased energy capture, improved reliability, and development of planning and operating practices for cost- effective wind integration.	**Expand Developable Areas.** Collaboration to increase market access to U.S. wind resources through improved power system flexibility and transmission expansion, technology development, stream- lined siting and permitting processes, and environmental impact and competing use of research.	**Increase Economic Value for the Nation** Collaboration to support a strong and self-sustaining domestic wind industry through job growth, improved competitiveness, and articulation of wind's benefits to improve decision making process.
Issues Addressed	Continuing declines in wind power costs and improved reliability are needed to improve market competition with other electricity sources.	Continued reduction of deployment barriers as well as enhanced mitigation strategies to responsibly improve market access to remote, low wind speed, offshore, and environmentally sensitive locations.	Capture the enduring value of wind power by analyzing job growth opportunities, evaluating existing and proposed policies, and disseminating credible information.
***Wind Vision Study Scenario* Linkages**	Levelized cost of electricity reduction trajectory of 24% by 2020, 33% by 2030, and 37% by 2050 for land- based wind power technology and 22% by 2020, 43% by 2030, and 51% by 2050 for offshore wind power technology to substantially reduce or eliminate the near- and mid- term incremental costs of the *Study Scenario*.	Wind deployment sufficient to enable national wind electricity generation shares of 10% by 2020, 20% by 2030, and 35% by 2050.	A sustainable and competitive regional and local wind industry supporting substantial domestic employment. Public benefits from reduced emissions and consumer energy cost savings are to be achieved.

2.9.6 Bibliographical References

1) Energy and Power Generation Handbook - Established and Emerging Technologies, edited by K. R. Rao, ASME Press, New York, 2011.
2) Wind Energy for Power Generation - Meeting the Challenge of Practical Implementation by K. R. Rao, Volumes 1

[111] https://www.energy.gov/sites/default/files/2015/03/f20/wv_full_report.pdf

CHAPTER

3

WIND ENERGY IN THE U.S.

Thomas Baldwin, Gary Seifert and K. R. Rao[1]

3.1 INTRODUCTION

Humans have been harnessing the energy in the wind for several thousand years. Early uses included sailboats and windmills used to pump water and grind grain into flour — hence the term wind *mill*. In the 1800s, settlers in the western United States used wooden windmills to pump water, and many are still standing. In the late 1800s, the windmill was connected to an electric generator to produce electricity — hence, wind *turbine*.

In the 1970s and 1980s, wind turbines were clustered into wind farms and connected to the electric grid in California, which marked the first commercial, utility-scale use of wind energy. The size of those wind turbines were 100 kW and smaller. In the following decades, wind turbine technology progressed quickly, and by 2010, grid-connected wind turbines were typically in the 2-MW range and turbines as large as 6-MW have been deployed [17].

Adequate wind speeds are essential to the success of wind energy facilities. The potential energy in wind is determined as a function of the cube of the speed. Given the cubic relationship between power and wind speed, when wind speed doubles there is eight times more power available.

In 2009, wind supplied about 1.8% of the electricity in the United States [1]. However, experience shows that wind can provide up to one fifth of a power system's electricity, and that amount can increase with optimized system designs. For example, wind power currently provides more than 20% of the electricity distributed in northern Spain and in Denmark. A goal is to increase wind generationin the United States to 20% by 2030 [2].

Utility transmission lines carry wind-generated electricity from vast and sparsely populated areas where the wind is most abundant, like the Great Plains, to large cities where demand for electricity is high. At the moment, there is insufficient transmission infrastructure connecting the windiest parts of the country with large cities. Enhancing the transmission line power capacity for utility-scale wind plants is a key issue that must be resolved in the coming years [35].

Most regions of the United States are served by "power pools" of utilities that join together to generate electricity and transmit it to where it is needed. The name "power pool" is descriptive of the electric power coming from many different sources (a coal-fired power plant, a hydro plant, and others) flows into a "pool" from which it is distributed to thousands of end users. A power pool can easily absorb the electric power from a wind plant and add it to the generation mixture up to a penetration level of around 20%. Wind penetration greater than 20% requires advancements in transmission capacity, forecasting accuracy, and energy storage as addressed in the Wind Energy Research section. Wind plants could be installed in many locations, providing income, jobs, and electricity for homes and businesses.

Wind energy is a particularly appealing way to generate electricity because it is essentially pollution-free once the energy used to manufacture the turbine is offset. More than half of all the electricity that is used in the United States is generated from burning coal, and in the process, large amounts of toxic metals, air pollutants, and greenhouse gases are emitted into the atmosphere. Developing just 10% of the wind potential in the ten windiest U.S. states would provide more than enough energy to displace emissions from many of the nation's fossil-fuel power plants and offset their growth for many years.

Wind farms and plants can also revitalize the economy of rural communities, providing steady income through lease or royalty payments to farmers and other landowners. Although leasing arrangements can vary widely, in 2010, a reasonable estimate for income to a landowner from a single utility-scale turbine was approximately $4,000 a year, depending on the wind resource, the size of the turbine, and other factors. For a typical farm, income from wind comes with little interruption in farming activities with about one acre removed from agricultural production per turbine. Farmers can grow crops or raise livestock next to the towers. While wind farms may extend over a large geographical area, their actual "footprint" covers only a very small portion of the land, making wind development a cooperative way for farmers to earn additional income.

[1]Thomas Baldwin and Gary Seifert are the original authors of this chapter that appeared as Chapter 8 in the "Energy and Power Generation Handbook - Established and Emerging Technologies", published in ASME in 2011. This chapter has been updated by Editor K. R. Rao in Section 3.16. K.R. Rao has provided relevant write-ups from on-line sources in the public domain covering "Scope of Wind Energy WIND ENERGY IN THE U.S since 2009 to date".

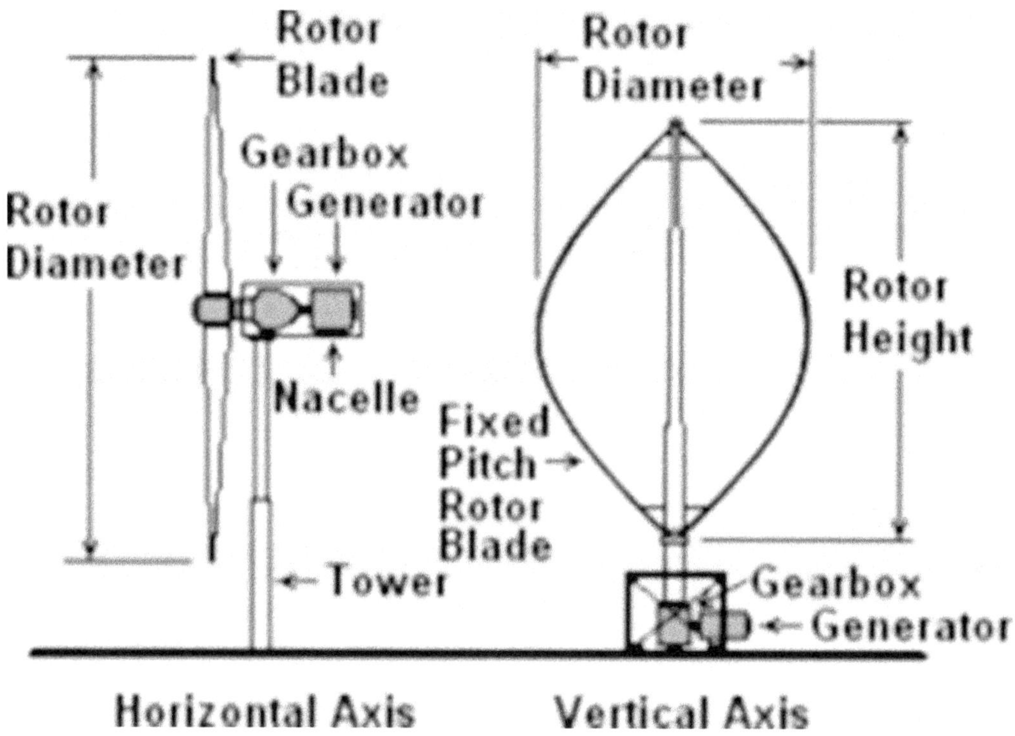

FIG. 3.1: TYPICAL WIND TURBINE CONFIGURATIONS

3.2 WIND TURBINE TECHNOLOGIES

Wind energy systems transform the kinetic energy of wind into mechanical and then into electrical energy, which is distributed for industrial, commercial, and residential use.

3.2.1 Types of Wind Turbines

There are two basic designs of wind electric turbines: horizontal-axis (propeller-style) machines and vertical-axis, or "egg-beater" style as shown in Fig. 3.1. The horizontal-axis wind turbines are commonly used in the United States, constituting nearly all of the "utility-scale" (100 kW, capacity and larger) turbines. The use of vertical turbines continues to lag at the utility scale because of the high cost to place the rotors into the higher wind speeds at elevations of 80 to 100 m above ground. However, use for residential units is gaining popularity with regular new entries in the market.

Wind turbines vary substantially in size. Table 3.1 depicts a variety of historical turbine sizes and the rated amount of electric power they are capable of generating. In 2010, offshore wind turbines operated at the 5-MW range with rotor diameters reaching 120 m. Technology developments foresee growth to the 10-MW range projected by 2015. Onshore turbines are operating up to the 3-MW range with 100 m rotor diameters.

3.3 WIND RESOURCES IN THE U.S.

There are three parts that make up a successful wind energy project. They includea wind resource, an energy sale contract, and a transmission/grid interconnection. Successful wind projects have tended to cluster along transmission grid corridors that have adequate wind. At the beginning of a project, a survey of wind resources must be conducted to ensure that a desired location has adequate energy supply. Wind resource maps, developed by the U.S. Department of Energy (DOE), provide general wind characteristics for areas of interest and are used to screen locations for suitable investigation. A review of these maps provides an initial decision point whether further exploration is warranted. Wind resource at the local level can vary significantly. Local wind measurements and evaluations should be conducted for the specific site of interest, a process known as micro-siting and validation. An evaluation characterizes the wind resource and predicts the turbine performance and economic feasibility.

The Wind Powering America organization also provides high-resolution wind maps and estimates of wind resource potential. The maps show the predicted mean annual wind speeds with a spatial resolution of 2.5 km for a height of 80-m above the terrain, the typical wind turbine hub height. Locations with annual average wind speeds greater than 6.5 m/s at the 80-m height are considered to have suitable wind resource for development of most areas.

The U.S. DOE's National Renewable Energy Laboratory (NREL) has identified practical locations for wind energy exploration. In the wind energy resource maps, wind velocity characteristics are categorized and presented in terms of wind classes. Table 3.2 lists the definitions of the wind classes, which range from class 1 (the lowest) to class 7 (the highest). Each class represents a range of mean wind power density or approximate mean wind speed at specified heights above the ground. Areas designated as class 4 or greater are suitable for most wind energy applications. Class 3 areas are transitional, where the power purchase costs become important. Class 2 areas are marginal and class 1 sites are generally not suitable for development.

TABLE 3.1 CHANGES IN WIND TURBINE ROTOR DIAMETERS, POWER RATINGS AND ESTIMATED ENERGY OUTPUT OVER TIME

Typical dimensions	1981	1985	1990	1996	1999	2000	2010
Rotor diameter (m)	10	17	27	40	50	71	120
Power rating (kW)	25	100	225	550	750	1,650	5000
Annual energy (MWh)	45	220	550	1,480	2,200	5,600	17,500

NREL conducted a preliminary review and validated the AWS Truepower's (formerly AWS Truewind) 80-m map estimates for 19 selected states (6 Western states, 6 Midwestern states, and 7 Eastern states) using tower measurements at heights of about 50 m and greater from more than 300 locations.

3.3.1 Wind Resource Potential

In a joint effort between NREL and AWS Truepower of Albany, New York, a national dataset was produced of estimated gross capacity factors. These resource maps identify potential energy resources, both on and offshore. The data is interpreted to a spatial resolution of 200 m for heights at 80 m and 100 m. Wind power is greatly impacted by the turbine height, as illustrated by Table 3.3.

Mean wind speed is based on the Rayleigh speed distribution of equivalent mean wind power density. The power available in the wind depends on air density as well as wind speed. In Table 3.3, wind speed is reported for standard sea-level conditions. To maintain the same power density at higher elevations, wind speed must increase 3% per 1000 m (5%/5000 ft) of elevation gain.

Each wind power class is designed to span a range of power densities. For example, Wind Power Class 3 represents a wind power density range between 150 W/m^2 and 200 W/m^2. The offset cells in the first column of Table 3.3 illustrate this concept.

TABLE 3.2 DEFINITIONS OF WIND POWER CLASSES

Wind class	Description
1	$V < 5.9$ m/s
2	5.9 $V < 6.9$ m/s
3	6.9 $V < 7.5$ m/s
4	7.5 $V < 8.1$ m/s
5	8.1 $V < 8.6$ m/s
6	8.6 $V < 9.4$ m/s
7	V 9.4 m/s

3.3.2 Potential Energy Supply for the U.S.

Useful wind energy resources for generating electricity can be found in nearly every state. In the 20% Wind Energy by 2030 report issued by the U.S. DOE in 2008, a scenario has been presented by which 20% of the nation's electricity could be produced by wind energy by the year 2030. [2] By comparison, wind energy accounted for less than 2% of the nation's electricity in 2009 [1].

U.S. wind resource potential is significantly greater than 20% of U.S. demand in 2009. North Dakota alone is theoretically capable of producing enough wind-generated energy to exceed 25% of the nation's 2009 electricity energy demand (3,950 TW/h [3]). In 2010, however, electric transmission constraints would prevent much of that electricity from being distributed to load centers throughout the county; transmission will be discussed later in this chapter. The theoretical potentials of the windiest states are shown in Table 3.4.

These wind potential estimates resulted from a collaborative project between the NREL and AWS Truepower. This is the first comprehensive update of the wind energy potential by state since 1993. NREL has worked with AWS Truepower for almost a decade updating wind resource maps for 36 states and producing validated maps for 50 m height above ground level. U.S. DOE's Wind Powering America project supported the mapping efforts, and the results are publicly available. Table 3.4 indicates that the wind resources available in the United States vastly exceed current electric energy demand. As discussed later, careful selection of sites is essential to ensure adequate power generation.

Wind power facilities tend to develop in areas with adequate local transmission infrastructure and capacity factors of 35% or greater. Figure 3.2 indicates that much more wind power capacity could be installed if sites with capacity factors as low as 25% were developed.

The two lines shown in Fig. 3.2 plot the installed power capacity versus capacity factor at two different hub heights. As Table 3.3 indicates, taller hub heights would result in greater power capacity. Turbine manufacturers tend to design turbines for sites with capacity factors of 30% or greater. Research and design of lower-wind speed turbines would support the development of 25% capacity factor sites.

3.3.2.1 Wind Maps NREL provides both a national wind resource assessment of the United States and high-resolution wind data as a service to the public. The national wind resource assessment was created for the U.S. DOE in 1986 by the Pacific Northwest Laboratory and is documented in the Wind Energy Resource Atlas of the United States, October 1986, and subsequently updated with modern 80 m maps available on NREL and DOE websites. According to NREL:

TABLE 3.3 CLASSES OF WIND POWER DENSITY AT 10-M AND 50-M; VERTICAL EXTRAPOLATION OF WIND SPEED BASED ON THE 1/7 POWER LAW

Wind power class	10 m (33 ft)		50 meter (164 feet)	
	Power density [W/m2]	Mean speed [m/s (mph)]	Power density [W/m2]	Mean speed [m/s (mph)]
1	0	0	0	0
2	100	4.4 (9.8)	200	5.6 (12.5)
3	015	5.1 (11.5)	300	6.4 (14.3)
4	200	5.6 (12.5)	400	7.0 (15.7)
5	250	6.0 (13.4)	500	7.5 (16.8)
6	300	6.4 (14.3)	600	8.0 (17.9)
7	400	7.0 (15.7)	800	8.8 (19.7)
	1000	9.4 (21.1)	2000	11.9 (26.6)

The wind resource assessment was based on surface wind data, coastal marine area data, and upper-air data, where applicable. In data-sparse areas, three qualitative indicators of wind speed or power were used when applicable: topographic/meteorological indicators (e.g., gorges, mountain summits, sheltered valleys); wind deformed vegetation; and eolian landforms (e.g., playas, sand dunes). The data was evaluated at a regional level to produce 12 regional wind resource assessments; the regional assessments were then incorporated into the national wind resource assessment.

The conterminous United States was divided into grid cells ¼ degree of latitude by ⅓ degree of longitude. Each grid cell was assigned a wind power class ranging from 1 to 6, with 6 being the windiest. The wind power density limits for each wind power class are shown in Table 3.1. Each grid cell contains sites of varying power class. The assigned wind power class is representative of the range of wind power densities likely to occur at exposed sites within the grid cell. Hilltops, ridge crests, mountain summits, large clearings, and other locations free of local obstruction to the wind will be well exposed to the wind. In contrast, locations in narrow valleys and canyons, downwind of hills or obstructions, or in forested or urban areas are likely to have poor wind exposure.

Wind classes are defined in Table 3.2. The degree of certainty with which the wind power class can be specified depends on three factors: the abundance and quality of wind data; the complexity of the terrain; and the geographical variability of the resource. A certainty rating was assigned to each grid cell based on these three factors, and is included in the Wind Energy Resource Atlas of the United States [36] (Fig. 3.3).

TABLE 3.4 THE TOP 20 STATES FOR WIND ENERGY POTENTIAL, MEASURED IN BILLIONS OF KWH, FACTORING IN ENVIRONMENTAL AND LAND USE EXCLUSIONS FOR WIND CLASSES OF 3 AND GREATER

Ranking	State	TWh/Yr	Ranking	State	TWh/Yr
1.	North Dakota	1,210	11.	Colorado	481
2.	Texas	1,190	12.	New Mexico	435
3.	Kansas	1,070	13.	Idaho	73
4.	South Dakota	1,030	14.	Michigan	65
5.	Montana	1,020	15.	New York	62
6.	Nebraska	868	16.	Illinois	61
7.	Wyoming	747	17.	California	59
8.	Oklahoma	725	18.	Wisconsin	58
9.	Minnesota	657	19.	Maine	56
10.	Iowa	551	20.	Missouri	52

Source: An Assessment of the Available Windy Land Area and Wind Energy Potential in the Contiguous United States, Pacific Northwest Laboratory, August 1991. PNL-7789.

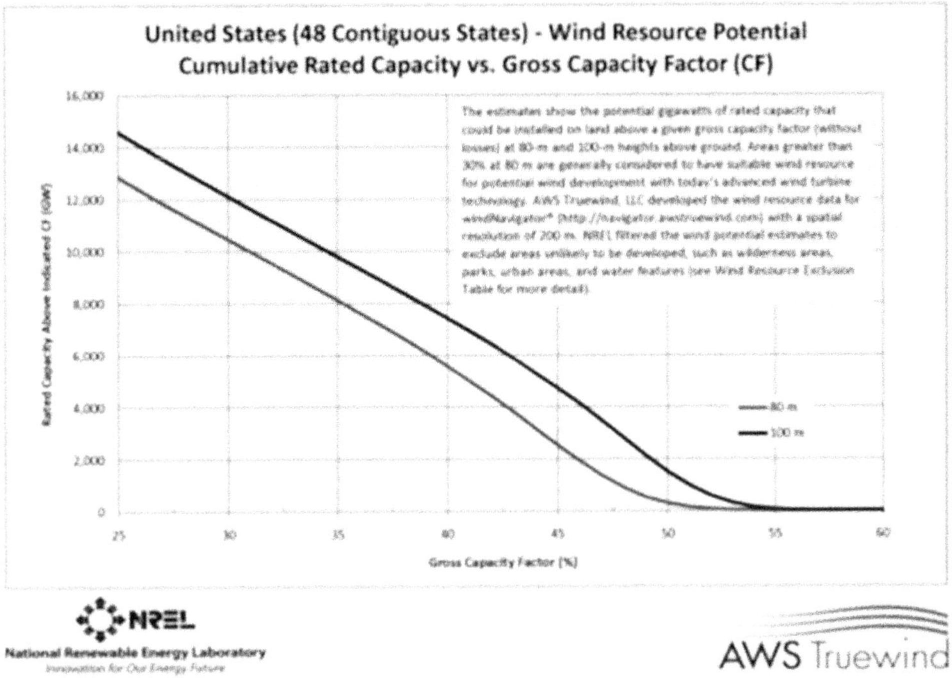

FIG. 3.2 U.S. CUMULATIVE RATED CAPACITY VS. GROSS CAPACITY FACTOR [5]

3.3.2.2 Local Terrain Effects When using public wind maps, it is important to consider the effect of local terrain, which may not be well represented on such data sources. A site's topography affects its power density (W/m2), which will impact both the number of turbines that can be placed on a site and turbine spacing (see the Wind Farm Development section of this chapter). In areas with relatively flat terrain, such as the Great Plains, power density tends to be greater than in mountainous areas. In areas with complex, mountainous terrain, wind turbine placement is typically limited to areas that are well exposed, such as ridgetops and tables. In heavily wooded areas, turbines are best placed in clear spaces with limited wind obstructions. In any region,

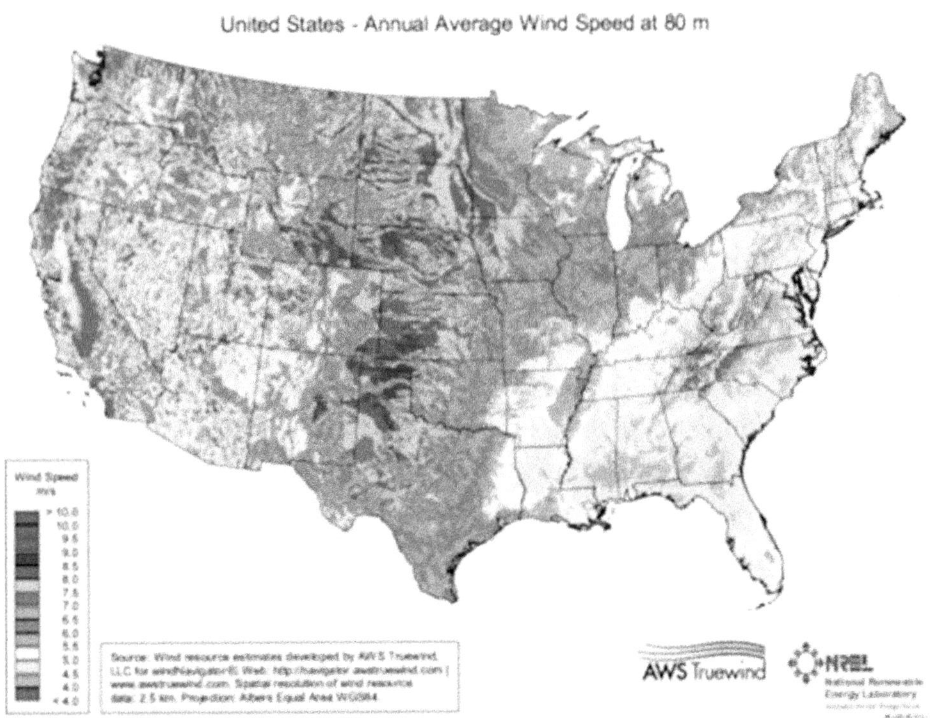

FIG. 3.3 MAP OF U.S. WIND RESOURCE AT 80-M HEIGHT ABOVE GROUND LEVEL [6]

measuring a potential site's *actual* wind resource (micro-siting) is an essential step when developing a wind farm and designing turbine arrays.

3.3.2.3 High-Resolution Wind Data High-resolution wind datasets are geographic shape-files generated from the original raster data. According to NREL, the original raster data varied in resolution from 200 m to 1000 m cell sizes and provide an estimate of annual average wind resource for specific states or regions. The data is separated into two distinct groups: NREL produced and AWS Truepower produced/NREL validated [36].

The NREL-produced map data only applies to areas of low surface roughness (i.e., grassy plains), and excludes areas with slopes greater than 20%. For areas of high surface roughness (i.e., forests), the values shown may need to be reduced by one or more power classes. The AWS Truepower-produced resource estimates factor surface roughness into their calculations and do not exclude areas with slopes greater than 20%. This data was produced in cooperation with U.S. DOE's Wind Powering America program and have been validated by NREL and other wind energy meteorological consultants. To download state wind resource maps, visit the Wind Powering America website and the Idaho National Laboratory's (INL) Virtual Renewable Energy Prospector (VREP) website. It is important to emphasize that all of these publicly available data is intended to be used as initial screening tools only. Site-specific measurements, as described in the Wind Farm Development section, are required to make sound decisions.

3.4 WIND PLANT ECONOMICS

From 1990 to 2010, the cost of electricity from utility-scale wind systems dropped by more than 80%. In the early 1980s, when the first utility-scale turbines were installed, wind-generated electricity cost as much as 30 cents per kilowatt-hour. Now, state-of-the-art wind power plants can generate electricity between 3 and 8 cents/ kWh with tax incentives, a price that is competitive with and often lower than new coal- or gas-fired power plants.

The DOE is working with the wind industry to develop a next generation of wind turbine technology. The products from this program are expected to generate electricity at prices that will be lower still and allow expansion to lower wind speed requirements around the nation.

The most important factors in determining the cost of wind-generated electricity from a wind farm are (1) size of the wind farm; (2) wind speed at the site; (3) cost of installing the turbines; and (4) grid interconnection and transmission costs. Each of these factors can have a major impact. Generally speaking:

- The larger the wind farm, all other factors being equal, the lower the cost of energy;
- The higher the wind speed, the lower the cost of energy;
- The less expensive construction costs are, the lower the cost of energy;
- The closer to transmission, the lower the cost of energy.

On isolated ridgelines, for example, wind farms are likely to be smaller and cost more to install than in the flat terrain of northern

(Billions of 2006 Dollars[1])

TYPE OF INCENTIVE	ENERGY SOURCE							SUMMARY FOR INCENTIVE TYPE	
	Oil	Nat. Gas	Coal	Hydro	Nuclear	Renewable[2]	Geothermal	Total	Share
Tax Policy	173	88	31	12		20	2	326	45%
Regulation	116	3	7	5	11			142	20%
R&D	7	6	32	1	67	19	3	135	19%
Market Activity	5	2	2	59		2	2	72	10%
Gov't Services	31	1	14	1	1	2		50	7%
Disbursements	3		8	2	-14	2		1	~0%
Total	**335**	**100**	**94**	**80**	**65**	**45**	**7**	**726**	
Share	46%	14%	13%	11%	9%	6%	1%		100%

[1] All estimates quoted are in constant 2006 dollars, unless otherwise noted, and refer to actual expenditures in the relevant fiscal year, rounded to the nearest billion. Totals and percentages may differ slightly due to independent rounding.

[2] Renewables are primarily wind and solar energy sources.

FIG. 3.4 SUMMARY OF FEDERAL ENERGY INCENTIVES FROM 1950 THROUGH 2006

Plains states. While larger wind farm power may cost less than 5 to 6 cents/kWh in the northern Plains, it may cost 6 to 7 cents/kWh in small sites.

In the case of offshore wind farms, the distance that power must be transmitted to shore and water depth are potentially significant cost elements, yet offshore farms can be significantly closer to population and load centers, changing transmission impacts.

3.4.1 Energy Subsidies and Incentives

In 2008, the Nuclear Energy Institute (NEI) asked Management Information Services Inc. (MISI) to prepare an independent assessment [7] of the amounts and types of federal incentives provided from 1950 to 2006 and the energy sources targeted with each type of incentive. As summarized in Fig. 3.4, the largest beneficiaries of federal energy incentives have been oil and gas, receiving more than half of all incentives provided since 1950. Carbon-emitting sources (oil, natural gas, and coal) combined account for approximately 73% of all incentives over this time period.

3.4.1.1 Wind Power Plants Cost Comparison to Other Renewable Energy Sources Wind is the lowest-cost renewable energy resource to develop in the beginning of the 21st century and consistently ranks low in utility integration resource planning.

3.4.1.2 Production Tax Credit for Wind Energy A 1.5-cent per kilowatt-hour [1] production tax credit (PTC) for wind energy was included in the Energy Policy Act of 1992. Passage of the credit reflected recognition of the important role that wind energy can and should play in the nation's energy mix. It also was intended to partially correct an existing bias of the federal energy tax code, which has historically favored conventional fossil energy technologies, such as oil and coal, as shown in Fig. 3.4.

Generally, the credit is a business credit that applies to electricity generated from wind plants for sale at "wholesale" (i.e., to a utility or other electricity supplier which then sells the electricity to customers at "retail"). It applies to electricity produced during the first 10 years of a wind plant's operation. The company that owns the wind plant subtracts the value of the credit from the business taxes that it would otherwise pay.

As part of the American Recovery and Reinvestment Act 2009 (ARRA), legislation (H.R. 1) expanded the eligibility criteria by removing the wind turbine size capacity limitation. This allowed all Renewable Electricity Production Tax Credit (PTC) eligible facilities to use the 30% Business Energy Investment Tax Credit (ITC) on any size of wind turbine project. The 30% credit is applied to the total project cost. Confirmation of the tax credits and their application to this specific project were confirmed by three methods. The following documents show the details of these different incentives:

Department of the Treasury Internal Revenue Service FORM 8835 [8] (Department of the Treasury, Internal Revenue Service, 2009),

Renewable Energy Production Tax, FORM 3468 [9] (Department of the Treasury, Internal Revenue Service, 2009),

Investment Credit Tax, FORM 3800 [10] (Department of the Treasury, Internal Revenue Service, 2009)

Under the ARRA program, the federal government will give a cash rebate for 30% of the cost of building a renewable energy facility, awarded 60 days after an application is approved, which is contingent upon the first energy date. Investors are also given valuable accelerated depreciation deductions, which help offset taxes. Critics have used the government subsidies as a way to draw negative attention to renewable energy, but in a report by the Environmental Law Institute (ELI) for the years 2002 to 2008 the government provided more subsidies to fossil fuels than to renewable energy sources. Approximately $72 billion dollars in fossil fuel subsidies represented a direct cost to tax paying Americans, whereas only $29 billion were given to the renewable fuels. The numbers explain the need for government subsidies for all types of fuels. Fossil fuels have been well established as the prime energy generation source, and the incentives are written in as permanent provisions. Renewable fuels incentives are mostly time-limited and are not as useful or effective as they could be if they were permanent. On a final note from the study done by ELI, more than half of the subsidies for renewables are for the processing of corn into ethanol (Environmental Law Institute, 2009) [11].

3.4.1.3 Green Pricing Programs There are several reasons for the cost premium (typically 2 to 3 cents per kilowatt-hour) that most green marketers charge for wind-generated electricity. Green pricing varies from state to state and is regulated by state public utility commissions.

Assume wind energy costs 2 cents more per kilowatt-hour (2 cents/kWh) than the rest of the electricity a utility is generating or buying — a conservative estimate. If the utility were to decide to use wind energy to generate 10% of its electricity (more than nearly all utilities in the United States), then the added cost would be 0.2 cents/kWh. An average U.S. home uses about 800 kWh per month, so a ratepayer would pay an extra $1.60 per month or about a nickel a day.

At times when the prices of natural gas, oil, and other fossil fuels are increasing, such as 2005, wind energy is more economical. A study of wind integration into the New York State electric power system, looking at a 10% addition of wind generation (3,300 MW of wind in a 34,000-MW system), projected a reduction in payments by electricity customers of $305 million in one year. [12] Natural gas prices are lower in 2010 than they were in 2005, but utility companies are still interested in adding wind.

3.4.1.4 Energy Payback Time for a Wind Turbine The "energy payback time" is a term used to measure the net energy value between production and the construction of a wind turbine or other power source. It is expressed in the number of years required for a plant to operate in order to generate the total energy required for its manufacture, construction, operation, and retirement. Several studies over the years have examined this quantity and have concluded that wind energy has one of the shortest energy payback times of any energy technology. Wind turbines typically take between 3 and 8 months, depending on the local average wind speeds to reach the payback point. From this point forward, the wind turbine will generate emission-free electricity. By comparison, conventional fossil-fuel power plants require several years to pay back the infrastructure's embedded energy and continue to generate emissions for their entire life.

3.4.1.5 Cost of Non-Dispatchable Energy Some critics claim that since wind power cannot be dispatched to address corresponding load demand, a utility should have available equal proportions of power producing capability from dispatchable fossil fuel generation as connected wind power. For example, 100 MW of fossil-fuel generation is needed for every 100 MW of wind turbines to allow for the times when the wind is calm.

Utilities use complicated statistical models to determine the value in added energy and power capacities that each new generating plant adds to the system. According to those models, the energy capacity value of a new wind plant can be equal to the product of its capacity and capacity factor. Thus, adding a 100-MW wind plant with an average capacity factor of 35% to the system is approximately the same as adding 40 MW of conventional fueled generating capacity with an 87.5% availability factor.

The exact answer depends on, among other factors, the correlation between the time that the wind blows and the time that the utility sees peak demand and diversity of multiple wind farms and rate structures and agreement determined by the PUC/PSC. Thus, wind farms whose output is highest in the spring months or early morning hours will generally have a lower capacity value than wind farms whose output is high on hot summer evenings or cold winter days.

3.4.2 Turbine Performance

3.4.2.1 Capacity Factor The Capacity Factor is a commonly used metric employed in measuring the productivity of a wind turbine or any other power production facility. It compares the plant's actual production over a given period of time with the amount of power the plant would have produced if it had run at full capacity for the same amount of time.

$$\text{Capacity factor} = \frac{A}{B} \tag{3.1}$$

Where:

A = Actual amount of power produced over time.

B = Power that would have been produced if the turbine operated at maximum output 100% of the time.

A conventional utility power plant uses fuel, so it will normally run much of the time unless it is idled by equipment problems or for maintenance. A capacity factor of 60% to 95% is typical for conventional fossil fuel and nuclear power plants and 30% to 50% for hydropower plants [13].

A wind turbine's prime mover is the wind, which blows steadily at times and not at all at other times. Modern utility-scale wind turbines typically operate 65% to 90% of the time, often at less than full capacity. Therefore, annual capacity factors ranging between 25% and 40% are common, although they may operate at much higher capacity factors during windy times.

It is important to note that while capacity factor is almost entirely a matter of reliability and market demand for a fueled power plant, it is not for a wind plant. For a wind plant, it is a matter of economical turbine design and wind resource. With a large rotor and a small generator, a wind turbine could run at full capacity whenever the wind blew and have a 60% to 80% capacity factor, but it would produce very little electricity compared to investment. The optimal electricity per dollar of investment is gained by using a larger generator with high wind resource and optimizing the capacity factor and revenue. Wind turbines are fundamentally different from fueled power plants in this respect and their inability to be dispatched by grid operators.

3.4.2.2 Availability Factor The Availability Factor is a measurement of the reliability of a wind turbine or other power plant. It refers to the percentage of time that a plant is ready to generate (i.e., not out of service for maintenance or repairs). Modern wind turbines have an availability of more than 95%, higher than most other types of power plant.

3.5 TECHNICAL ISSUES

3.5.1 Grid Operations with Wind Power

Modern wind turbine systems can provide very good voltage regulations, but the output from the power plant cannot be controlled by grid operators, rather it is dependent on the environment.

Wind power fluctuations are stochastic in nature, but not completely random. Since the wind does not blow all the time a wind farm's contribution to a utility's generating capacity is the subject of debate.

Utilities must maintain enough power plant capacity to meet expected customer electricity demand at all times, plus an additional reserve margin. They schedule many resources, including wind, to meet those loads. All other things being equal, utilities generally prefer plants that can generate as needed (that is, conventional plants) to plants that cannot (such as wind plants).

However, despite the fact that the wind is variable and sometimes does not blow at all, wind plants do increase the overall statistical probability that a utility system will be able to meet demand requirements while reducing the cost of fossil fuel input. A rough rule of thumb is that the capacity value of adding a wind plant to a utility system is about the same as the wind plant's capacity factor multiplied by its capacity. Thus, a 100-MW wind plant with a capacity factor of 35% would be similar in capacity value to a 35MW conventional generator. For example, in 2001 the Colorado Public Utility Commission found the capacity value of a proposed 162-MW wind plant in eastern Colorado (with a 30% capacity factor) to be approximately 48 MW [14].

The exact amount of capacity value that a given wind project provides depends on a number of factors, including average wind speeds at the site and the match between wind patterns and utility load (demand) requirements. It also depends on how dispersed geographically wind plants on a utility system are, and how well-connected the utility is with neighboring systems that may also have wind generators. The broader the wind plants are scattered geographically, the greater the chance that some of them will be producing power at any given time, increasing wind power's contribution to the electric grid.

3.6 ENVIRONMENTAL ISSUES

3.6.1 Wildlife Considerations

Birds occasionally collide with wind turbines, as they do with other tall structures such as buildings. Avian deaths have become a concern at Altamont Pass in California, which is an area of extensive wind development and also high year-round raptor use. Detailed studies and monitoring following construction at other wind development areas indicate that this is a site-specific issue that will not be a problem at most potential wind sites. However, it is an important issue which should be considered during planning and layout with site specific studies. Overall, wind turbines' impact on birds is low compared with other human-related sources of avian mortality. See "Avian Collisions with Wind Turbines," for more information. Figure 3.5 presents results of that study.

No matter how extensively wind is developed in the future, bird deaths from wind energy are unlikely to ever reach as high as 1% of those from other human-related sources such as hunters, house cats,

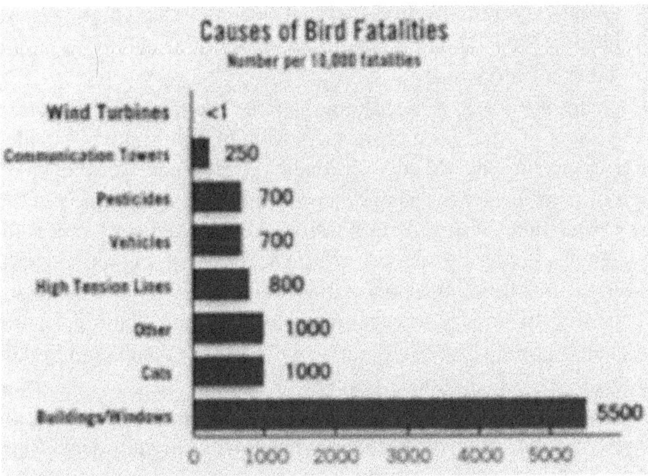

FIG. 3.5 CAUSES OF BIRD FATALITIES [15]

buildings, and autos (house cats, for example, are believed to kill 1 billion birds annually in the United States alone). Wind turbine causes are proportionally insignificant. Areas that are commonly used by threatened or endangered bird species should be regarded as less suitable for wind development. The wind industry is working with environmental groups, federal regulators, and other interested parties to develop methods of measuring and mitigating wind energy's effect on birds and accommodating special issues.

Wind energy can also negatively impact birds and other wildlife by fragmenting habitat, both through installation and operation of wind turbines themselves and through the roads and power lines that may be needed. This has been raised as an issue in areas with unbroken stretches of prairie grasslands or of forests. More research is needed to better understand these impacts.

Bat collisions at wind plants generally tend to be low in number and to involve common species which are quite numerous. Human disturbance of hibernating bats in caves is a far greater threat to species of concern. Still, a surprisingly high number of bat kills at a new wind plant in West Virginia in the fall of 2003 has raised concerns, and research at that plant and another in Pennsylvania in 2004 suggests that the problem may be a regional one. The wind industry has joined with the U.S. Fish and Wildlife Service, the U.S. DOE's NREL, and Bat Conservation International to form the Bats and Wind Energy Cooperative (BWEC), which funded the 2004 research program and is continuing to explore ways to avoid or reduce bat kills.

3.6.1.1 Additional Information Sources

- Comparative Impacts of Wind and Other Energy Sources on Wildlife
- Avian Collisions with Wind Turbines: A Summary of Existing Studies and Comparisons to Other Sources of Avian Collision Mortality in the United States
- Wind Turbine Interactions with Birds and Bats: A Summary of Research Results and Remaining Questions, National Wind Coordinating Committee.

3.7 RADAR IMPACTS

In the first decade of the 21st Century, deployment of wind power systems around the world grew significantly. This growth has led to many interaction issues between wind farms, military, aviation and weather radar systems. The UK RAF issued three studies in 2005, which illustrated new interactions and the need for additional review. The U.S. government is also concerned about the same issues and initiated a military readiness review in 2006, and issued interim policies that had the unfortunate effect of delaying over 4 billion dollars of wind projects in 2005 and over 600 billion since. Diligent efforts reduced this impact significantly and established new and reasonable policies.

Mitigation and coexistence is the goal for wind energy developers, yet defense entities are focused on eliminating wind radar interaction, creating impasses in many locations. Many mitigation methods have been successfully implemented, including wind farm/turbine site adjustments, radar (software or hardware) upgrades, and improved assessments. Early communication between affected parties in a wind farm development is a key to identifying possible problem areas and solutions to those problems.

The use of wind farms in the United States to generate electricity has increased dramatically from 2000 to 2010 [17]. Such installations can have over 100 turbines with blade-tip heights over 120 m (400 ft) above ground level (AGL). Blade-tip heights of over 180 m (600 ft) AGL are expected within a few years. Building a wind farm involves many considerations, including consultation with various aviation interests, both civil and military, and also with weather radar installations. These parties may raise objections to a proposed wind farm for a variety of reasons.

Data from the national network of Weather Surveillance Radar1988, Doppler (WSR-88D) systems are a key component in the decision making process of issuing weather forecasts and severe weather warnings, and supporting air traffic safety. Experience has shown that when wind farms are located "close" to weather radar systems, the turbine towers, rotating blades, and the wake turbulence induced by the blades negatively impact data quality and can degrade the performance of radar algorithms particularly due to Doppler return from the turbine blades. Another example, the wind turbines might appear as echoes on the display of radar used in air traffic control (ATC) and reduce radar sensitivity near and over wind farms. These echoes may distract the air traffic controller from the aircraft echoes which are his main interest and can reduce the effectiveness of the radar by masking genuine aircraft returns.

Recognizing the radar concerns relating to the Wind Radar Issue is the first step. Supporting a resolve to such concerns as quickly as possible is the next step. Protecting our borders, to provide air traffic support, forecasting the weather, and preceding with rapid development of domestic sustainable energy resources are all important. They are all vital to national security. The wind industry is in a confusing and uncertain position because it is unclear how many projects will be affected by Wind Radar. The broad-brush approach being taken by some stops all development over a multi-state region pending outcome of a study and often downplays industry and military operational experience. A number of U.S. government installations have both wind turbines and functional radar co-existing, and the British military has a track record of successfully addressing this challenge. Policies on wind turbine impacts to radar should explore the solutions used elsewhere in the world and look at ways of mitigating the problem and developing new technologies, rather than limiting wind development in large areas. Wind turbines are installed at U.S. Air Force bases and near airports in the United States and elsewhere, and the experience at those sites demonstrate this is a solvable challenge. Several radar manufacturers are confident that the majority of radar systems can perform well in presence of wind turbines, albeit with some optimization, modifications, and upgrades. Impacts, costs of mitigation technologies and DOD and FAA guidelines should be able to help streamline the development and integration processes and integrate correction technologies. Systems updated toASR-9 have had good experience mitigating wind systems and others are having good luck with their ASR-11 radar in presence of turbines.

The DHS and DOD have contested many proposed wind turbines, which stalled development of several thousands of MW of wind energy nationwide (billions of dollars of investment). The large number of denials is a serious impediment to the nation's growth of sustainable energy.

Technology can help, as Radar is basically designed to filter out stationary objects and display moving ones, and moving wind turbine blades create radar interference. It is possible to modify a radar installation to eliminate this problem, according to a consulting firm that has studied it for the British government [16]. That study concludes that " . . . radars can be modified to ensure that air safety is maintained in the presence of wind turbine farms. Individual circumstances will dictate the degree and cost of modification required, some installations may require no change at all while others may require significant modification."

If a wind project is proposed near an airport or military airfield, this issue will likely require further technical investigation. The interference is generally limited to objects (airplanes) that are physically obscured over and nearby turbines. Radar manufacturers are aggressively upgrading the radar system to allow discrimination between airplanes and turbines and foster co-existence.

3.8 LOCAL IMPACTS

Local opposition to proposed wind farms sometimes arises because some people perceive that the development will spoil the view that they are accustomed to. It is true that a large wind farm can be a significant change, but while some people express concern about the effect wind turbines have on the beauty of our landscape, others see them as elegant and beautiful, or symbols of a better, less polluted future.

The visual effect of wind farms is a subjective issue. Most of the other criticisms made about wind energy today are exaggerated or untrue and simply reflect attempts by groups to discredit the technology, worry local communities, and turn them against proposed projects. In the electronic age, myths and misinformation about wind power spread at lightning speed and due diligence are needed to dispassionately inform decision makers of the facts.

3.8.1 Visual Impacts

Visual impacts can be minimized through careful design of a wind power plant. Using turbines of the same size and type and spacing them uniformly generally results in a wind plant that satisfies most aesthetic concerns. Computer simulation is helpful in evaluating and demonstrating visual impacts before construction begins. Public opinion polls show that the vast majority of people favor wind energy, and support for wind plants often increases after they are actually installed and operating. The bibliography references [30, 31], and [32] provide more information on public attitudes toward wind.

Shadow flicker is occasionally raised as an issue by close neighbors of wind farm projects. A wind turbine's moving blades can cast a moving shadow on a nearby residence, depending on the time of the year (which determines how low the sun is in the sky) and time of day. It is possible to calculate very precisely whether a flickering shadow will in fact fall on a given location near a wind farm, and how many hours in a year it will do so. Therefore, it is easy to determine the time and duration of the flicker. Normally, it should not be a problem in the United States, because at U.S. latitudes (except in Alaska) the sun's angle is not very low in the sky. Further, the appropriate setbacks for noise will be sufficient to prevent shadow flicker problems from becoming longer-term impacts.

3.8.2 Acoustic Noise

Impacts and levels of audible noise emanating from a wind turbine depend on its surroundings and proximity to civilization.

Noise was an issue with some early wind turbine designs, but it has been largely eliminated as a problem through improved engineering and through appropriate use of setbacks from nearby residences. Aerodynamic noise has been reduced by changing the thickness of the blades' trailing edges, adjusted tip speed and by making machines "upwind" rather than "downwind" so that the wind flows throughthe rotor blades first, then the tower (on turbines with the rotor "downwind" of the tower, a thumping noise occurs each time a blade passes behind the tower). A small amount of noise is generated by the mechanical components of the turbine. To put this into perspective, a wind turbine 300 m away is no noisier than the reading room of a library.

3.9 ADDRESSING NEEDS FOR WIND TO REACH ITS FULL POTENTIAL IN THE U.S.

3.9.1 Consistent Policy Support

Since 1999, the federal production tax credit has been extended four times, but three times (2000, 2002, and 2004) Congress allowed the credit to expire before acting, and then only approved short durations. Lapses in the PTC have a significant effect on wind turbine installations as evidenced in Fig. 3.6. These expiration-and-extension cycles inflict a high cost on the industry, cause large lay-offs, and hold up investments. Long-term, consistent policy support would help unleash the industry's pent-up potential.

3.9.2 Electric Transmission

3.9.2.1 Nondiscriminatory Access to Transmission Lines To provide a stable electric power grid, system operators must maintain the balance between power consumption by customer loads and power production by the generation facilities. For each hour of the day, the operators dispatch a sufficient amount of generation to meet the expected load. Transmission grid operators typically charge generation providers substantial penalty fees if they fail to deliver electricity when it is scheduled to be transmitted. One

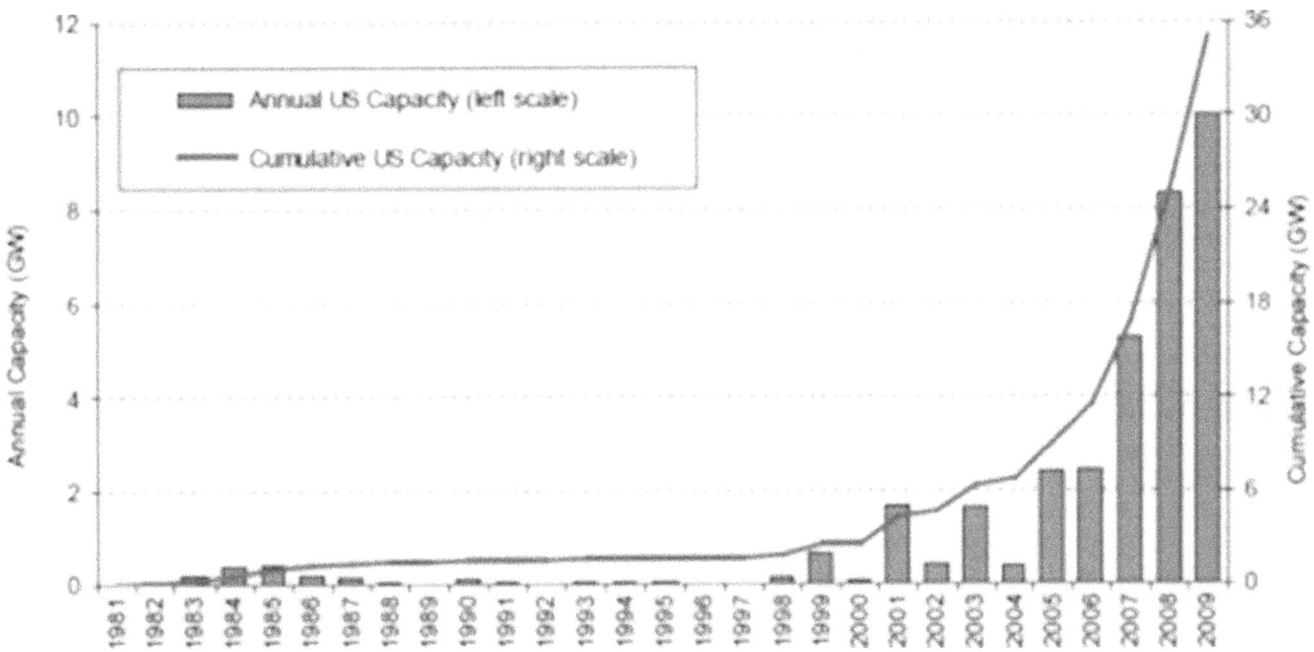

FIG. 3.6 ANNUAL AND CUMULATIVE GROWTH IN U.S. WIND POWER CAPACITY [17]

purpose of these fees is to deter power providers from using transmission scheduling as a "gaming" technique to raise spot prices or gain advantage against competitors.

Because the wind is variable, wind plant owners naturally cannot guarantee delivery of electric power for transmission at a scheduled time. Wind energy needs a modified policy that recognizes the different nature of wind plants and allows these plants to compete on a fair basis.

3.9.2.2 New Transmission Lines The transmission grid of the wind-rich Great Plains, which covers the central one third of the United States, was designed to meet the power needs of a less densely populated region. In order to capture and transmit the wind energy, extensive grid develop is required. At present, this system consists mostly of smaller-power rural transmission service. A series of extra-high-voltage transmission lines is needed to transport power from wind plants to population centers. Such a redevelopment will be expensive, but it will also benefit consumers and national security by making the electrical transmission system more reliable, reducing dependence on fossil energy, and reducing shortages, sensitivity and price volatility due to natural gas and other fossil fuels. National energy policy addressing transmission grids will be a key component for the development of the wind industry's future over the next two decades.

Wind is an intermittent energy source and therefore adds complexity for grid operators. At current levels of use, wind integration has not yet become a significant issue on most utility systems. Current best practice is to permit wind production to operate up to the point where wind generates about 10% of the electric power that the system is delivering in any given hour of the day. At this operating level, there is enough flexibility in the form of spinning reserve built into the system for generation reserve backup, varying loads, etc. Studies show, that there is effectively little difference between a system with 10% wind production and a system with 0% wind generation. Observations of actual wind generation at 10% indicate that the power variations introduced by wind are smaller than routine variations in customer load demand.

At the point where wind is generating 10% to 20% of the power it can become a larger concern for grid operators. Significant improvement can be achieved with wind forecasting and spatial diversity. Both are the subjects of current research and involve the application of statistical tools. Concepts of an energy grid overlay to the power grid are possible solutions to creating a virtual super-wind plant where the power production across vast distances is integrated together to form very predictable wind power generation.

Once wind is generating more than about 20% of the electricity, the system operator can incur significant additional expense because of the need to procure additional equipment that is solely related to the system's increased variability and/or managing load differently.

These figures assume that the utility system has an "average" resource portfolio that is complementary to wind's variability (e.g., hydroelectric dams and natural gas peaking plants) and an "average" amount of load that can vary quickly (e.g., electric arc furnace steel mills). Actual utility systems can vary quite widely in their ability to handle as-available output resources like wind farms, especially when they are part of a larger grid system. However, as wholesale electricity markets grow, fewer, larger utility systems are emerging. Therefore, over time, more and more utility systems will be able to integrate larger ratios of renewable power.

For detailed information on this topic, see "Grid Impacts of Wind Power: A Summary of Recent Studies in the United States," Milligan et al, NREL [33].

Industry and utilities note that one of the biggest constraints to the expanded growth of wind power in the United States will be the ability of the transmission grid to deliver large amounts of wind energy to customers. AWEA, DOE, and WPA work together to transform electric industry practices, including the area of transmission, through technology, planning advocacy and outreach.

Transmission is an important factor in the development of wind energy because some of the best wind resources in the country are located in areas remote from the large electric markets. By expanding and upgrading transmission systems, the nation could have better access to wind energy, which could be more easily moved from distant areas to population centers where electricity demand is greatest. Moreover, by facilitating the expansion and geographical dispersion of wind power across a wide area, an upgraded transmission grid improves both the reliability and capacity of wind. When the wind is not blowing at one location, it is usually blowing somewhere else. Thus, dispersed wind power compensates for short-term fluctuations and increases capacity value.

Electric utilities originally developed their electric transmission grids over time to connect local load with local generation, and then gradually those local utility grids were interconnected in furtherance of reliability benefits, into a "national electric grid." That national grid was never intended to function, and never has functioned, to move truly large amounts of power from one region of the country to another. America's electric grid could be designed to function much like the interstate highway system functions to efficiently transport goods across the country. This "electric superhighway" would allow low-priced clean energy to reach consumers across the country. Such a national grid would require state, regional, and national rethinking in terms of transmission line construction processes, yet provide vastly improved transfer capability.

3.10 WIND FARM DEVELOPMENT

To be viable, a candidate wind plant site needs to have the following components:

- Favorable economics
  - Ability to generate income
    - Sufficient wind resource
    - Buyer/PPA
  - Reasonable construction costs
    - Roads
    - Cost effective turbine
    - Interconnection fees
- Transmission/path to market
  - Proximity to transmission lines
  - Cost effective interfacing substation
- To be legally permissible
  - Siting/permitting
    - Environmental impacts
    - Radar impacts
    - Local concerns
- Access to financing

Typically, mechanical engineers are not involved in all aspects of these steps, but need to be aware of them in order to properly assess a candidate site's viability. A given site may have an excellent wind resource, but other factors (i.e., if it is near an airport or lacks adequate transmission) may prevent its project development. The following sections address the project development steps typically handled by engineers.

3.11 WIND RESOURCE ASSESSMENT

A site's wind resource is a key factor in determining the ability to generate income. There are many steps involved in resource assessment, but the end goal is estimating how much annual energy (MWh) can be generated at a site. Wind resource metrics include average wind speed at a certain height, frequency distribution of wind speeds at the site, and wind shear. Each metric will be discussed individually.

3.11.1 Power Available in the Wind

Average wind speed is one of the most frequently discussed metrics of a farm or plant site. Because it is a single number, it is easy to discuss and is the metric used to define wind classes (refer to Table 3.1). However, relying on a site's average wind speed can be very misleading because of the cubic relationship between instantaneous wind speed, instantaneous power available, and the total annual energy production. It is important to understand the nature of the power available in the wind.

The following equation describes the amount of power, P, available in the wind as a function of air density, cross-sectional or "swept" area, A, of the rotor, and the wind speed, v.

$$P = \frac{1}{2} r A v^3 \tag{3.2}$$

Note that this equation does not describe the power delivered to the grid. To calculate grid power, one must account for generator efficiency and line losses. The total available energy is obtained from the time integral of the power.

$$E = \int P(t)dt \tag{3.3}$$

Consider for example the impact of the wind characteristics on the annual energy available at two sites having the same annual average wind speed, but very different distribution of wind speeds. It is assumed that each project has four discrete wind speeds evenly distributed over the year (a quarter of the time is equal to 2190 hours). The site data is provided in Table 3.5. For both sites, the simplified wind speed distributions yield the same annual average wind speed of 10 (m/s). However, Site A's average cubic wind speed, which is proportional to power, is 2020 $(m/s)^3$, nearly double that of Site B. If one assumes the turbines placed at each site have equal rotor diameters (i.e., equal swept area) and each site experiences the same air density over the course of a year, then site A will generate approximately double the energy that Site B generates. This over-simplified example demonstrates the importance of a site's wind speed frequency distribution in addition to its average velocity and the importance of having actual long-term wind speed data to assist in decisions and optimization.

3.11.2 Wind Speed Frequency Distribution

The best way to determine a site's wind speed frequency distribution is to measure the wind speed over the course of a full year at a minimum. In professional practice, many investors are requiring wind sampling over the course of 2 years to gain more confidence in a site's wind resource. Anemometers placed on meteorological towers have been the most common instrument used to collect wind speeds up to the early part of this decade. Sonar and laser type instruments capable of measuring wind speed and direction at

TABLE 3.5 AVERAGE WIND SPEED

Site A			Site B		
Time (hours/yr)	**Wind speed (m/s)**	**Wind speed cubed (m/s^3)**	**Time (hrs/yr)**	**Wind speed (m/s)**	**Wind speed cubed $(m/s)^3$**
2190	2	8	2190	8	512
2190	8	512	2190	9	729
2190	12	1728	2190	11	1331
2190	18	5832	2190	12	1728
Averages	10	2020	Averages	10	1075

various heights without use of tower (SODARs and LIDARs) have become more widely used since 2005.

Regardless of which instrument is used to measure wind speed and direction, data collection is a fairly uniform process. Most data loggers used in 2010 sample wind speed and direction every 2 seconds and report wind speed and direction in 10-minute averages. Wind resource engineers use this data to develop wind speed frequency distribution curves similar to the one shown in Fig. 3.7.

The wind speed frequency distribution can be used to estimate the annual energy (kWh) production as plotted in the figure. Although 12 mph wind is the most frequent, occurring at 6% of the time, the 22 mph winds generate the greatest amount of energy annually. This reflects the cubic relationship between wind speed and power.

3.11.3 Wind Shear

As with other flows, wind blowing across land has velocity profiles. In general, wind nearest the ground has the least velocity and wind at higher levels above the ground has the greatest velocity (i.e., free stream). The variation of wind speeds at different heights is known as wind shear.

Wind shear is an important factor in determining the optimal tower height for wind turbine performance. To optimize tower height, two competing factors are considered. First, in general, wind speeds increase with height above the ground, which can lead to greater energy production and ultimately greater income. Conversely, taller towers increase a project's capital and operations and maintenance (O&M) costs. Increased capital costs include the cost of the towers themselves, transportation costs to the site, taller cranes required to erect the turbines, stronger/larger foundations, and roads suitable to carry the loads. O&M costs increase because of the increased nacelle height, which will affect everything from routine maintenance (due to the increased time and energy required for technicians to gain access) to replacing large components, especially blades and gearboxes.

In order to optimize tower height, a clear understanding of the site's wind shear is crucial. Experience has shown that wind shear varies daily, seasonally and with changes in wind direction (especially for sites with mountainous topography). The best way to understand wind shear is to directly measure wind speed and direction at multiple heights, preferably including the hub height and rotor diameter. Figure 3.8 shows the hub height, top and bottom of the rotor of a Suzlon S-88 wind turbine with an 80-m tower and 88-m rotor diameter.

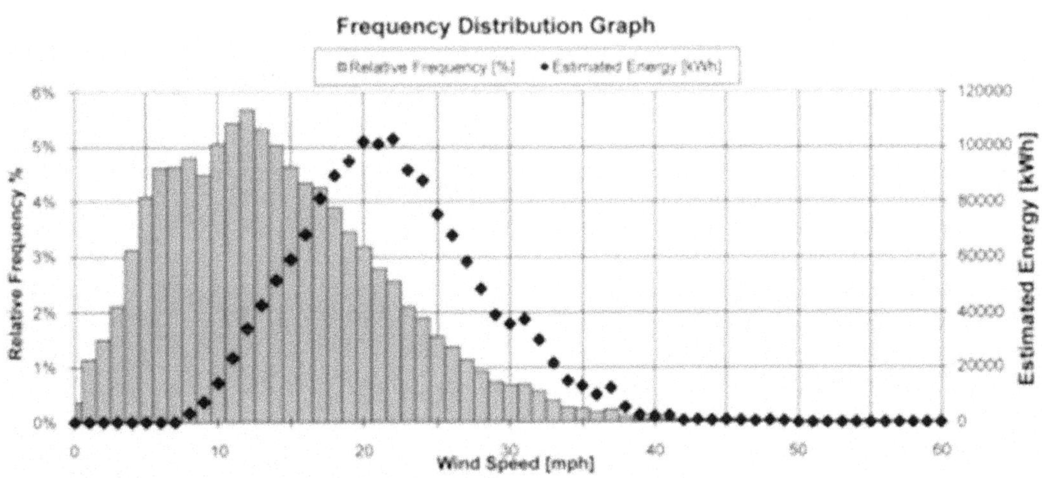

FIG. 3.7 FREQUENCY DISTRIBUTION GRAPH OF WIND SPEEDS COLLECTED OVER A 5-MONTH PERIOD IN 2009 FROM A SITE IN IDAHO. THE X-AXIS INDICATES WIND SPEED. THE BARS INDICATE PERCENT OF TIME THAT A GIVEN WIND SPEED OCCURRED DURING THE PERIOD (LEFT VERTICAL AXIS). THE BLACK DIAMONDS INDICATE ENERGY (RIGHT VERTICAL AXIS) [18]

FIG. 3.8 WIND TURBINE NEAR MOUNTAIN HOME, ID (PHOTO COURTESY OF TODD HAYNES, 2008)

In practice, few meteorological towers are as tall as wind turbine towers (i.e., they cannot measure wind speed at hub height). As a result, wind speeds are measured at various heights (typically 10-m intervals) along the met-towers. The measured wind speeds are used to create a velocity profile which can be extrapolated to the turbine's hub height and maximum rotor height. Meteorologists have developed relationships to estimate wind speed at one height given a known wind speed at another height. In North America, the power law is most often used.

$$\frac{v}{v_0} = \left(\frac{h}{h_0}\right)^a \quad (3.4)$$

In the power law, v_0 is a known velocity at height v_0 and v is an unknown velocity at height h. The exponent is the wind shear exponent. Under very precise meteorological conditions, the wind shear exponent has been shown to be $a = 1/7$. In practice, these conditions are not often present. Inexperienced wind resource engineers and developers will assume that $a = 1/7$ in the absence of measured data. This can lead to inaccurate estimates of annual energy generation, and thus inaccurate estimates of income potential. Wind resource engineers should exercise caution when estimating wind shear, and reminded that for a given site, wind shear will vary under different atmospheric conditions. The best way to know a site's wind shear is to measure it. A discussion of various instruments available follows in the Micro-siting section.

3.11.4 Prospecting

Given the costs associated with installing met-towers or other instrumentation, wind farms developers typically engage in "prospecting" sites before actually measuring data. Prospecting involves collecting cursory information in a low cost manner to determine if a site warrants a more detailed resource assessment.

During the 1990s and 2000s, the amount and quality of public data available for prospecting has increased considerably. Low-tech prospecting may be as simple as observing tree growth. If the wind blows consistently from one direction, a tree will "flag" as it grows. That is, the trunk will bend away from the prevailing wind, and the branches will tend to grow on the downwind side as shown in the photograph of Fig. 3.9.

There are many sources of public information that can be useful for prospecting. The U.S. DOE's Wind Powering America (DOE–WPA) program has produced publicly available wind maps for the nation (refer to Fig. 3.3) and many individual states.

In addition to wind maps, DOE-WPA initiated a state anemometer loan program in the early 2000s. In this program, DOE-WPA provided dozens of met-towers to participating states. The met-towers are loaned on an annual basis to interested landowners — both public and private. If the site shows promise, it is possible for the tower to remain a second year and continue collecting data. If not, the tower is removed and installed at another location in the state. DOE-WPA does charge the landowners to participate, but requires that the data collected be made publicly available for the common good. More than 20 states participate in the anemometer loan program.

Much of this data has been collected on 20-m and 30-m towers, which are considered short by 2010 wind resource assessment standards. These data are not meant to be relied upon for wind project development. Instead, they are meant to be indicators of sites which show sufficient promise that a developer or other interested party might invest in installing taller met-towers or other instrumentation. State of the art wind resource instrumentation is discussed in the Micro-siting/Data Collection section. Links to the publicly available met-data can be found by clicking on the respective states at the website as shown in Fig. 3.10.

For example, by clicking on the Idaho link on the website leads to websites hosted by the INL and Boise State University, both of which contain further links to wind data collected at individual

FIG. 3.9 GERALD FLEISCHMAN, WIND ENERGY ENGINEER AT THE IDAHO OFFICE OF ENERGY RESOURCES, STANDS NEAR A FLAGGING TREE (PHOTO COURTESY OF KURT MYERS, 2002)

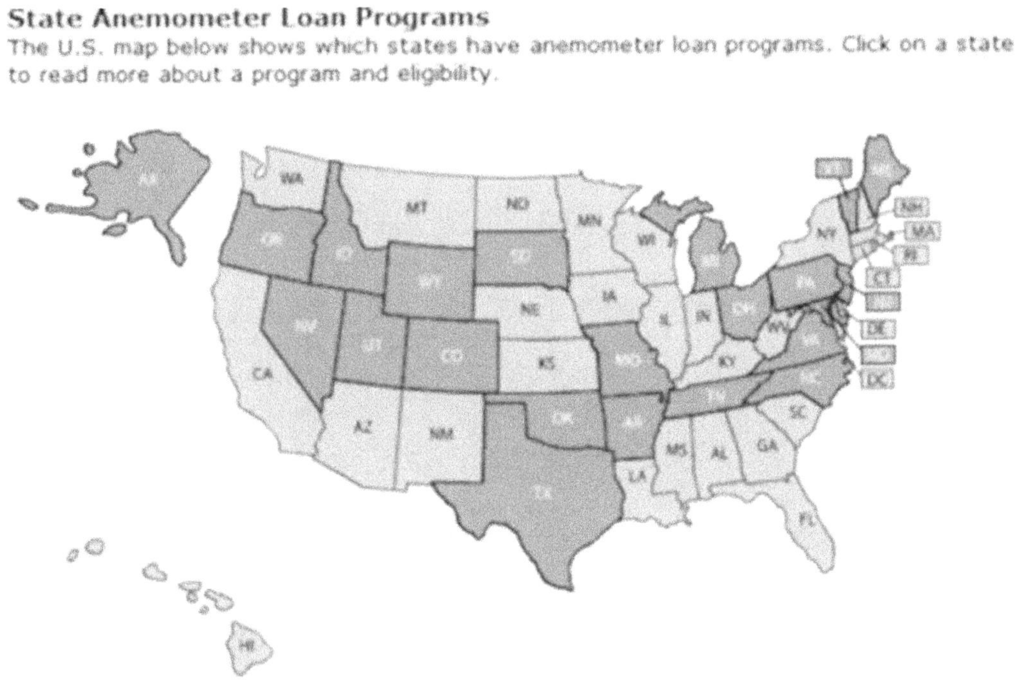

FIG. 3.10 MAP OF STATES WHICH PARTICIPATE IN THE DOE-WPA ANEMOMETER LOAN PROGRAM [19]

FIG. 3.11 MAP OF INDIVIDUAL ANEMOMETER LOAN PROGRAM SITES IN IDAHO. IDAHO NATIONAL LABORATORY AND BOISE STATE UNIVERSITY ARE PARTNERS WHICH ANALYZE AND POST THESE DATA [20]

sites in Idaho. Other states with anemometer loan programs (shown in Fig. 3.10) have similar public data sources. The Idaho National Lab map contains data going back to the program's inception in 2002; see Fig. 3.11. The Boise State University College of Engineering became involved in 2008 and contains data collected and analyzed since that time. By clicking on one of the dots on the map at the website, the user will find raw data and summary data for that site, similar to the data shown in Fig. 3.12.

3.11.5 Micro-siting

Once a candidate site has been identified through the prospecting process, the next step is to collect "investment grade" data. This data is typically collected with a tall tower (60-m or greater) or a sensor which measures multiple heights without a tower (SODAR/LIDAR).

3.11.5.1 Met-Towers In 2010, the most commonly installed met-towers are 60-m tall. Taller towers are available, but it can be difficult to obtain permits for towers taller than 200 ft (60 m) because of FAA regulations regarding lights and the can be cost prohibitive. Regardless of height, met-towers typically contain multiple instruments. This can include anemometers at 10-m height intervals, a direction vane and temperature sensor.

Most counties in the U.S. require that met-towers taller than 33 ft (10-m) obtain conditional use or special use permits. Additionally, in 2010, the wind industry was made aware of concerns from crop dusters over met-towers being installed on or near farmland. These issues typically do not prevent met-towers from being installed. They are addressed here to alert the reader to be cognizant of the need to inform appropriate officials, especially since most met-towers do not have lights at the top.

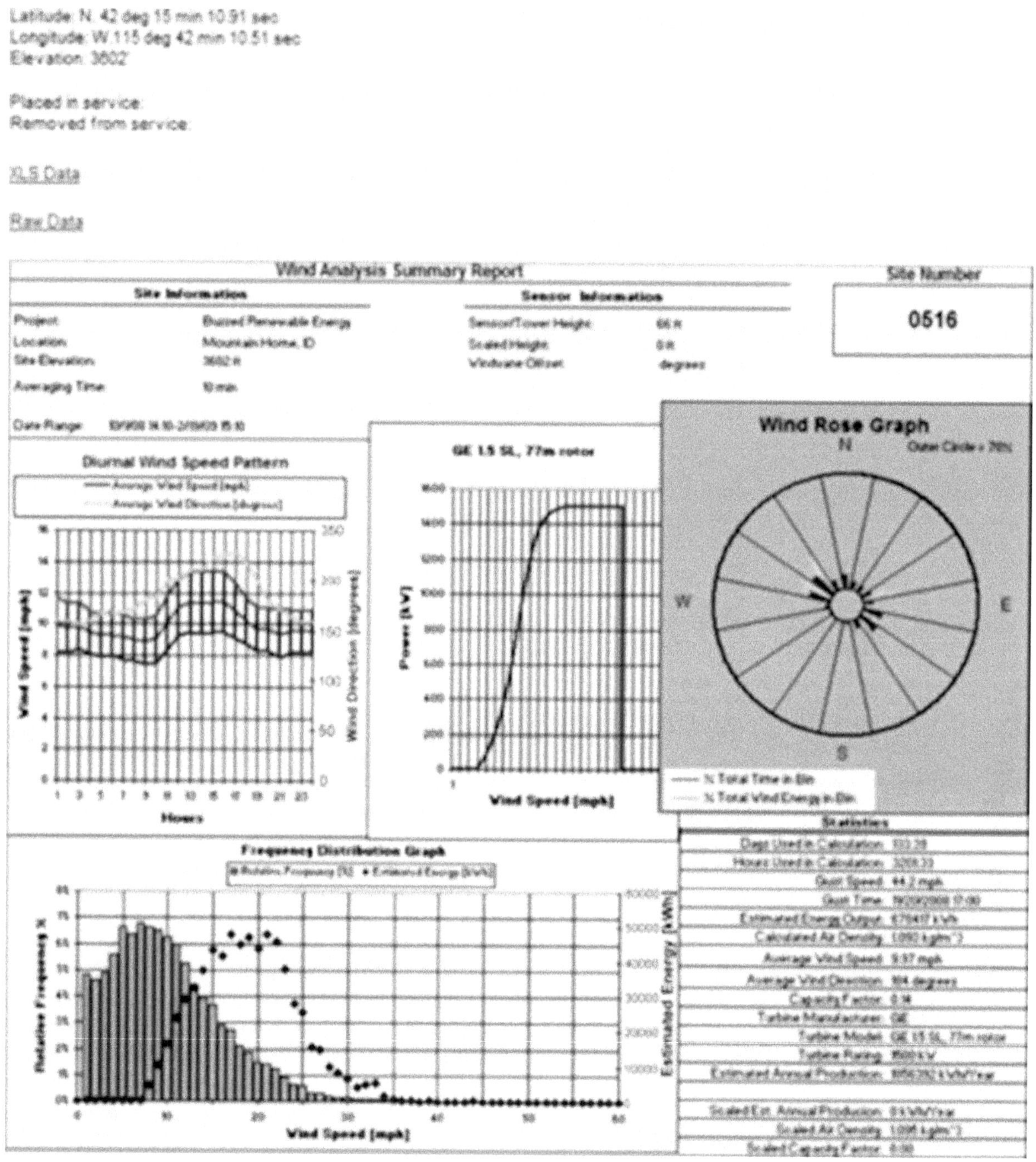

FIG. 3.12 EXAMPLE OF THE SUMMARY SHEET AND LINKS TO ANALYZED (XLS) AND RAW DATA AVAILABLE FROM THE ANEMOMETER LOAN PROGRAM [21]

Many companies manufacturer met-towers in North America. Typical packages include the tower, appropriate instrumentation and a choice of data loggers. A typical met-tower is shown in Fig. 3.13.

Depending on the geographic footprint of the project being considered, more than one met-tower may be necessary. This is especially true in complex (i.e., mountainous) terrain. When placing multiple met-towers, it is advisable to locate some towers at points expected to have good wind (i.e., ridge tops or other high points) and others at places which are expected to have less wind. This will enable the wind resource engineer to create a more realistic model of the wind speeds across the site than would be the case if measurements were only taken on ridge tops.

3.11.5.2 SODARs and LIDARS In addition to MET-towers, other instruments have gained popularity in the wind resource assessment field over the last decade. Sonic Detection and Ranging (SO-DAR) instruments use sound to measure wind speed and direction, taking advantage of the doppler shift. Light Detection and Ranging (LIDAR) instruments use light to measure wind speed and direction. Both SODARs and LIDARs are able to measure wind speed and direction at multiple height intervals without using a tower. This allows an engineer to collect measurements higher above the ground than can be done with most towers, which can lead to more accurate estimates of wind shear at a given site. In practice, SO-DARs and LIDARs are often used in conjunction

FIG. 3.13 A 50-M MET-TOWER IN SOUTHERN IDAHO (PHOTO COURTESY OF TODD HAYNES, 2009)

with met-towers to create very high resolution wind maps of a potential site. Figure 3.14 shows a typical SODAR setup.

SODARs and LIDARs have both advantages and disadvantages as compared to met-towers. Advantages include the ability to measure at more heights as already discussed. Another advantage is the fact that they can be moved to different locations much easier than towers. One disadvantage to these technologies is that they tend to cost more than towers. However, the costs are coming down and nearing convergence with tall towers in 2010. Another disadvantage to SODARs is the noise emitted. They are generally not suitable for placement near residences or places of business. This is often not an issue with wind resource assessment, because wind farms also tend to be located away from homes and businesses.

3.11.6 Data Analysis

Data is typically collected at 2-second sample rates and averaged over 10-minute intervals. This data is used to create frequency distribution curves (similar to Fig. 3.7), calculate a sites average wind speed, and also to create wind roses. A wind rose is helpful in determining the orientation of a wind turbine array. It indicates the direction of the prevailing wind(s), the percentage of time that wind comes from any particular direction and the percentage of energy associated with a wind direction. The site represented in Fig. 3.15 has two prevailing wind directions: the northwest and the southeast. It is interesting to note that although the wind comes from the northwest much more frequently than the southeast (the green slice), there is only slightly more energy from the northwest than from the southeast. This suggests that southeasterly winds are typically higher velocity than those from the northwest.

3.11.6.1 Data Analysis Tools There are many software packages available to analyze wind data and create frequency distribution curves, wind roses and the like. Some are offered by the met-tower manufacturers. Others are offered as services along with data loggers. Some are similar to Computational Fluid Dynamics (CFD) software codes in that they use measured wind speeds at one or more locations on a site to create a high resolution wind map of the candidate site. Some create visualizations of how various wind turbines will look from a distance when placed at a given site. The sophistication of these packages in increasing rapidly in 2010 and wind developers and engineers are encouraged to become familiar with several options to determine which will best suit a project's needs.

The U.S. DOE's INL has created free data analysis software tools which are based in MS Excel. This software allows a user to input wind data which has been measured at a site (regardless of the instrument). It contains power curves for a variety of commercial wind turbines. The user is able to compare the performance of a variety of wind turbines for a given site. Figure 3.12 was created using the INL software, which is available for download at bibliographic reference [34].

3.12 WIND FARM DESIGN

After a candidate site is deemed viable in the wind resource assessment process, the design process may be initiated. There are many steps involved in the design of a wind power facility. The steps below are typically handled by a combination of mechanical, electrical and civil engineers.

3.12.1 Turbine Selection

At any given site, different turbines will perform differently. It is typically the job of a mechanical engineer to select an optimal turbine given a site's wind speed and turbulence characteristics. The selection process consist of (i) the selection of the number and type of turbine, including the manufacturer and power capacity, (ii) the tower height, and (iii) the blade length. The cost optimization process considers both income from energy generation and the costs for construction, operation, and maintenance.

3.12.2 Turbine Placement and Layout

The layout of a turbine array is a critical step in wind farm design. Proper spacing between turbines is necessary to prevent one turbine from being in another turbines wake, which can decrease energy output, increase turbulence, and increase O&M costs. Although each site should be carefully modeled to precisely space turbines, some rules of thumb or best practices exist. Side-to-side spacing of turbines, perpendicular to the prevailing wind direction, should be separated by 3 to 4 rotor diameter lengths. For front to back spacing, parallel to prevailing wind direction, turbines should be spaced 8 to 12 rotor diameters apart. In sites with a singular or predominant prevailing wind (such as indicated in the rose graph of Fig. 3.15), the low-end of the range can be used. In sites with frequent winds from multiple directions and having a more balanced wind rose pattern, the greater spacing distances should be used. Turbine layouts can be optimized using computer models.

One advantage to wind energy is that it allows a site to continue with its traditional land use. For wind turbines placed on active farmland, it may be necessary to consider the land's current use when creating a turbine array. For example, a landowner may prefer that some income from energy be forfeited in order to continue farming the majority of the land. In other words, rather than placing all turbines in the optimal spots for energy production, they may be moved to areas not farmed, such as irrigation pivot corners.

3.12.3 Access Road Design

Access roads must have adequate width and bases to support trucks transporting tower sections and blades as well as the cranes necessary to erect the turbines. When roads are built, curves must be of sufficient radius to accommodate the long blades of modern turbines, as shown in Fig. 3.16. Crane pads must be sufficient to support cranes as shown in Fig. 3.17. Civil engineers typically design a wind farm's roads and crane pads.

FIG. 3.14 BOISE STATE UNIVERSITY ENGINEERING STU-DENT KEN FUKUMOTO STANDS NEXT TO A TRITON™ SODAR NEAR MOUNTAIN HOME, ID (PHOTO COURTESY OF TODD HAYNES, 2009)

3.12.4 Electrical Collection System

Typical wind farm design in the United States includes many electrical power systems aspects. Wind farm engineers need to be cognizant of electrical grid and transmission issues when looking for suitable locations and planning their developments. Available transmission capacities, voltage levels, proximity to loads, and many other factors need to be considered. The voltage level of the proposed interconnect with the utility's transmission will have a direct impact on the cost of the interconnection equipment and the economics of the wind project. Most wind projects above 20 MW in size typically attempt to tie in to transmission voltage levels between 69kV and 230 kV, while only very large projects above 500 MW attempt interconnects at voltages above 230 kV due to substation costs.

Once an interconnect application is filed with a utility, both the utility and the wind developer's electrical engineers start more in-depth analyses of the power system from both perspectives. The utility will model its system and the wind farm addition with software programs addressing items such as load flow, system voltages and voltage support equipment, potential harmonics, system dynamics and stability, and any other power quality areas of concern. The wind developer's engineers will analyze the wind system with similar modeling tools and calculations and will refine the wind farm electrical design based on this interaction with the utility and other design factors. Layout and spacing of the wind turbines drives much of the design for the electric power collector system, along with other aspects such as soil thermal conductivity, electrical resistivity, and underground soil and rock characteristics. Therefore, the electrical engineers must work closely with the meteorologists doing the wind turbine layouts to put together an appropriate design.

Transformers and cables are sized to keep losses and voltage drops-below acceptable levels (typically less than 5%) for the project and to keep cables from being damaged due to overheating. The latter concern is related to the soil thermal conductivity determined in testing and number of cables in parallel, and designs should consider appropriate cable sizing and deratings based on applicable electrical codes and good design practices. Wind farm collector systems are typically run underground to reduce potential bird and environmental impacts, and once the electrical system is far enough away from the turbines or back to the substation, it usually transitions back to overhead power lines to keep costs down. Typical collector system voltages are from the 25 kV and 35 kV voltage classes with 34.5 kV being a common nominal voltage. Some smaller projects may have systems at 25 kV or lower distribution voltages. Typical wind turbine output voltages are 575 V to 690 V, so step-up transformers are located either in the turbine nacelle or on the ground next to the turbine to increase the voltage to the collection voltage level and reduce transmission losses. Grounding of the wind turbine and electrical system is also a significant component of the electrical design, and is related to the electrical resistivity tests and wind turbine grounding requirements that must be part of every wind farm design.

Wind turbines are designed with circuit breakers and/or contactors and fault protection devices that monitor voltages, frequency, current and other parameters in order to protect the turbines.

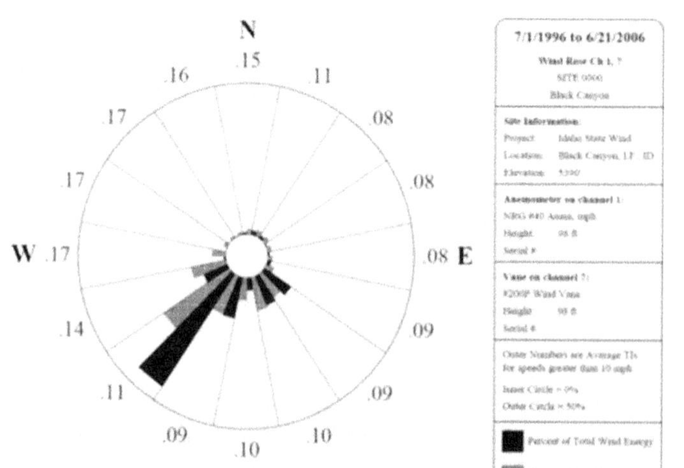

FIG. 3.15 EXAMPLE OF A WIND ROSE GRAPH (TOTAL 10-MINUTE INTERVALS: 524592. INTERVALS USED IN CALCULA-TIONS: 217906. PERCENT DATA USED: 41.5) [20]

FIG. 3.16 TRUCK TRANSPORTING A 44-M BLADE TO A WIND FARM NEAR BLISS, ID (PHOTO COURTESY OF TODD HAY-NES, 008)

Additional fault protective systems to protect the wind farm transformers, cables and other equipment should be designed in the electrical system and substation at the point of interconnect. These systems usually include high and low side transformer breakers and switches, electronic protective relays, feeder breakers and relays, switchgear, grounding transformers, current and potential transformers, and related equipments. Other items that may be incorporated in the substation include metering equipment, control building and backup batteries, voltage support equipment, such as switched capacitor banks or reactors, and other equipment determined necessary during the project design.

3.12.5 Permitting

Engineers typically support the permitting process steps by providing:

- topographical maps describing the proposed turbine array,
- visual simulations of the turbines from a distance (such as how they might be visible from a nearby city or town),
- estimates of sound levels at the property boundary, and
- technical details about the proposed turbines.

During the permitting process, wind farm developers must consider potential environmental impacts, radar proximity, and local ordinances regarding sound, set back, and flicker.

3.13 WIND ENERGY RESEARCH

From 1998 through 2009, installed wind power capacity grew very rapidly. At the end of 2009, the cumulative total capacity of wind power in the United States was greater than 35,000 MW [17]. Given wind power's intermittent nature, there is evidence that this rapid growth is adding complexity to grid operators and electric utility companies. Grid operators must constantly balance load demand and generation supply on the electric grids. Since typical users do not schedule their demand ahead of time, grid operators must forecast demand ahead of time and adjust generation to meet the forecast. Forecasts and scheduling are done hourly in many markets and sub-hourly in other markets. The grid is kept in balance by adjusting generation power levels. Output from most traditional energy sources thermal generators and hydropower can be controlled, or "dispatched," by grid operators to meet the load forecasts. Intermittent energy sources, such as wind, cannot be controlled by grid operators. In fact, the power produced by intermittent sources is accounted as negative loading on the grid. As currently configured, this fact increases the complexity of keeping electric grids in balance.

FIG. 3.17 CRANES USED TO ERECT TOWER AND ROTOR SECTIONS AT A WIND FARM NEAR BLISS, ID (PHOTO COUR-TESY OF TODD HAYNES, 008)

3.13.1 Grid Integration

During some periods, the output from wind plants is increasing or decreasing at the same time as load, which helps keep the grid in balance. At times when wind is increasing and load is decreasing, or vice-versa, the wind output makes balancing the grid much more difficult.

The Bonneville Power Administration's (BPA) balancing control area in the Pacific Northwest has the largest penetration of wind generation in the nation at approximately 30% of peak load (3000 MW of 10,500 MW) [22]. The U.S. DOE has a stated goal of 20% wind energy nationally by the year 2030 [1], and BPA was well ahead of this goal in 2009. Its experiences are considered a leading indicator of wind power's effects (both positive and negative) on electric transmission grids. BPA has created a Wind Power section on its website which contains many resources regarding wind power's grid impacts (http://www.bpa.gov/corporate/ WindPower/).

Some of the data is near real-time, such as Fig. 3.18, which plots BPA's total load and total generation from several sources, including wind, over a rolling 7-day period. The red line indicates the daily load cycles; the blue, green and brown lines indicate generation contributions from wind, hydro and thermal energy sources respectively. During this 7-day period, the contribution from wind was near-zero, which is not unusual during the winter season. There was around 1000MW contribution from wind on December 9, 2010.

While there is often little wind generation during the winter months, the spring and fall months tend to have much more significant contribution. Figure 3.19, showing a 7-day period in March 2010, indicates wind energy contributed 2000 MW or greater on March 12, 13, 14, and 17. The figure also demonstrates the phenomena known as "wind ramps," dramatic increases or decreases in wind power over a short period of time. An extreme wind ramp occurred on March 17, when wind power generation increased from 0 MW around noon to greater than 2000 MW around 4 p.m. This up-ramp was followed by a down-ramp, which was nearly as extreme. From approximately 8 p.m. March 17 to 3 a.m. March 18, wind power generation decreased from around 2500 MW

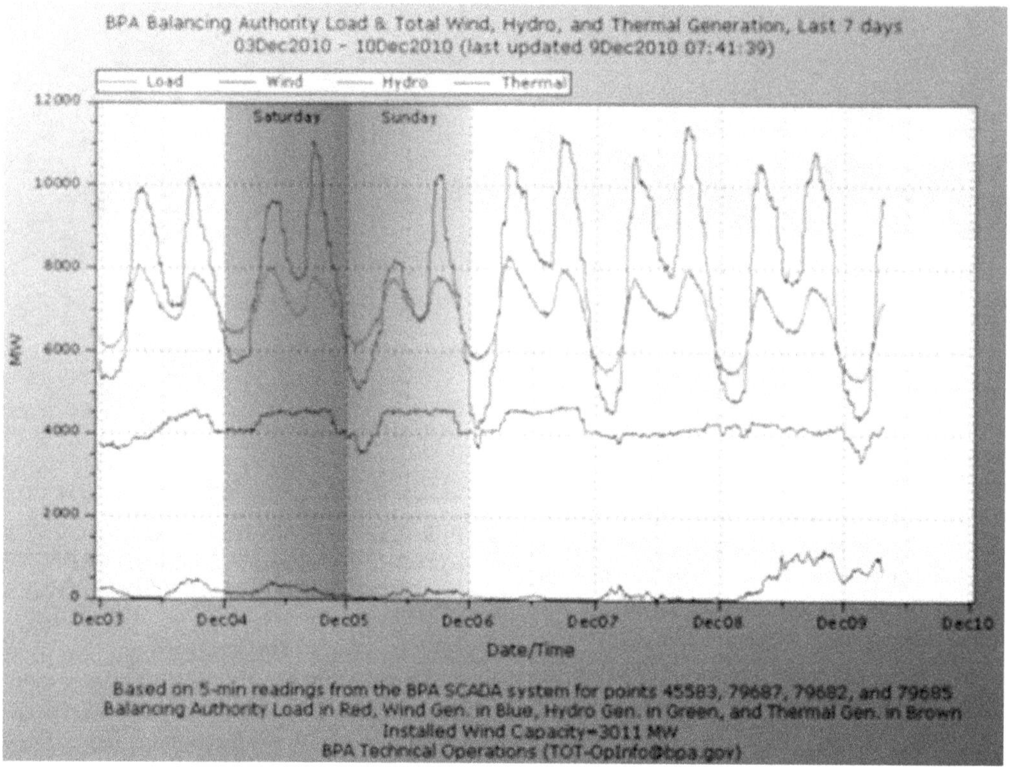

FIG. 3.18 PLOT OF 7-DAY PERIOD IN DECEMBER 2010, SHOWING LOAD (RED LINE) AND GENERATION CONTRIBUTIONS FROM SEVERAL ENERGY SOURCES ON THE BONNEVILLE POWER ADMINSTRATION GRID [23]

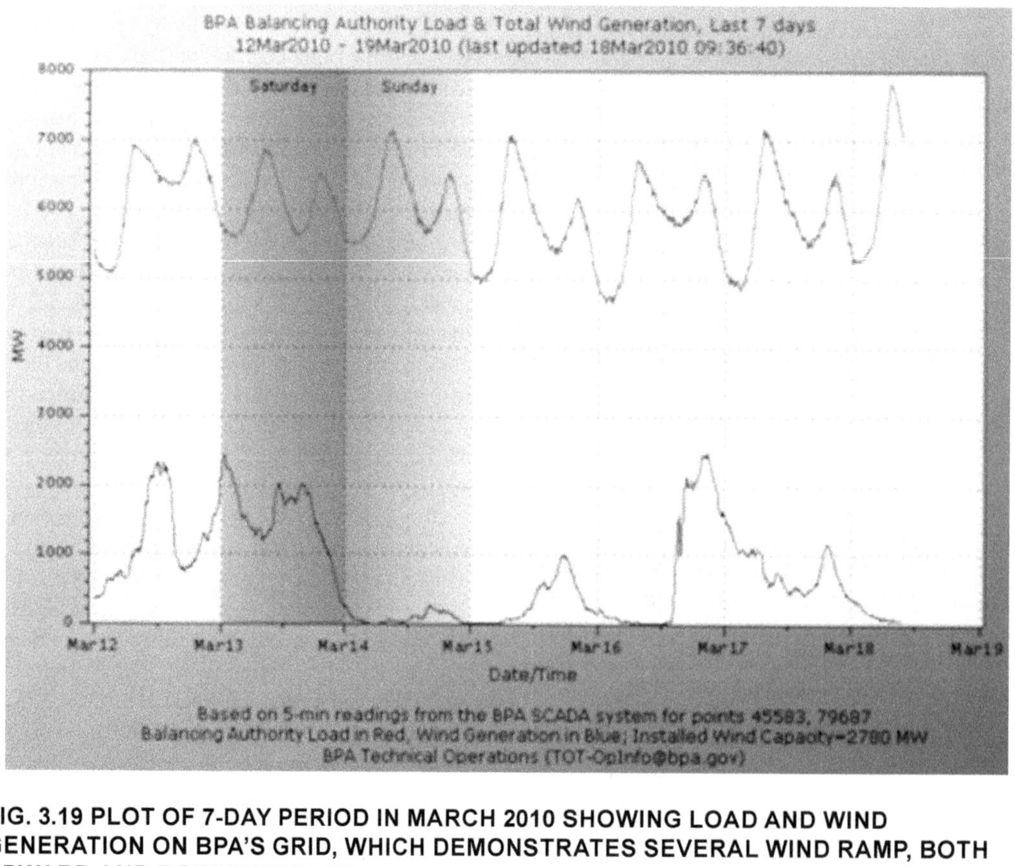

FIG. 3.19 PLOT OF 7-DAY PERIOD IN MARCH 2010 SHOWING LOAD AND WIND GENERATION ON BPA'S GRID, WHICH DEMONSTRATES SEVERAL WIND RAMP, BOTH UPWARD AND DOWNWARD [22]

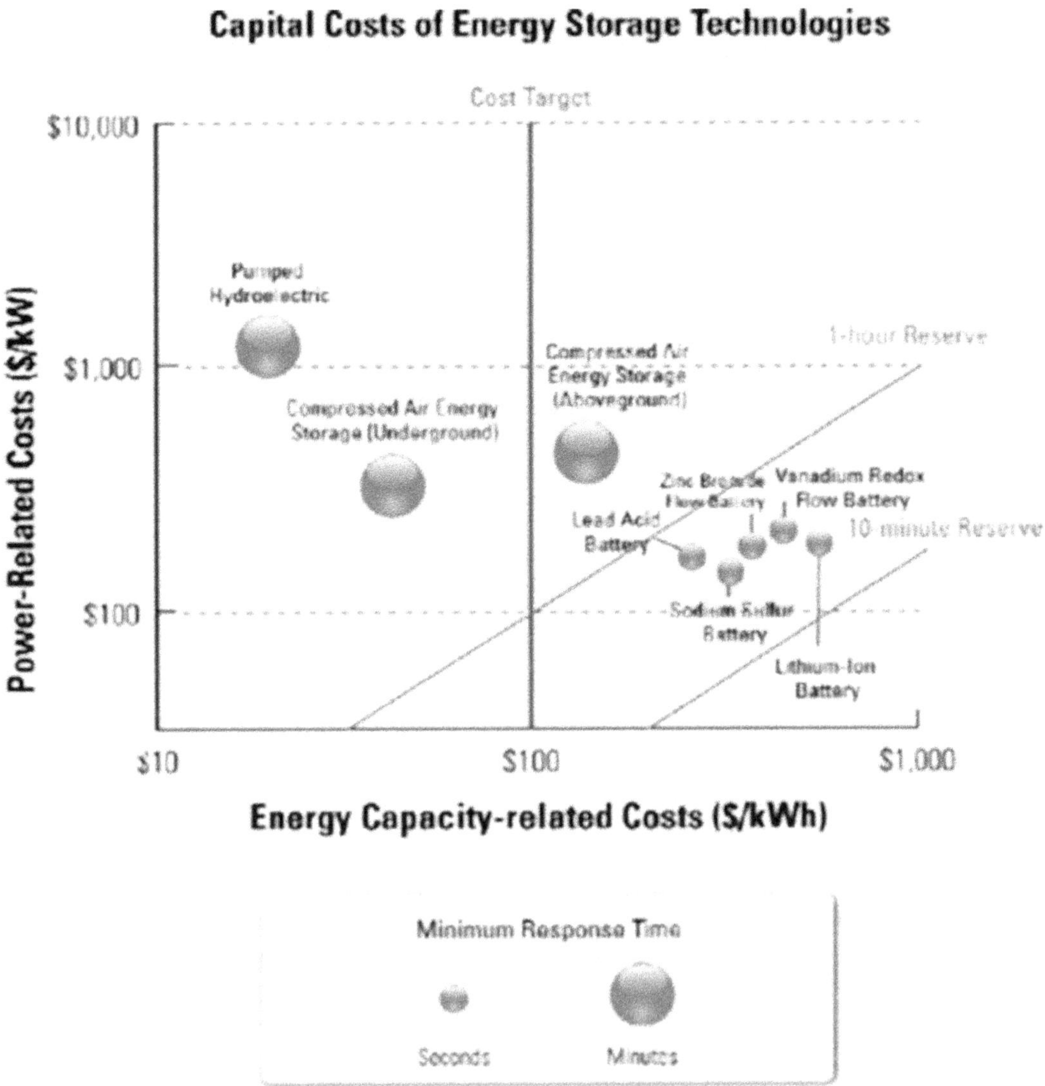

FIG. 3.20 POWER COSTS AND ENERGY COSTS ARE DISPLAYED FOR A RANGE OF ENERGY STORAGE TECHNOLOGIES [27]

to around 1000 MW. While these ramps are interesting to study alone, it is important to view them in relationship to changes in load (red line). The March 17 up-ramp provided a more difficult situation than the down-ramp. This is because most of the up-ramp was coincident with a down-ramp in load. The down ramp beginning around 8 p.m. was coincident with a down-ramp in load; this had the effect of assisting grid operators in balancing the grid.

Wind ramps such as these demonstrate some of the most pressing research needs for continued growth in wind energy. Ongoing research in grid integration, specifically forecasting, storage and development of hybrid energy systems, is intended to allow greater percentages of intermittent generation on electric grids.

3.13.1.1 Forecasting Grid operators balance the grid by forecasting load and adjusting generation output accordingly. Since utility companies cannot directly dispatch wind, the ability to forecast wind output would be beneficial by allowing them to properly adjust output from other generation — to account for changes in load and wind generation simultaneously.

State-of-the-art monitoring and statistical computation technologies allow wind farm developers and operators to fairly accurately predict how much energy can be generated annually at a facility. However, knowing when that energy will be available is a much more difficult problem. Utility companies are continually working to keep the grid in balance, and rely on short- and medium-term load forecasts, ranging from hour-ahead to 24-hours-ahead. The ability to accurately forecast wind plant outputs in those time frames has been identified by BPA as a high priority [24]. Similarly, the U.S. DOE Energy Efficiency and Renewable Energy Office (EERE) provided funding in 2010 to two research projects with the goal of improving short-term wind forecasting. The research teams include collaborators from private companies, universities, the DOE, and National Oceanic and Atmospheric Administration (NOAA) [25].

3.13.1.2 Storage To allow high percentages of intermittent energy generation on transmission grids, economical grid-scale energy storage is arguably more important than forecasting. Such storage would allow energy from wind, solar, and other intermittent sources to be generated when available and stored until it was needed to meet load. Although many forms of energy storage are readily available (batteries, reservoirs, capacitors, etc.), few are economical,

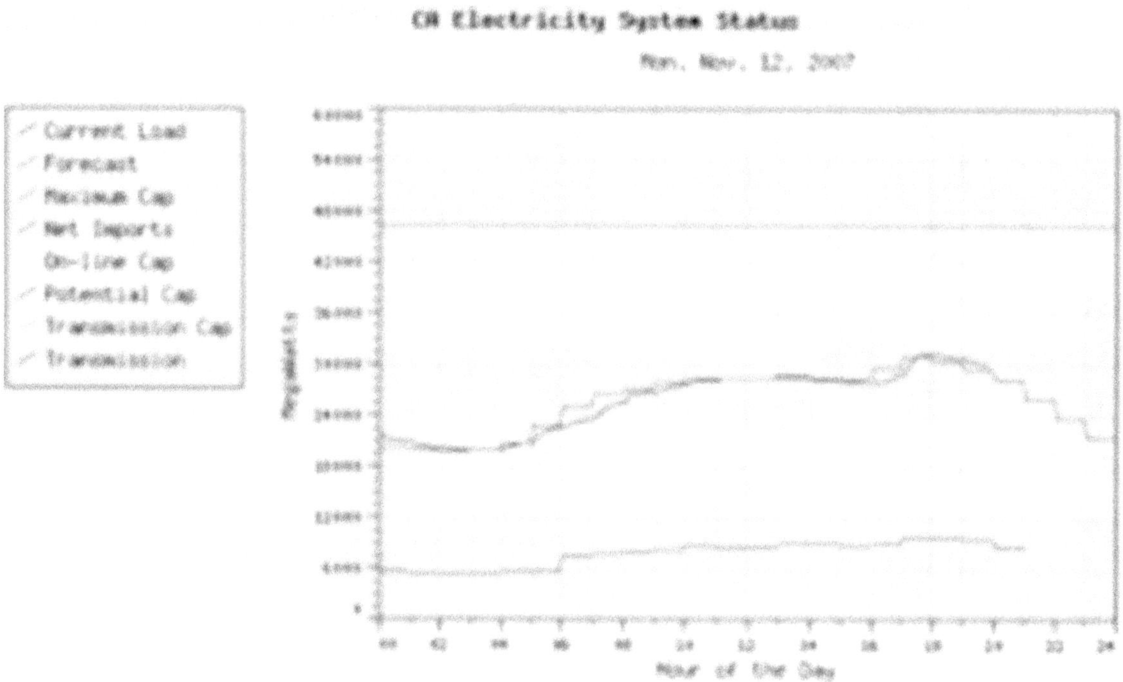

FIG. 3.21 EXAMPLE OF A 24-HOUR LOAD CYCLE ON THE CALIFORNIA ELECTRICAL SYSTEM [29]

large-scale and ubiquitous. In 2010, pumped-hydro storage is both large-scale and relatively cost effective, but it is not ubiquitous; that is, the number of candidate sites is limited by topography, availability of water and environmental considerations. Batteries (including chemical and flow) are ubiquitous, but are not economical on large scale. Capacitors and ultra-capacitors are ubiquitous and relatively economical, but cannot store large quantities of energy over time.

The U.S. DOE recognizes the need to develop larger scale energy storage for improving grid integration. Through the Advanced Research Projects Agency-Energy (DOE ARPA-E) 12 grid-scale, rampable energy storage research projects were initiated in 2010 [26]. The project teams include private companies, universities and national laboratories. Figure 3.20 indicates some of the ARPA-E targets, with respect to cost, energy and power, for the research.

3.13.1.3 Hybrid Energy Systems While wind, solar, and other renewable energy sources tend to be intermittent, other energy sources, such as nuclear, tend to operate most efficiently at steady state. Electric load tends to be cyclic in nature, varying daily and seasonally. Figure 3.21 shows the electric demand (load) in California on November 12, 2007. The load is lightest at night and in the morning, and peaks in mid/late afternoon. The shape of such curves will vary regionally and seasonally, but all have peaks and troughs.

Neither intermittent sources nor steady-state base-loading energy sources naturally lend themselves to match such cyclic load patterns. Hybrid Energy Systems (HES) offer a possible solution by combining steady-state and intermittent energy sources to meet electrical demand when needed and create liquid fuels at times of low-demand. The INL is actively researching HES. The following excerpt introduces the HES concept.

> *Broadly described, HES are energy product production plants that take two or more energy resource inputs (typically includes both carbon and non-carbon based sources) and produce two or more energy products (e.g., electricity, liquid transportation fuels, industrial chemicals) in an integrated plant. Nuclear energy integration into HES offers intriguing potential, particularly if smaller (<300 MWe) reactors are available. Although the concept of using nuclear energy in a variety of non-electrical process applications is certainly not new, renewed interest in more tightly coupled energy product plants (such as HES) that meet the objectives outline above have gained additional interest recently, an interest likely sparked by sharpening energy security concerns. Studies have shown that non-nuclear integrated (hybrid) energy systems can have appealing attributes in terms of overall process efficiency, enhanced electric grid stability, renewable energy integration, and economic performance, and lifecycle greenhouse gas emissions. These attributes seem to be sufficiently compelling that several significant commercial investments in fossil-renewable HES are being made in the United States while the U.S. Defense Advanced Research Projects Agency (DARPA) has openly solicited information regarding nuclear energy integration schemes. In testimony before the U.S. Senate, a senior researcher at Rand Corporation summed up the potential value of hybrid systems well, stating ". . . the combined use of fossil and solar or nuclear technologies may make for cost-effective and environmentally superior approaches" [28].*

3.14 ACRONYMS

AGL	Above ground level
ARRA	American Recovery and Reinvestment Act of 2009
ASR-11	Raytheon Terminal Radar ASR-9 Grumman Terminal Radar
ATC	Air Traffic Control

BPA	Bonneville Power Administration BWEC Bats and Wind Energy Cooperative
CFD	Computational Fluid Dynamics
DARPA	Defense Advanced Research Projects Agency
DHS	Department of Homeland Security
DOD	Department of Defense
DOE	Department of Energy
DOE-WPA	Department of Energy Wind Powering America
FAA	Federal Aviation Administration
HES	Hybrid Energy Systems
INL	Idaho National Laboratory
ITC	Investment Tax Credit
kVa	Kilovolt amps
kW	Kilowatts
kWh	Kilowatt-hours
LIDAR	Light Detection and Ranging
m/s	Meters per second
MISI	Management Information Services Inc.
MS Excel	Microsoft Excel
MW	Megawatts
NEI	Nuclear Energy Institute
NREL	National Renewable Energy Laboratory
O&M	Operations and Maintenance
PPA	Power Purchase Agreement
PTC	Production Tax Credit
SODAR	Sonic Detection and Ranging
USDOE	United States Department of Energy
VREP	Virtual Renewable Energy Prospector
W/m^2	Watts per meter squared
WSR-88D	Weather Surveillance Radar-1988, Doppler

3.15 REFERENCES

1) (U.S. Energy Information Administration, http://www.eia.doe.gov/cneaf/electricity/epm/table1_1.html,http://www.eia.doe.gov/cneaf/electricity/epm/table1_1_a.html, accessed 12/13/10)
2) 20% Wind Energy by 2030, US Department of Energy Office of Energy Efficiency and Renewable Energy (DOE-EERE) Wind and Hydropower Technologies Program (WHTP), July 2008
3) (U.S. Energy Information Administration, http://www.eia.doe.gov/cneaf/electricity/epa/epa_sum.html, accessed 12/3/10)
4) Source: An Assessment of the Available Windy Land Area and Wind Energy Potential in the Contiguous United States, Pacific Northwest Laboratory, August 1991. PNL-7789
5) http://www.windpoweringamerica.gov/pdfs/wind_maps/us_wind_potential_chart.pdf
6) http://www.windpoweringamerica.gov/pdfs/wind_maps/us_windmap_ 80meters.pdf
7) Analysis of Federal Expenditures for Energy Development, Management Information Services, Inc., September 2008.
8) Department of the Treasury, Internal Revenue Service. (2009). Form 8835
9) Department of the Treasury, Internal Revenue Service. (2009). Investment Credit, Form 3468
10) Department of the Treasury, Internal Revenue Service. (2009). General Business Credit, Form 3800
11) Environmental Law Institute. (2009). Estimating U.S. Government Subsidies to Energy Sources: 2002-2008.
12) "The Effects of Integrating Wind Power on Transmission System Planning, Reliability, and Operations: Report on Phase 2: System Performance Evaluation," Executive Summary, p. 2.13.
13) http://www.jcmiras.net/surge/p130.htm
14) http://www.nrel.gov/docs/fy01osti/30551.pdf
15) Source: Erickson, et. al, 2002. Summary of Anthropogenic Causes of Bird Mortality
16) http://www.bwea.com/aviation/ams_report.html
17) Wiser, Ryan and Bolinger, Mark, "2009 WindTechnologies Market Report", August 2010, pp 3. http://www.windpoweringamerica.gov/filter_detail.asp?itemid=2788
18) This frequency distribution graph was taken from publicly available wind data posted on the Idaho National Laboratory website.http://www.inl.gov/wind/idaho/
19) http://www.windpoweringamerica.gov/anemometer_loans.asp
20) http://www.inl.gov/wind/idaho/
21) http://coen.boisestate.edu/WindEnergy/WindData/index.asp
22) DOE Bonneville Power Administration, Wind Generation Capacity http://www.transmission.bpa.gov/business/operations/Wind/WIND_InstalledCapacity_current.xls
23) http://transmission.bpa.gov/Business/Operations/Wind/baltwg.aspx
24) Haynes, Todd; Dawson, Paul; and Nuss, Kevin, Forecasting for Wind Energy, February 2010, prepared for Bonneville Power Administration.
25) http://www1.eere.energy.gov/windandhydro/news_detail.html?news_ id=16316
26) http://arpa-e.energy.gov/ProgramsProjects/GRIDS.aspx
27) U.S. Department of Energy Advanced Research Projects Agency-Energy (DOE ARPA-E), Grid-Scale Rampable Intermittent Dispatchable Storage Funding Opportunity Announcement (DE-FOA-0000290), page 12, March 2010.
28) Aumeier, Steven, Cherry, Robert, Boardman, Richard and Smith, Joseph, Nuclear Hybrid Energy Systems: Imperatives, Prospects, and Challenges, pp. 2, October 2010
29) http://currentenergy.lbl.gov/ca/index.php
30) S. Krohn and S. Damborg, "On public attitude towards wind power," Renewable Energy, vol. 16, no. 1/4, pp. 954-960, 1999, Pergamon Press, Inc.
31) M. Wolsink, "Wind power implementation: The nature of public attitudes: Equity and fairness instead of backyard motives," Renewable and Substanable Energy Reviews, vol. 11, no. 6, pp. 1188-1207, Aug. 2007, Elsevier Ltd.
32) I. Bishop "Determination of thresholds of visual impact: the case of wind turbines" Environment and Planning B: Planning and Design, vol. 29, no. 5, pp. 707–718, 2002, Pion Ltd.
33) http://www.nrel.gov/docs/fy03osti/34318.pdf
34) http://www.inl.gov/wind

3.16 SCOPE OF WIND ENERGY IN THE US SINCE 2011

3.16.1 Introduction

The original authors of chapter 8 that appeared in the "Energy and Power Generation Handbook" published by ASME in 2011 [1] were invited to provide an update for their contributions of the original publication of 2011. It is quite understandable that they were unavailable for updating their original contribution since over a period of a decade their jobs may have taken different directions and their professional interests had shifted to newer and different avenues. Given that perspective and that they were unavailable for the current publication, Dr. K. R. Rao, editor-in-charge of this Renewable Energy Series decided to provide the update reflecting the present scenario of the subject addressed by Thomas Baldwin and Gary Seifert in Chapter 8 titled "WIND ENERGY IN THE U.S". Dr. KR Rao with his classic 2volume book, nearly

2,000 pages on "Wind Energy for Power Generation Meeting The Challenge of Practical Implementation" published by Springer in 2019 [2] makes him credible to provide the update for Baldwin/Seifert Chapter. This chapter 8 update has been renumbered as chapter 3 in the current publication. Most of the content and graphics including references of the original Baldwin / Seifert effort have been retained by the editor considering they are still germane for addressing the subject matter, even today. However, considering several advances which have occurred over a period of one decade, editor who provided this update retained the topical outline which is still pertinent today as initially published and the updates are provided against each of the topics mentioned in the 2011, ASME Publication [1].

Wind power in the United States has expanded rapidly over the recent decades, and January through December 2020, 337.5 terrawatt-hours were generated by wind power, or 8.42% of all generated electrical energy in the United States. In 2019, wind power surpassed hydroelectric power as the largest renewable energy source generated in the U.S. As of January 2021, the total installed wind power nameplate generating capacity in the United States was 122,478 megawatts (MW). This capacity is exceeded only by China and the European Union. Thus far, wind power's largest growth in capacity was in 2020, when 16,913 MW of wind power was installed and surpassed by leaps and bounds what was achieved in 2012, which saw the addition of 11,895 MW, representing 26.5% of the new power capacity installed during that year. By September 2019, 19 states had over 1,000 MW of installed capacity with 5 states (Texas, Iowa, Oklahoma, Kansas, and California) generating over half of all wind energy in the nation. Texas, with 28,843 MW of capacity, about 16.8% of the state's electricity usage, had the most installed wind power capacity of any U.S. state at the end of 2019. The state generating the highest percentage of energy from wind power is Iowa at over 57% of total energy production, while North Dakota has the most per capita wind generation. The Alta Wind Energy Center in California is the largest wind farm in the United States with a capacity of 1,548 MW. GE Power is the largest domestic wind turbine manufacturer.[2]

3.16.2 Wind Turbine Technologies

3.16.2.1 Types of Wind Turbines

Apart from the wind turbines described in chapters 1 and 2 and shown in figure 1[3] other types which have been developed since 2010.

3.16.2.2 Types of Unconventional types of Turbines[5]

Unconventional wind turbines are those that differ significantly from the most common types in use. As of 2012, the most common type of wind turbine is the three-bladed upwind horizontal-axis

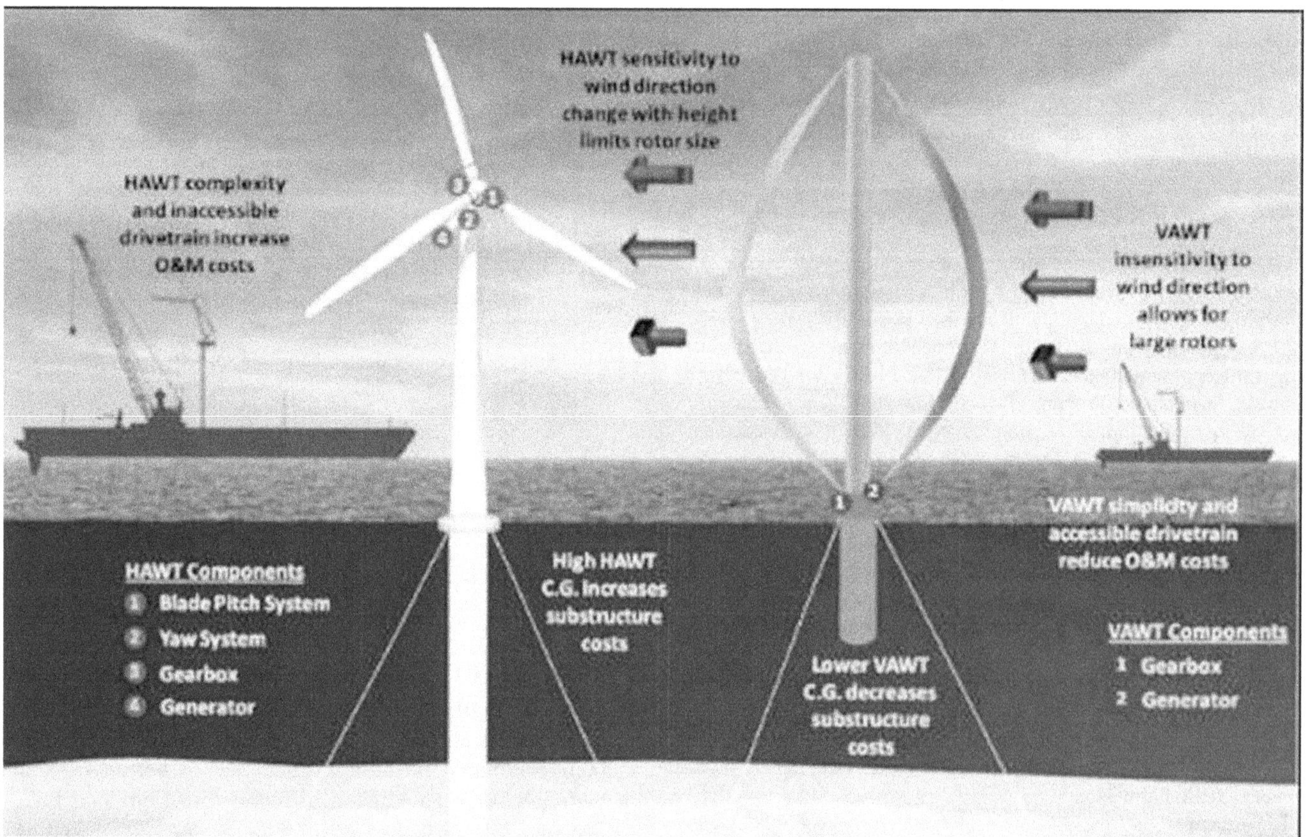

FIG. 3.22 TYPES OF TURBINES (SOURCE: [4])

[2] https://en.wikipedia.org/wiki/Wind_power_in_the_United_States
[3] https://www.ablison.com/types-of-wind-turbines/
[4] https://www.ablison.com/types-of-wind-turbines/
[5] https://en.wikipedia.org/wiki/Unconventional_wind_turbines_10-24-21

wind turbine (HAWT), a second major unit type is the vertical-axis wind turbine (VAWT). Some unconventional designs have entered commercial use, while others have only been demonstrated or are only theoretical concepts. Unconventional designs cover a wide gamut of innovations, including different rotor types, basic functionalities, supporting structures and form-factors which are listed below:

- Crosswind kite generator with fast motion transfer.
- Downwind rotor: Nearly all wind turbines place the rotor in front of the nacelle when the wind is blowing (upwind design). Some turbines place the rotor behind the nacelle (downwind design).
- Ducted rotor consists of a turbine inside a duct that flares at the back. They are also referred as Diffuser- Augmented Wind Turbines (i.e. DAWT). Its main advantage is that it can operate in a wide range of winds and generate a higher power per unit of rotor area. Another advantage is that the generator operates at a high rotation rate, so it doesn't require a bulky gearbox, allowing the mechanical portion to be smaller and lighter.
- Co-axial, multi-rotor; Two or more rotors may be mounted to a single driveshaft, with their combined co-rotation together turning the same generator: Power was multiplied several times using coaxial, multiple rotors.
- Counter-rotating horizontal-axis. The placement of a second rotor behind the first takes advantage of the disturbed airflow and can gain up to 40% more energy from a given swept area as compared with a single rotor.
- Furling tail and twisting blades. In addition to variable pitch blades, furling tails and twisting blades are other improvements on wind turbines. Similar to the variable pitch blades, they may also greatly increase efficiency.
- Wind-mill style is with absence of bearings and gears and thus eliminates frictional inefficiencies and noise.

FIG. 3.23 COUNTER ROTATING WIND TURBINE, SOURCE [6]

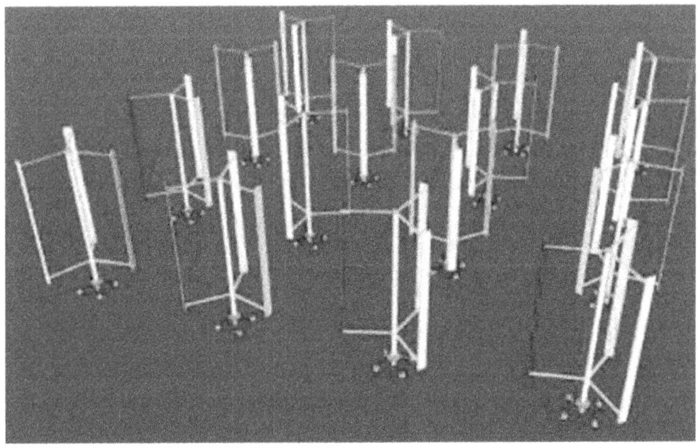

FIG. 3.24 VERTICAL AXIS WIND TURBINES OFFSHORE; (SOURCE: [7])

FIG. 3.25 HIGHWAY WIND TURBINE (SOURCE: [8])

---

[6] https://en.wikipedia.org/wiki/Unconventional_wind_turbines_10-24-21
[7] https://en.wikipedia.org/wiki/Unconventional_wind_turbines_10-24-21
[8] https://en.wikipedia.org/wiki/Unconventional_wind_turbines_10-24-21

The generator can operate in low-light environments, costs are low due to low cost components and simple construction.

- Vertical axis wind turbines offshore:
- Rooftop Wind-turbines can be installed on building roofs. Examples include Marthalen Landi-Silo in Switzerland, Council House 2 in Melbourne, Australia. Ridgeblade in the UK is a vertical wind turbine on its side mounted on the apex of a pitched roof. Discovery Tower is an office building in Houston, Texas, that incorporates 10 roof top wind turbines.
- Traffic-driven wind turbines generating power from the energy in the draft created by traffic.

3.16.2.3 Turbine Monitoring and Diagnostics:

Data transmission problems, structural health monitoring of winter turbines is performed by several accelerometers and strain gauges attached to the nacelle to monitor the gearbox and equipment. Currently digital image correlation and stereo photogrammetry are used to measure dynamics of wind turbine blades. These methods measure displacement and strain to identify location of defects. Dynamic characteristics of non-rotating wind turbines are measured using digital image correlation and photogrammetry. Three-dimensional point tracking is used to measure rotating dynamics of wind turbines.[9]

3.16.2.4 Recent Developments in Turbine Technology:

Wind turbine technology developments are made to increase efficiency. This requires blades to be stiff, strong, light and resistant to fatigue. Materials with these properties are composites such as polyester and epoxy, while glass fiber and carbon fiber have been used for reinforcing. Construction uses this manual layup are injection molding.[10]

Diffuser-augmented Wind Turbine is modified with cone-shaped wind diffusers to increase the effectiveness of converting wind power into electricity. The diffuser increases the speed; thus, the efficiency is increased. In the traditional bare turbines, rotor blades are mounted vertically on the support shaft or tower. In the Diffuser-augmented wind turbine, rotor blades are installed in the diffuser. It is then placed at the top of a shaft or support tower. To increase the efficiency of the device, more modifications have been carried out with the diffuser[11]

Size of offshore wind turbines over the next decade is expected to continually increase with a two-bladed downwind turbine system with higher capacity factors redesigned with appropriate aerodynamics, structures, and controls.[12]

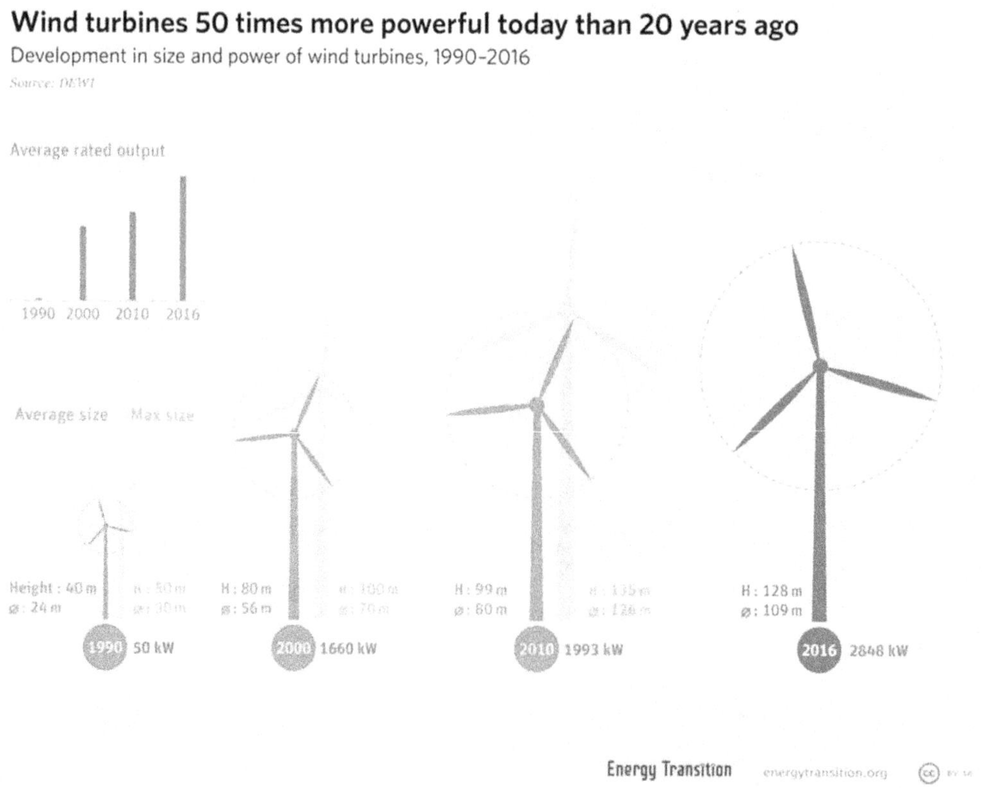

FIGURE 3.26: WIND TURBINES 1990 THROUGH 2016 (SOURCE: [13])

[9] https://en.wikipedia.org/wiki/Unconventional_wind_turbines_10-24-21

[10] https://en.wikipedia.org/wiki/Unconventional_wind_turbines_10-24-21

[11] https://en.wikipedia.org/wiki/Diffuser-augmented_wind_turbine

[12] https://www.sciencedirect.com/science/article/abs/pii/S0960148120305711

[13] https://en.wikipedia.org/wiki/Wind_turbine#/media/File:Growing_size_of_wind_turbines.png; (Source: Heinrich-Böll-Stiftung-https://www. flickr.com/photos/boellstiftung/26614518499/in/album-72157632531168908/, CC BY-SA 2.0, https://commons.wikimedia.org/w/index.php?curid=79053862)

There is also pseudo direct mechanism which has some advantages over the permanent magnet direct drive mechanisms. Turbines used in wind farms are usually 3 bladed with low torque ripple which contributes to reliability. The size and height of the turbines increase year after year for offshore wind turbines and are 8 MW today with a blade length up to 260 feet. Turbine designs for 2018 are 15 MW and those with 15+MW have 387 feet blades planned to be constructed by 2022 and multi-MW turbines have tubular steel towers with a height of 70 meters to 120 meters and in extreme up 260 meters.[14]

Nano-engineered polymers and composites are additions of small amount (0.5% weight) of nano reinforcement (carbon nanotubes or nano clay) in the polymer matrix of composites, fiber sizing or interlaminar layers can improve fatigue resistance, shear or compressive strength, and fracture toughness of the composites by 30% to 80%. Research has also shown that incorporating small amounts of carbon nanotubes (CNT) can increase the lifetime up to 1500%.[15]

3.16.2.5 Wind Turbines 50 Times More Powerful

3.16.3 Wind Resources In The U.S.

As mentioned in the original write-up by Seinfield and Baldwin the wind resources in the US comprise of three parts that make up a successful wind energy project, wind resource, an energy sale contract, and a transmission/grid interconnection. Wind resource in the US since publication of the initial chapter in 2011 are examined below. The following figures 3.27 and 3.28 illustrate U.S. Wind Power Resource at 100-meter Hub Height and the installed wind generating capacity for U.S. states at end of 2018, these make it possible to compare what is available in terms of wind speeds to what is achieved.

FIGURE 3.27: U.S. WIND POWER RESOURCE AT 100-METER HUB HEIGHT (SOURCE: [16])

[14] "https://en.wikipedia.org/wiki/Wind_turbine" \l "Recent_developments_in_technology"
[15] https://en.m.wikipedia.org/wiki/Wind_turbine_10-24-21
[16] https://en.wikipedia.org/wiki/Wind_power_in_the_United_States_10-24-21

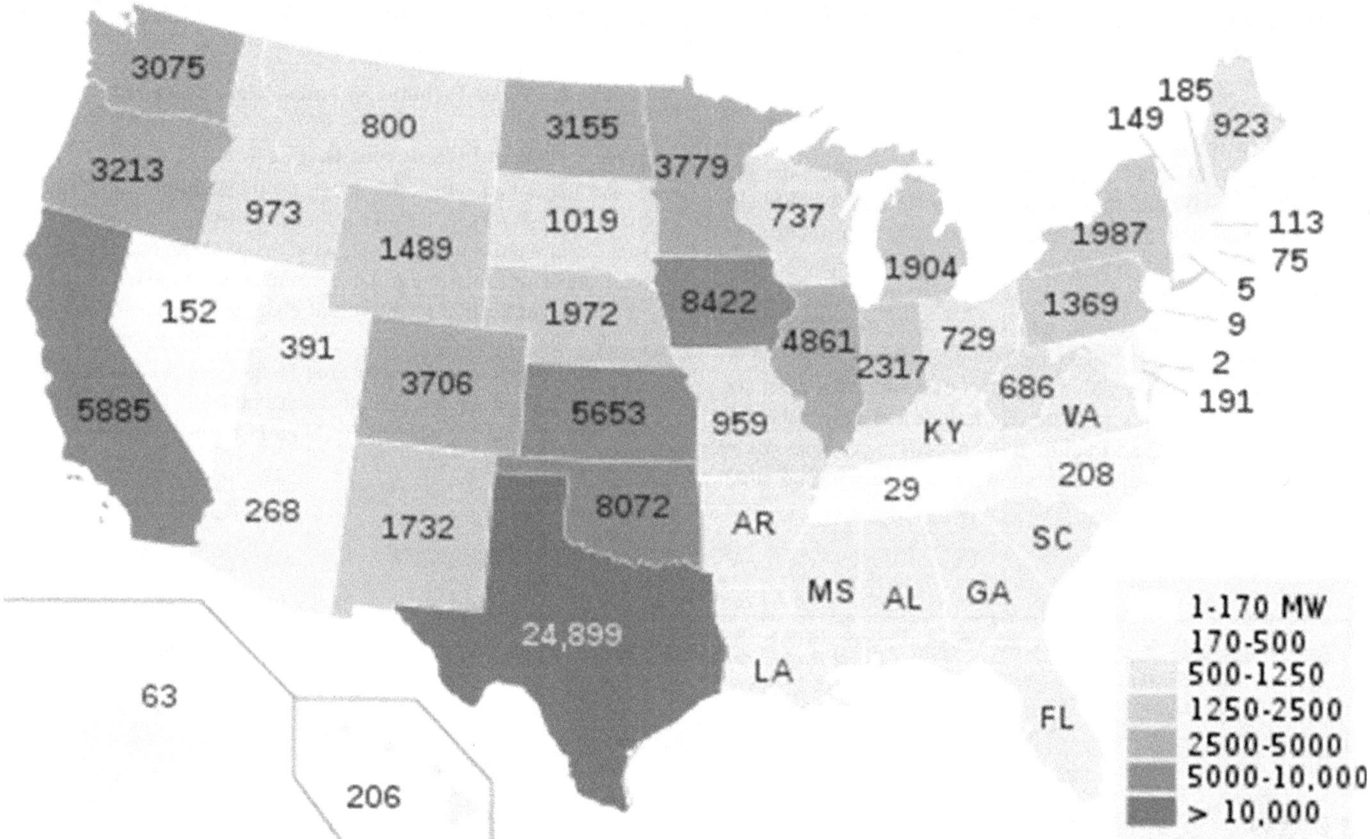

FIG. 3.28 U.S. INSTALLED WIND POWER CAPACITY 2021. (SOURCE: [17]) CURRENTLY, THE LARGEST WIND FARMS IN THE UNITED STATES ARE LISTED IN TABLE 3.6.

TABLE 3.6: LARGEST WIND FARMS CURRENTLY OPERATING IN THE US. (SOURCE: [18])

Project	Capacity (MW)	State
Alta Wind Energy Center	1548	California
The Traverse Wind Energy Center	999 MW	Oklahoma
Los Vientos Wind Farm	912	Texas
Shepherds Flat Wind Farm	845	Oregon
Meadow Lake Wind Farm	801	Indiana
Roscoe Wind Farm	781	Texas
Horse Hollow Wind Energy Center	736	Texas
Tehachapi Pass Wind Farm	705	California
Capricorn Ridge Wind Farm	662	Texas
San Gorgonio Pass Wind Farm	619	California
Limon Wind Energy Center	601	Colorado

[17] https://en.wikipedia.org/wiki/Wind_power_in_the_United_States
[18] https://en.wikipedia.org/wiki/Wind_power_in_the_United_States_10-24-21

3.16.3.1 Key Findings of the Wind Resources in the U.S. Since 2011[19]

- Turbine scaling is significantly boosting wind project performance, while the installed cost of wind projects has declined.
- Wind power sales prices and levelized cost of energy are at all-time lows, enabling economic competitiveness despite low gas prices.

Key findings of the wind resources are documented in source[20]:

- Wind Power Additions Continued at a Robust Pace in 2018, with 7,588 MW of New Capacity, Bringing the Total to 96,433 MW, see figure 3.29.
- The Geographic Spread of Wind Power Projects Across the United States Is Broad, with the Exception of the Southeast, see figure 3.30.
- The Domestic Supply Chain for Wind Equipment is Diverse, see figure 3.31.
- Domestic Manufacturing Capability for Nacelle Assembly, Towers, & Blades Reasonably Well Balanced Against Historical Demand, see Figure 3.32.
- Domestic Manufacturing Content is Strong for Nacelle Assembly, Towers, and Blades, but not Equipment Internal to the Nacelle, see Table 3.7.
- Turbine Capacity, Rotor Diameter, and Hub Height Have All Increased Significantly Over the Long Term, and in 2018, see figure 3.33.
- Growth in Rotor Diameter and Nameplate Capacity Have Outpaced Growth in Hub Height over the Last Two Decades, see figure 3.34.
- Turbines Originally Designed for Lower Wind Speed Sites Dominate the Market, see figure 3.35.
- Wind Turbines Continued to be Deployed in Somewhat Lower Wind-Speed Sites in 2018, see figure 3.36.
- Low Specific Power Turbines Are Deployed in Low & High Wind Speeds; Taller Towers More Common in Great Lakes & Northeast, see figure 3.37.
- The Interior Region Features the Highest Capacity Factors, in Part Reflecting the Strength of the Wind Resource, see figure 3.38.
- Wind Turbine Prices Remained Well Below the Levels Seen a Decade Ago, see figure 3.39:

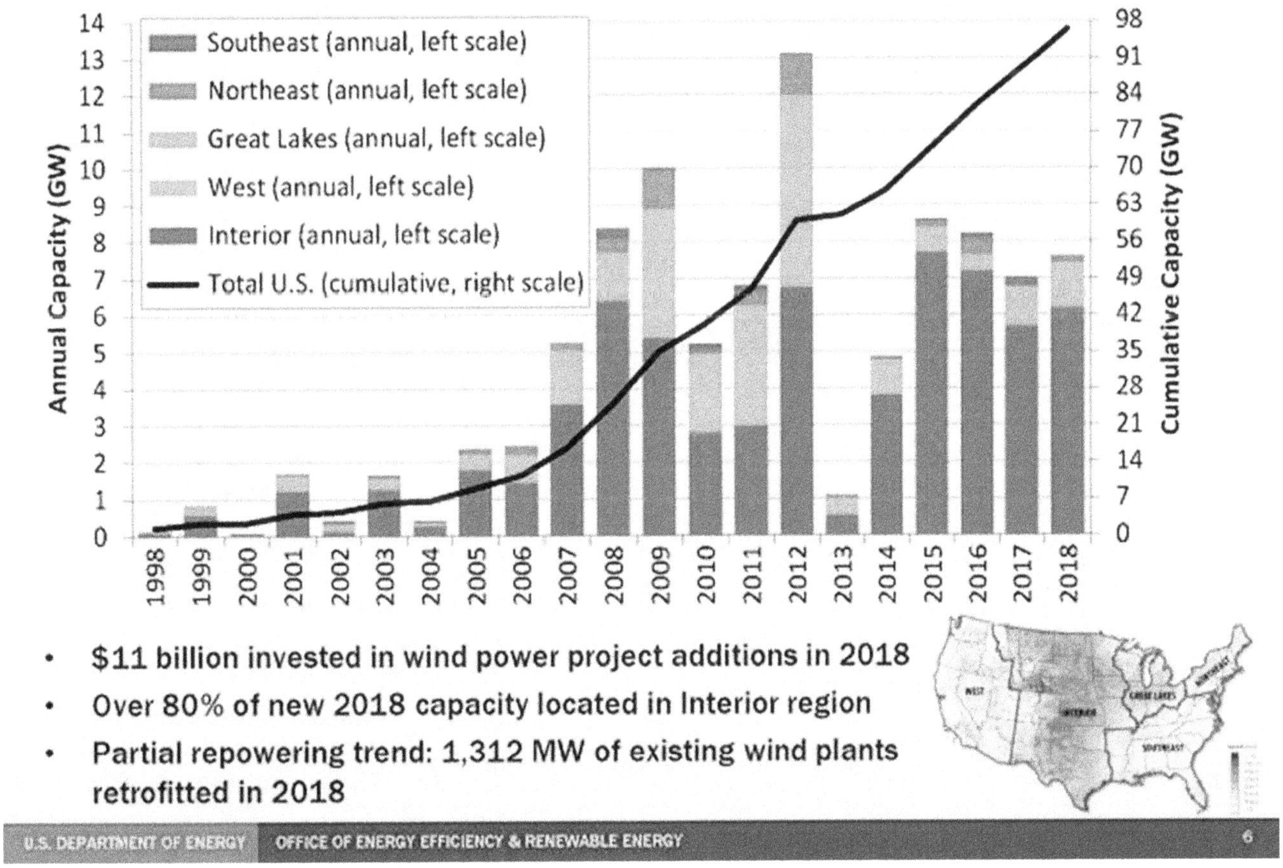

FIG. 3.29 GROWTH OF WIND POWER ADDITIONS IN THE US THROUGH 2018 WITH 7588 MEGAWATTS OF NEW CAPACITY, (SOURCE: [21])

---

[19] https://en.wikipedia.org/wiki/Wind_power_in_the_United_States_10-24-21

[20] https://www.energy.gov/eere/wind/downloads/2018-wind- technologies-market-report

[21] https://www.energy.gov/eere/wind/downloads/2018-wind- technologies-market-report

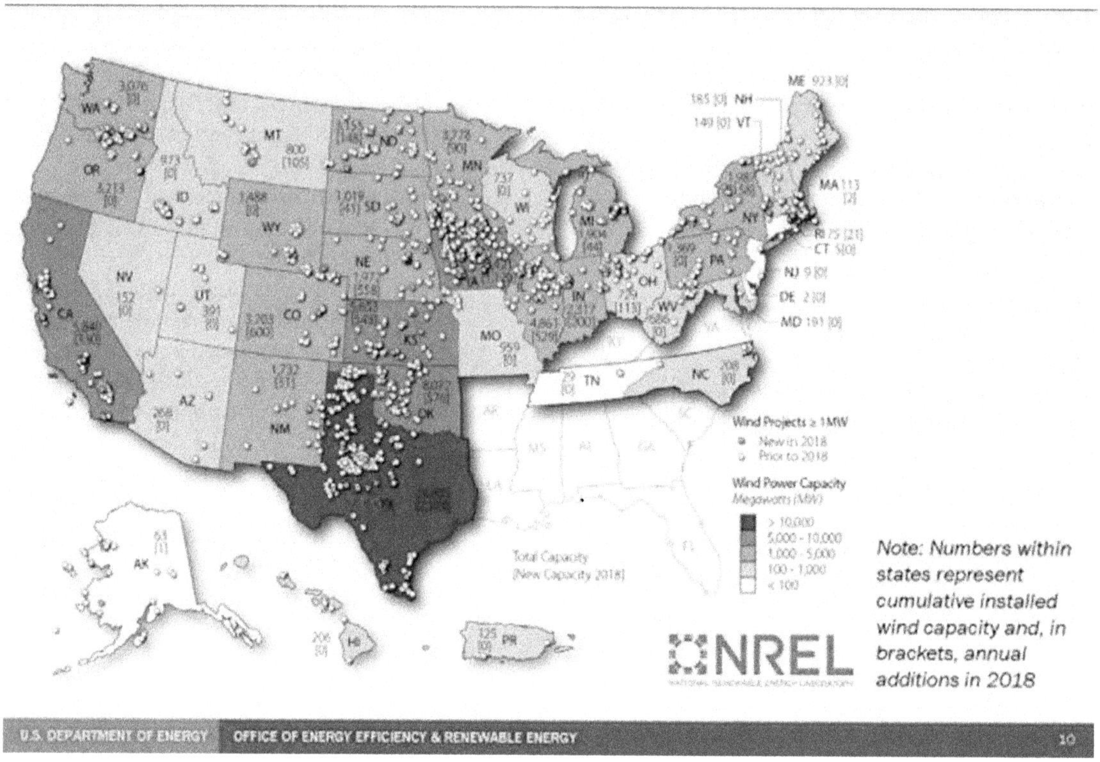

FIG. 3.30 THE GEOGRAPHIC SPREAD OF WIND POWER PROJECTS ACROSS THE UNITED STATES IS BROAD, WITH THE EXCEPTION OF THE SOUTHEAST, (SOURCE: [22])

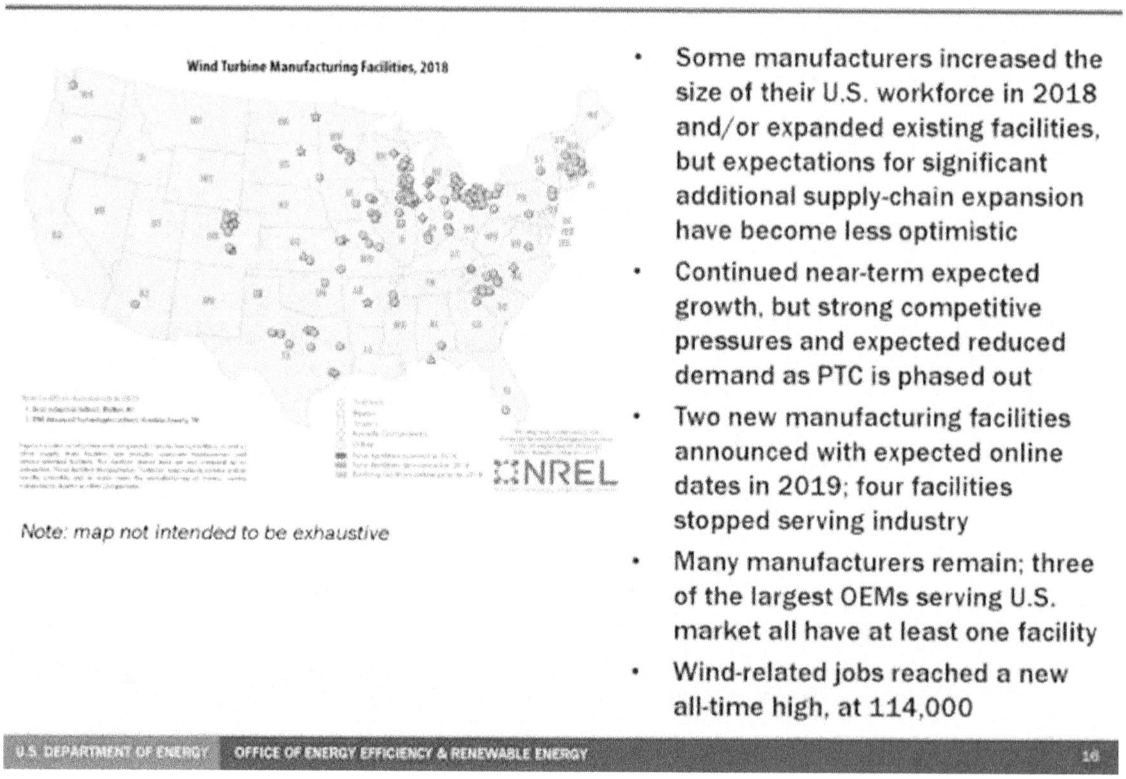

FIG. 3.31 THE DOMESTIC SUPPLY CHAIN FOR WIND EQUIPMENT IS DIVERSE, (SOURCE [23])

[22] https://www.energy.gov/eere/wind/downloads/2018-wind- technologies-market-report

[23] https://www.energy.gov/eere/wind/downloads/2018-wind- technologies-market-report

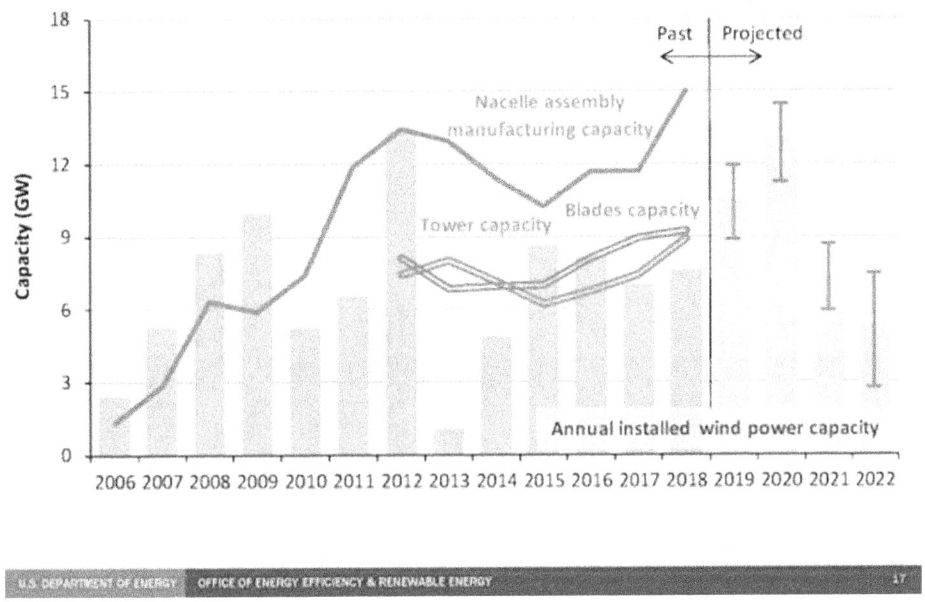

FIG. 3.32 DOMESTIC MANUFACTURING CAPABILITY FOR NACELLE ASSEMBLY, TOWERS, & BLADES REASONABLY WELL BALANCED AGAINST HISTORICAL DEMAND, (SOURCE [24])

TABLE 3.7: DOMESTIC MANUFACTURING CONTENT IS STRONG FOR NACELLE ASSEMBLY, TOWERS, AND BLADES, BUT NOT EQUIPMENT INTERNAL TO THE NACELLE [25]

Domestic Content for 2018 Turbine Installations in the United States		
Towers	**Blades & Hubs**	**Nacelle Assembly**
75–90%	50–70%	> 85%; Imports occur in untracked trade categories, including many nacelle internals; nacelle internals generally have domestic content of < 20%

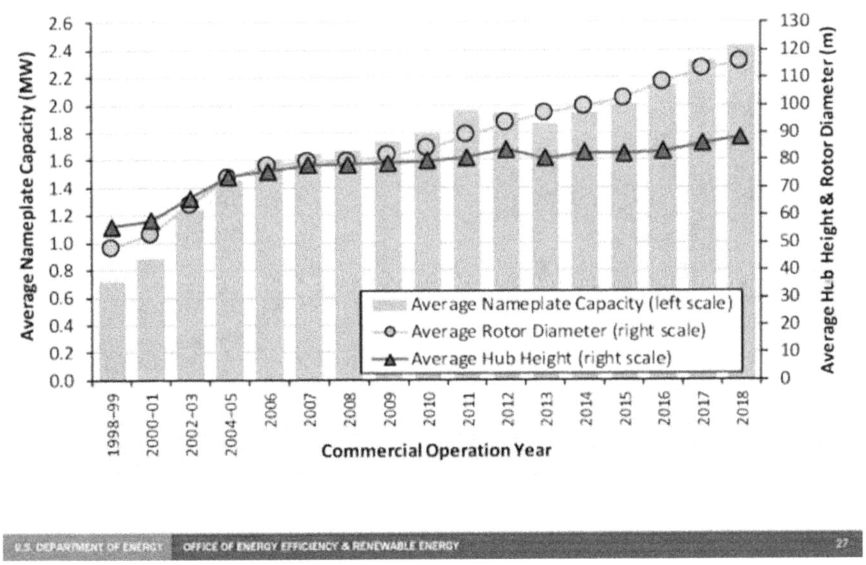

FIG. 3.33 TURBINE CAPACITY, ROTOR DIAMETER, AND HUB HEIGHT HAVE ALL INCREASED SIGNIFICANTLY OVER THE LONG TERM, AND IN 2018, (SOURCE: [26])

[24] https://www.energy.gov/eere/wind/downloads/2018-wind- technologies-market-report

[25] https://www.energy.gov/eere/wind/downloads/2018-wind- technologies-market-report

[26] https://www.energy.gov/eere/wind/downloads/2018-wind- technologies-market-report

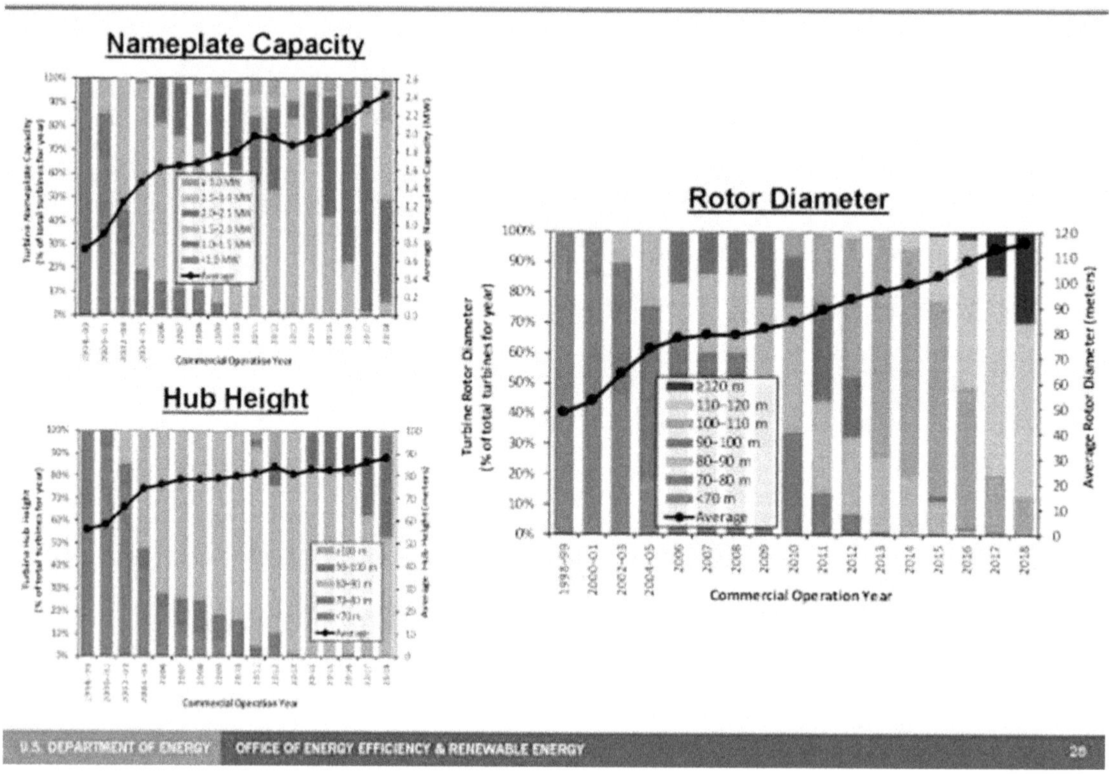

FIG. 3.34 GROWTH IN ROTOR DIAMETER AND NAMEPLATE CAPACITY HAVE OUTPACED GROWTH IN HUB HEIGHT OVER THE LAST TWO DECADES, (SOURCE: [27])

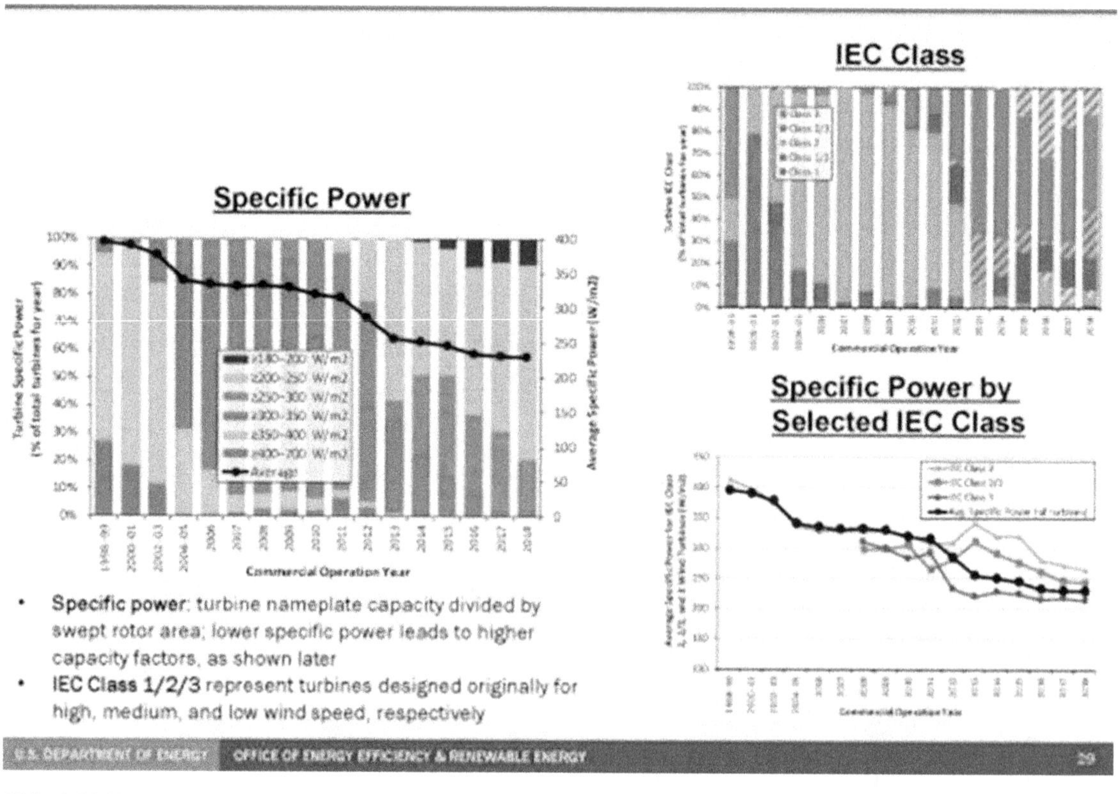

FIG. 3.35 TURBINES ORIGINALLY DESIGNED FOR LOWER WIND SPEED SITES DOMINATE THE MARKET, (SOURCE: [28])

[27] https://www.energy.gov/eere/wind/downloads/2018-wind- technologies-market-report

[28] https://www.energy.gov/eere/wind/downloads/2018-wind- technologies-market-report

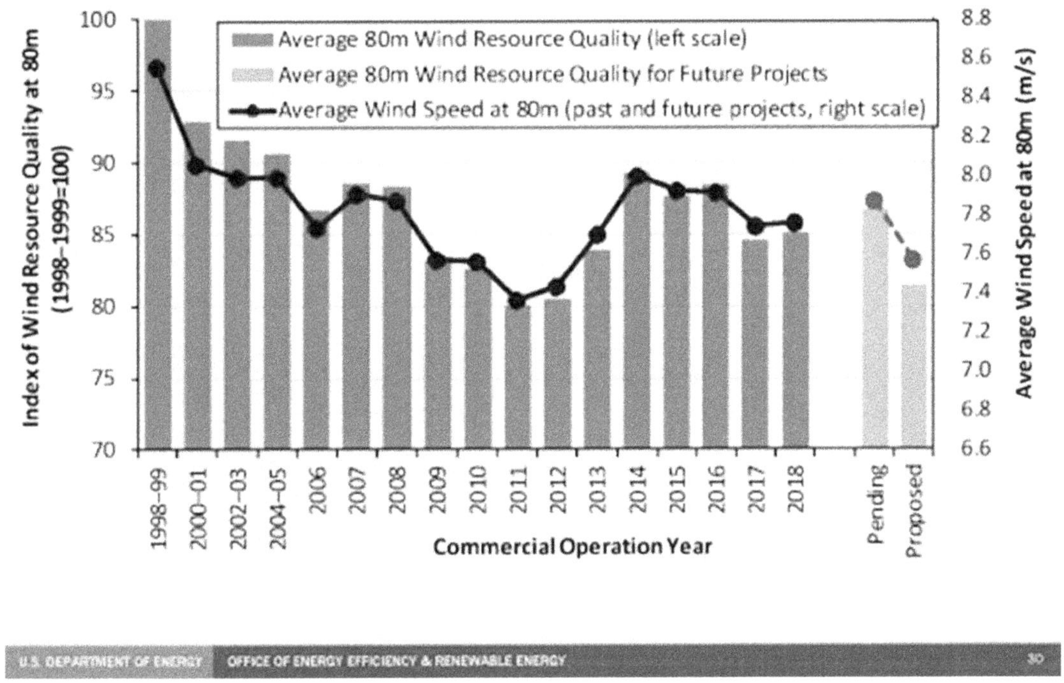

FIG. 3.36 WIND TURBINES CONTINUED TO BE DEPLOYED IN SOMEWHAT LOWER WIND-SPEED SITES IN 2018, (SOURCE: [29])

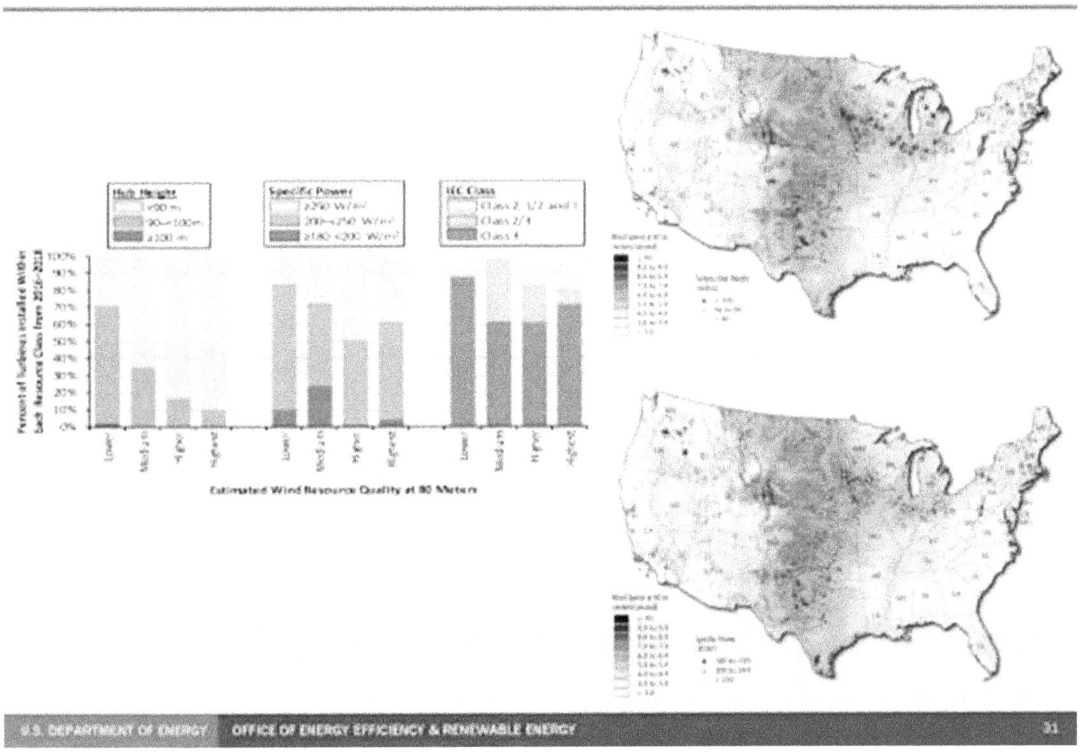

FIG. 3.37 LOW SPECIFIC POWER TURBINES ARE DEPLOYED IN LOW & HIGH WIND SPEEDS; TALLER TOWERS MORE COMMON IN GREAT LAKES & NORTHEAST, (SOURCE: [30])

[29] https://www.energy.gov/eere/wind/downloads/2018-wind- technologies-market-report

[30] https://www.energy.gov/eere/wind/downloads/2018-wind- technologies-market-report

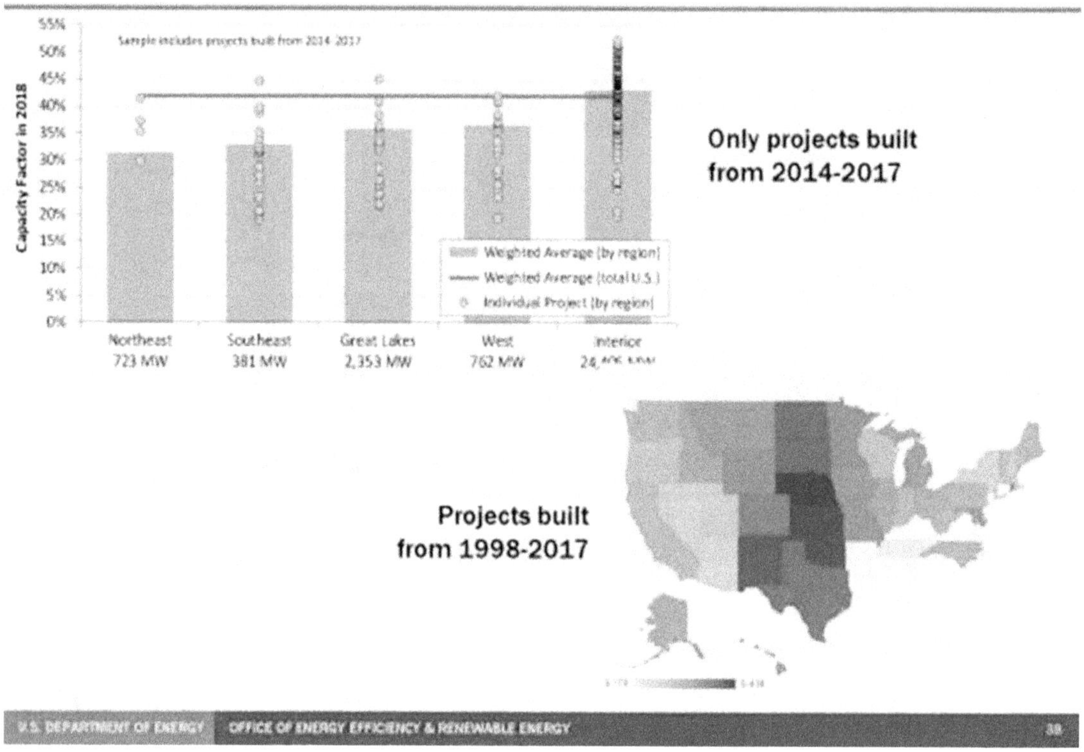

FIG. 3.38 THE INTERIOR REGION FEATURES THE HIGHEST CAPACITY FACTORS, IN PART REFLECTING THE STRENGTH OF THE WIND RESOURCE, (SOURCE: [31])

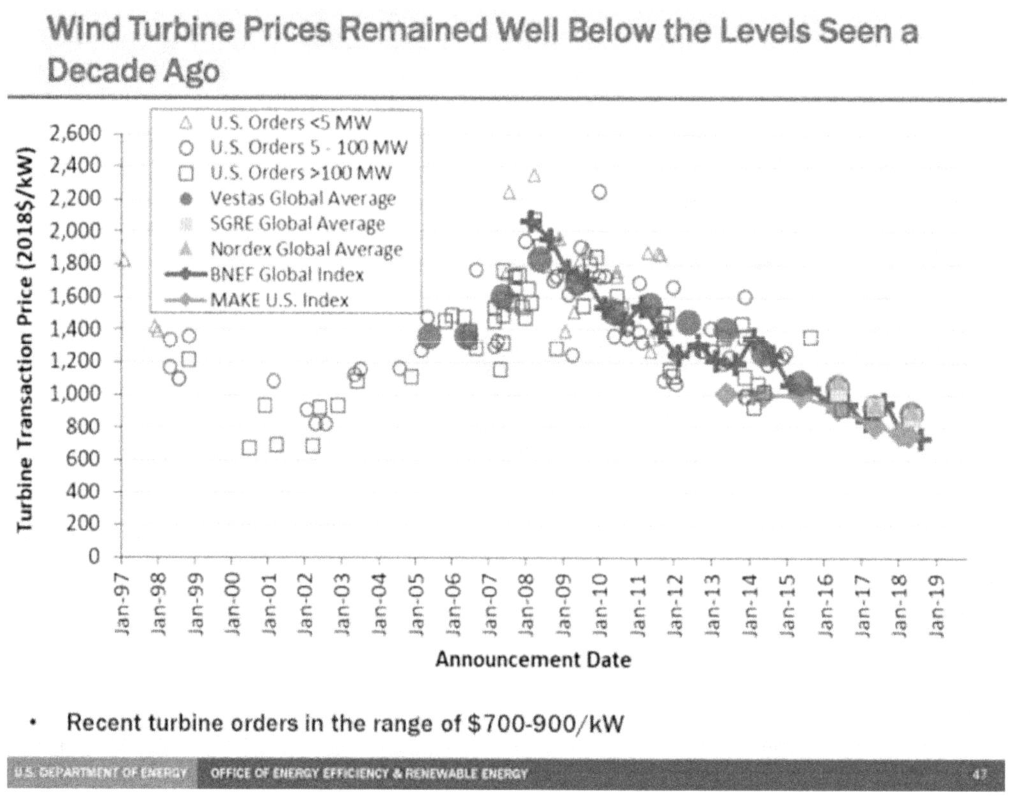

FIG. 3.39 WIND TURBINE PRICES REMAINED WELL BELOW THE LEVELS SEEN A DECADE AGO SOURCE [32]

[31] https://www.energy.gov/eere/wind/downloads/2018-wind- technologies-market-report

[32] https://www.energy.gov/eere/wind/downloads/2018-wind- technologies-market-report

3.16.4 Wind Plant Economics

The 2018 Wind Technologies Market Report, in its thirteenth consecutive year, provides an overview of developments and trends in the U.S. wind power market, with a particular focus on the year 2018. Wind capacity additions continued at a robust pace, with 7,588 megawatts (MW) added in 2018 and significant new builds anticipated in the near-term. Wind energy supplies more than 10% of total electricity generation in 14 states and more than 30% in three of those states, see figure 3.28. The U.S. supply chain is diverse and multifaceted, with strong domestic content for nacelle assembly, towers, and blades, see figure 3.31. Wind sector employment reached a new high of 114,000 full-time workers. Wind turbines continue to increase in size; the average generating capacity of newly installed wind turbines in the United States in 2018 was 2.4 MW and the average rotor diameter was 116 meters. Permit applications suggest that much taller turbines are on the way, see figure 3.33. Low wind turbine pricing continued to push down installed project costs, see figure 3.39. Wind energy prices are at historic lows, below 2 cents per kilowatt-hour. Recently signed wind energy contracts compared favorably to projections of the fuel costs of gas-fired generation. The levelized cost of wind energy, which allows the comparison of technologies with different lifespans, sizes, and initial capital costs also hit an all-time low, averaging $38 per megawatt-hour. While significant U.S. wind capacity additions are expected to continue for the next couple years growth in the medium-term remains uncertain as the Production Tax Credit is phased out. As of 2019, a wind turbine may cost around $1 million per megawatt. The following figures illustrate the economic impact of wind energy since 2011 to date:[33]

- Capacity Factors Have Increased Significantly Over Time, by Online Date (i.e., Commercial Online Date, COD), see figure 3.40.
- Economies of Scale Are Apparent, especially when Moving from Small- to Medium-Sized Projects, see figure 3.40.
- Most Projects, and All of the Low-Cost Projects, are Located in the Interior; Other Regions Have Higher Costs, see figure 3.41.
- O&M Costs are Lower for More-Recent Projects, and Increase with Age for the Older Projects, see figure 3.42.
- Wind PPA Prices Are at Historical Lows, see figure 3.44.
- A Smoother Look at the Time Trend Shows a Steep Decline in Pricing Since 2009; Prices Below $20/MWh in Interior Region, see figure 3.45.
- The Levelized Cost of Wind Energy Is at an All-Time Low: Nationwide Average of $38/MWh for Projects Built in 2018, see figure 3.46.

American Wind Energy Association (AWEA) market report of 11/3/2020 shows The U.S. wind industry installed nearly 2,000 MW of new wind power capacity in Q3 2020, bringing U.S. wind capacity to 112,000 MW. Development and construction activity remains strong despite the pandemic, as per 2020 third-quarter report and addendum to its 2019 annual report. Ten new wind projects totaling 1,934 MW started operations across nine states during the third quarter, enough to power more than 600,000 American homes. Texas led the country with 687 MW of new wind projects installed, followed by Colorado (496 MW), Illinois (200 MW), Iowa (168 MW) and Indiana (147 MW). The two largest single-phase wind projects in U.S. history came online during the third quarter of 2020: the 525 MW Aviator Wind project in Texas and the 496 MW Cheyenne Ridge project in Colorado. There are now in the United States,

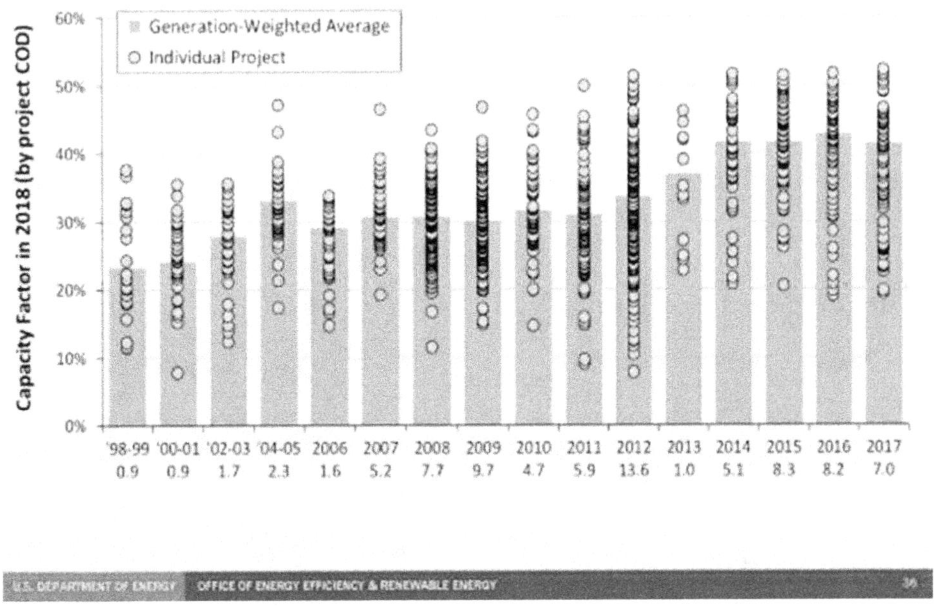

FIG. 3.40 CAPACITY FACTORS HAVE INCREASED SIGNIFICANTLY OVER TIME, BY ONLINE DATE (I.E., COMMERCIAL ONLINE DATE, COD), SOURCE [34]

[33] https://www.energy.gov/eere/wind/downloads/2018-wind- technologies-market-report

[34] https://www.energy.gov/eere/wind/downloads/2018-wind-technologies-market-report

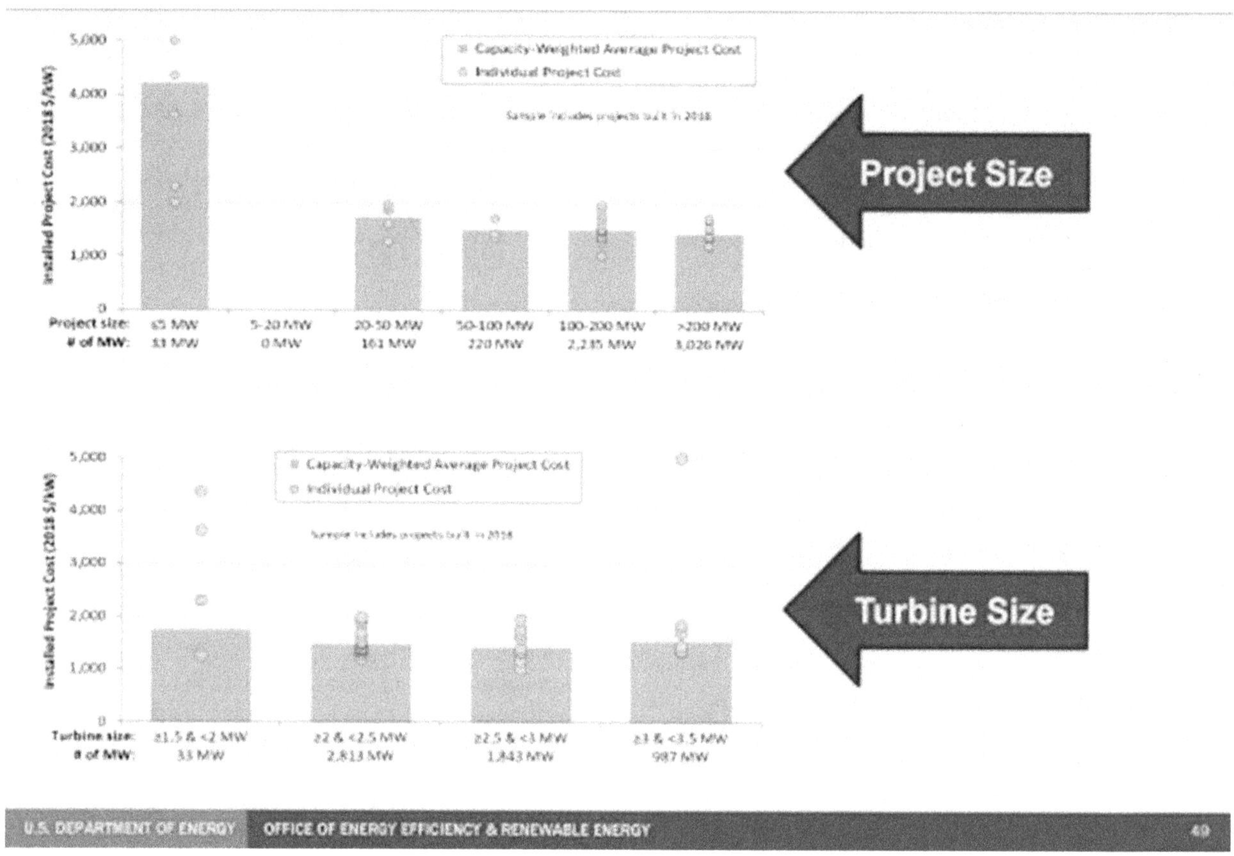

FIG. 3.41 ECONOMIES OF SCALE ARE APPARENT, ESPECIALLY WHEN MOVING FROM SMALL- TO MEDIUM-SIZED PROJECTS, SOURCE [35]

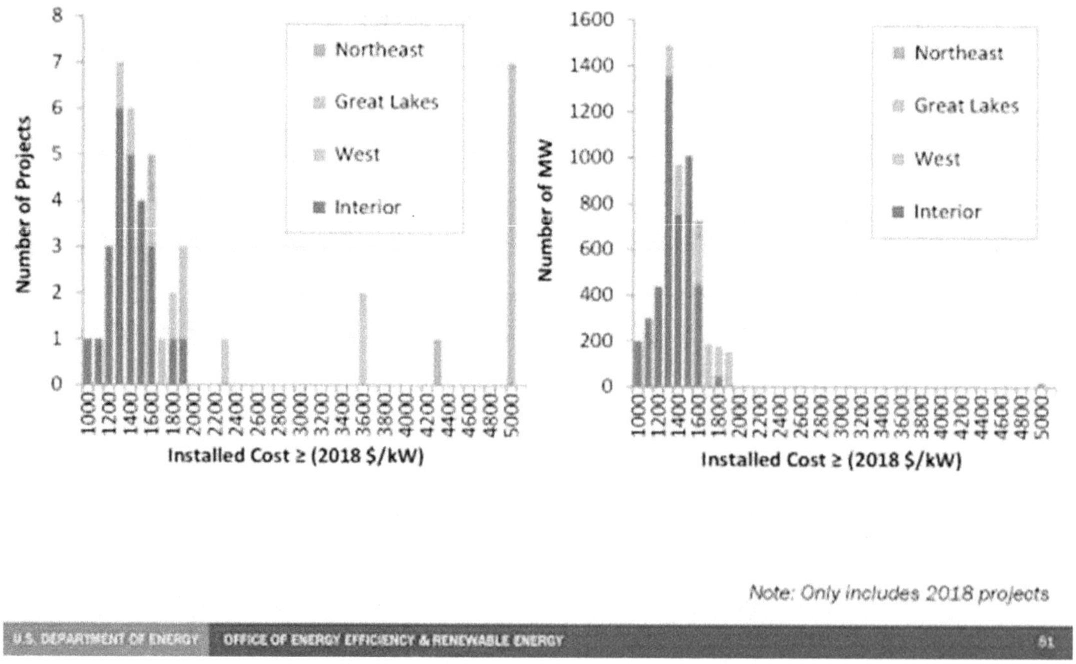

FIG. 3.42 MOST PROJECTS, AND ALL OF THE LOW-COST PROJECTS, ARE LOCATED IN THE INTERIOR; OTHER REGIONS HAVE HIGHER COSTS, SOURCE [36]

[35] https://www.energy.gov/eere/wind/downloads/2018-wind-technologies-market-report

[36] https://www.energy.gov/eere/wind/downloads/2018-wind-technologies-market-report

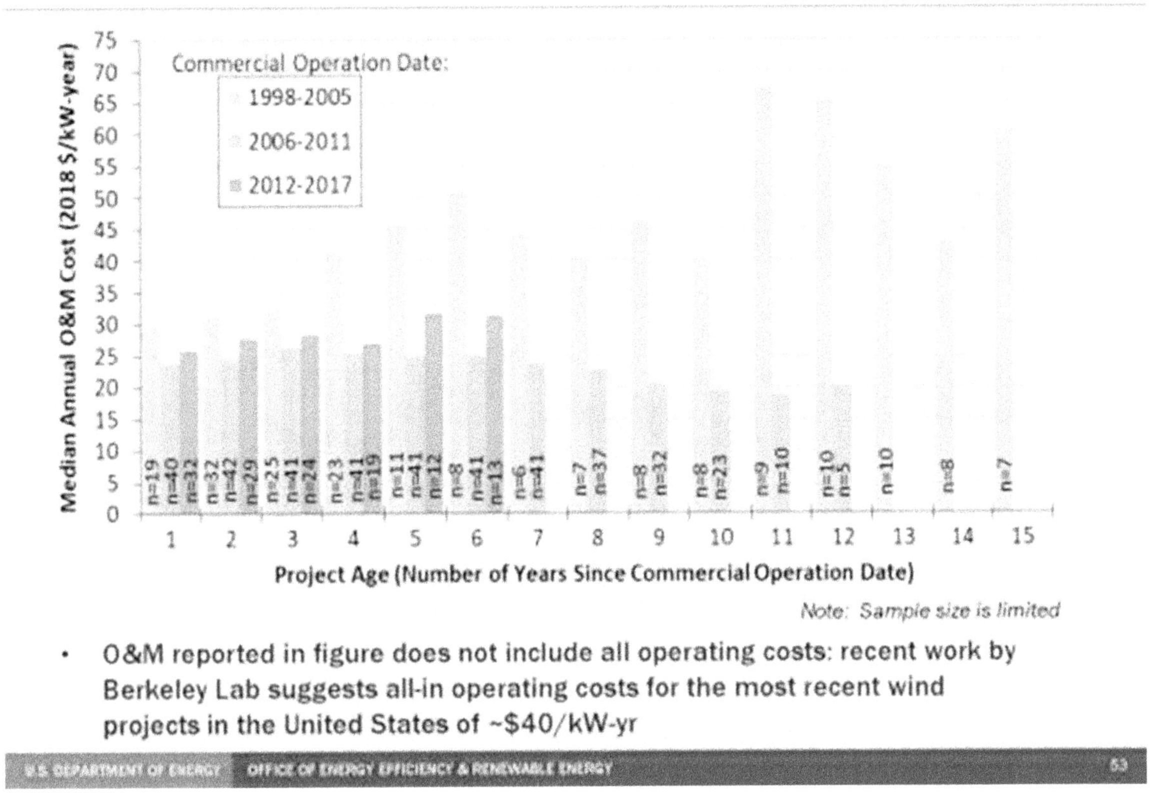

FIG. 3.43 O&M COSTS ARE LOWER FOR MORE-RECENT PROJECTS, AND INCREASE WITH AGE FOR THE OLDER PROJECTS, SOURCE [37]

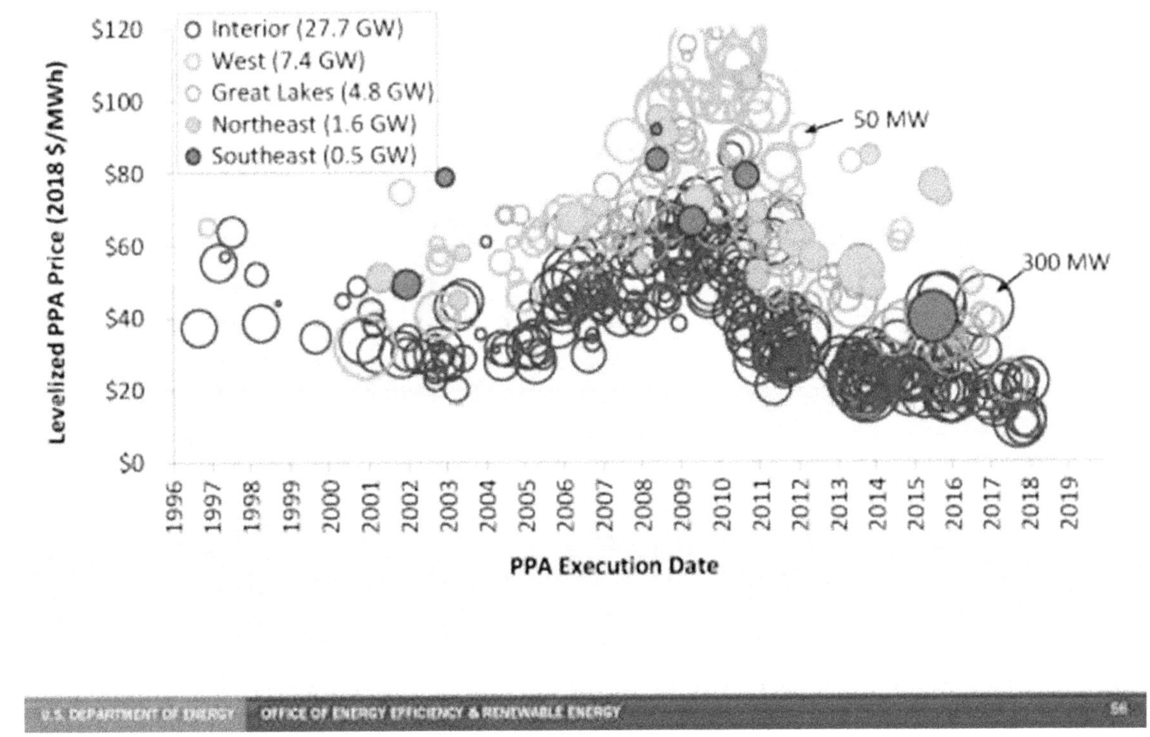

FIG. 3.44 WIND PPA PRICES ARE AT HISTORICAL LOWS, SOURCE [38]

[37] https://www.energy.gov/eere/wind/downloads/2018-wind-technologies-market-report

[38] https://www.energy.gov/eere/wind/downloads/2018-wind-technologies-market-report

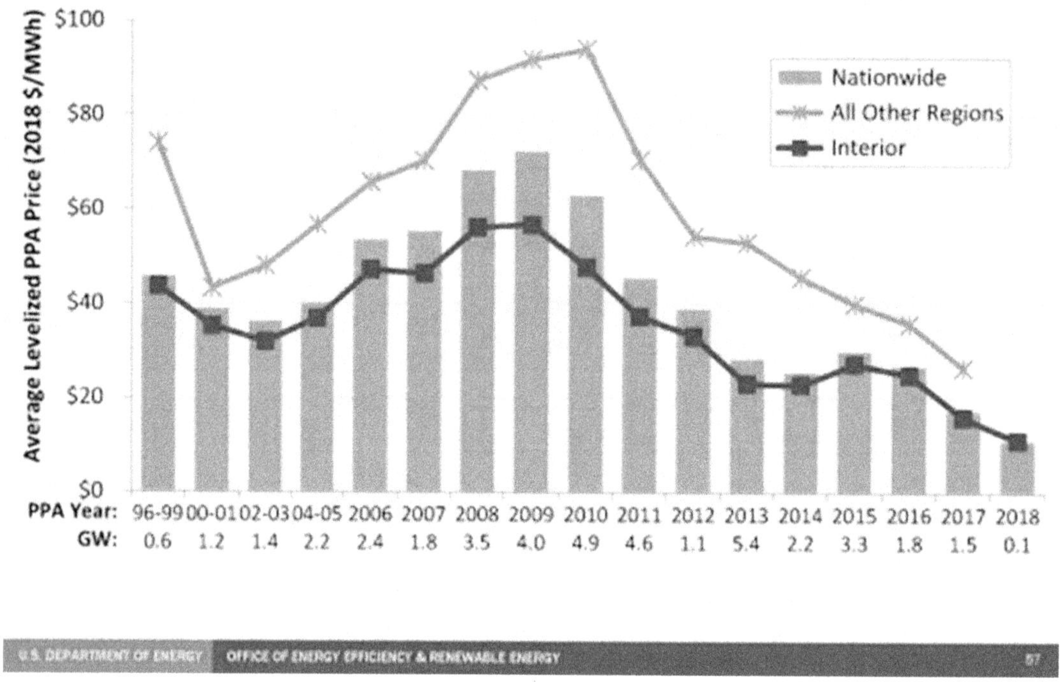

FIG. 3.45 A SMOOTHER LOOK AT THE TIME TREND SHOWS A STEEP DECLINE IN PRICING SINCE 2009; PRICES BELOW $20/MWH IN INTERIOR REGION, SOURCE: [39]

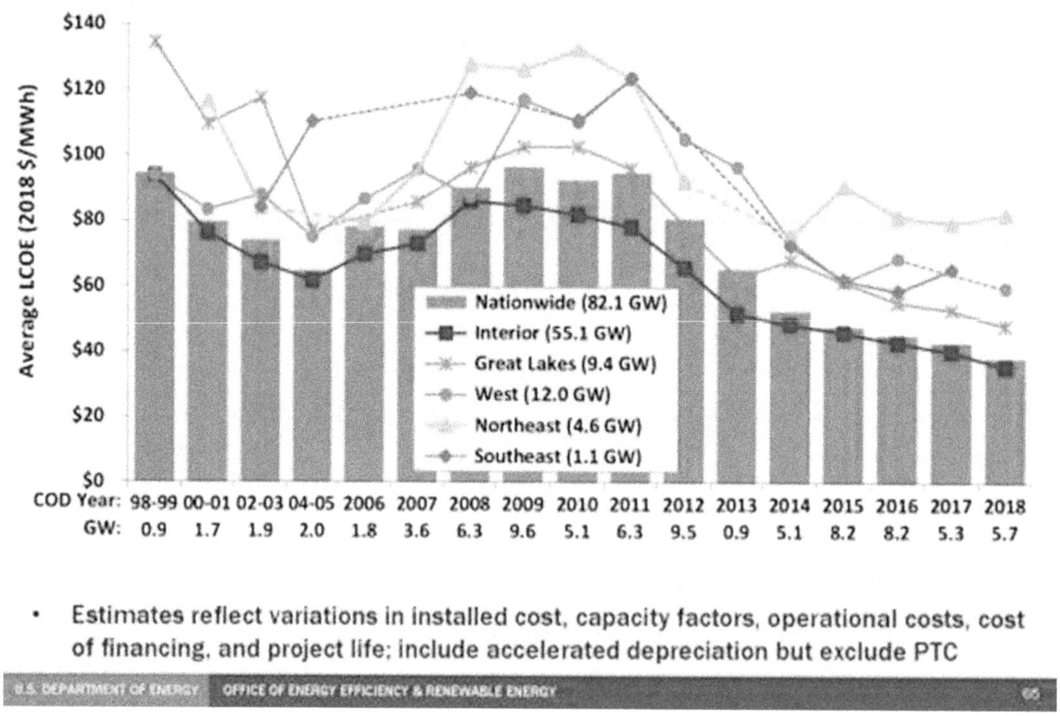

FIG. 3.46 THE LEVELIZED COST OF WIND ENERGY IS AT AN ALL-TIME LOW: NATIONWIDE AVERAGE OF $38/MWH FOR PROJECTS BUILT IN 2018, SOURCE [40]

[39] https://www.energy.gov/eere/wind/downloads/2018-wind-technologies-market-report

[40] https://www.energy.gov/eere/wind/downloads/2018-wind-technologies-market-report

enough for over 34 million American homes. Twenty states have more than 1,000 MW of installed capacity. Despite disruptions from COVID-19, project developers announced 2,420 MW in combination with new development activity, with projects totaling 972 MW starting construction and an additional 1,448 MW entering advanced development. At the end of September 2020, there were 43,575 MW of wind power capacity in the near-term pipeline, including 24,355 MW under construction and 19,220 MW in advanced development. Flexibility regarding tax credit qualification from the U.S. Treasury and Internal Revenue Service (IRS) has also been crucial for the American wind industry. Leadership at the local, state, and federal level, including forward-thinking policies will be vital as American wind power works to expand its support for local communities and to deliver clean, reliable, and affordable power to U.S communities. This policy leadership is especially important for the offshore wind industry, America's newest major energy source, which is poised to deliver up to 83,000 jobs and $25 billion in annual economic investments within the next decade. Market forecast for cost and price for American wind power posits strong growth over the next decade, with forecasters estimating nearly 60 gigawatts (GW) of new wind power installations over the next five years and 94 GW over the next ten years. Offshore wind is expected to set sail in the coming years, with forecasters anticipating at least 20 GW of operational offshore wind power by 2029 . Wind energy costs are expected to fall, with project costs declining 16% over just the past five years. The average installed cost of a wind project in 2019 was $1,436 per kilowatt (kW) and wind turbines last year priced from $700 $900 per kilowatt, a 50% decline from 2008. These falling costs have contributed to wind power purchase agreements pricing at roughly $20 per megawatt-hour (MWh), the lowest cost of power generation in many parts of the country.[41]

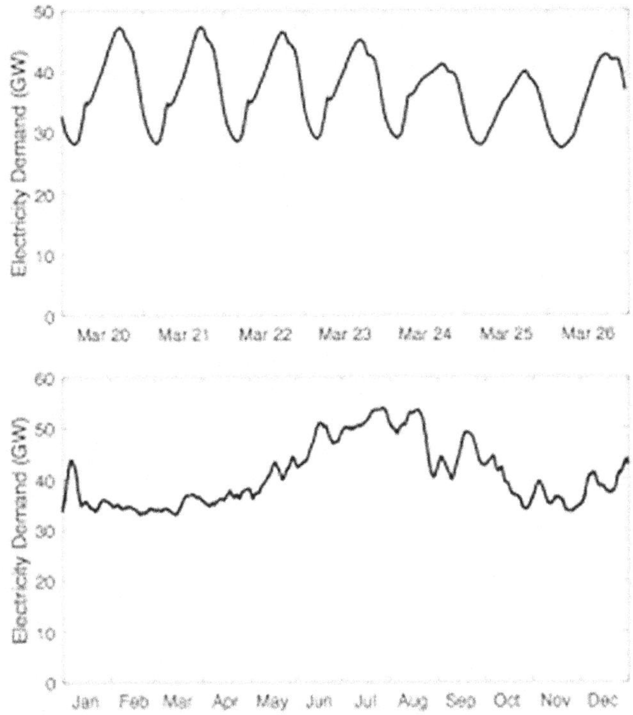

FIG. 3.47 ELECTRICITY DEMAND IN TEXAS ON TWO DIFFERENT TIMESCALES. TOP: HOURLY DEMAND FOR ONE WEEK, FROM MARCH 20-26, 2017. BOTTOM: DEMAND FOR ALL OF 2017, SHOWN AS A 168 HOUR (1 WEEK) RUNNING AVERAGE OF HOURLY. (SOURCE: [43])

3.16.5 Technical Issues

3.16.5.1 Grid Operations with Wind Power

Managing a grid with Variable Wind is prominent issue. As wind becomes a more prominent contributor to electricity supply, its variability complicates grid operation on ranges of timescales, from seconds to days to months. The duration of the mismatch between supply and demand (the length of the lull) determines the optimal response, which is some mix of flexible power from other sources, access to more distant winds, energy storage, and demand-side management. Prominent wind power also creates requirements for wind turbines to be able to help reduce the consequences of unforeseen grid disruptions; wind turbines are becoming increasingly helpful. Thus, integrating variable wind is an important factor of wind power generation. To illustrate the potential for mismatches between variable supply and variable demand an example is the Texas grid which has the highest installed wind capacity of any state in the US at the end of 2017. 21,000 megawatts were installed whereas wind turbines produced 17 percent of the state's electricity, and natural gas power plants produced 39 percent.[42]

Consider that all the excess wind power in the idealized energy system were stored and then used during the times of under-supply. The wind power input would meet demand exactly throughout the week. To be sure, this result requires the storage system to operate with no energy losses; in fact, there are always energy losses whenever a storage system acquires or discharges energy. For an energy system that even slightly resembles the one shown in Figure 3.48 in its prospective periods of excess and shortfall in energy supply, energy storage would be only one of many strategies to rebalance the system. Two countries where wind power already accounts for a large fraction of annual electricity production are Ireland (21 percent in 2016) and Denmark (44 percent in 2017). Like Texas, the Ireland grid is relatively isolated, with only a 9 percent interconnection capacity. Also, like Texas, Ireland achieves wind integration primarily with natural gas, which accounts for about 44 percent of Ireland's electricity. Denmark, by contrast, has relatively little gas generation to balance its high penetration of wind; its second largest electricity source is coal, which provides 25 percent of its electricity.

---

[41] https://www.compositesworld.com/news/awea-market-report-indicates-record-third-quarter

[42] https://acee.princeton.edu/wp-content/uploads/2019/04/AndlingerDistillate_Article8.pdf

[43] http://www.ercot.com/gridinfo/generation, cited in https://acee.princeton.edu/wp- content/uploads/2019/04/AndlingerDistillate_Article8.pdf.

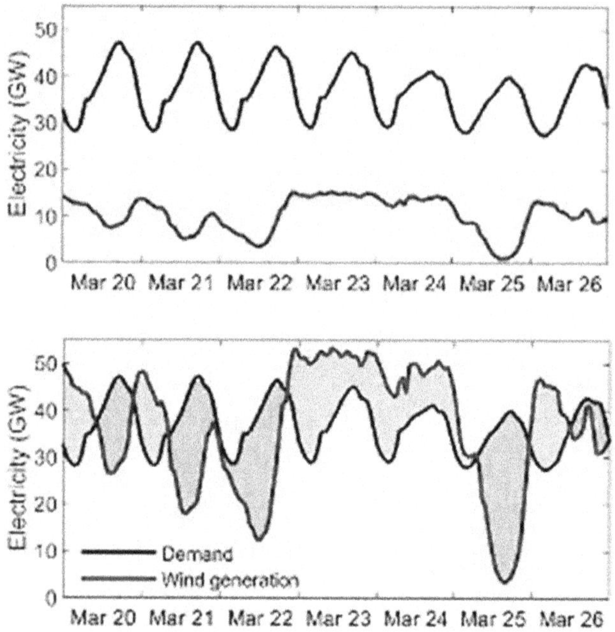

FIG. 3.48 TOP: ACTUAL ELECTRICITY DEMAND AND WIND GENERATION IN TEXAS, MARCH 20-26, 2017. BOTTOM: WIND CAPACITY, RESCALED TO CREATE THE COUNTERFACTUAL SITUATION WHERE THE TOTAL WIND GENERATION THAT WEEK EQUALS THE TOTAL ELECTRICITY DEMAND, AND NO OTHER CHANGES ARE MADE. GREEN AND RED SHADED AREAS REPRESENT EXCESS WIND AND A DEFICIT OF WIND, RESPECTIVELY. (SOURCE: [44])

3.16.5.2 Grid Flexibility

The principal way a current grid deals with threatened mismatches between supply and demand, when the mismatch is for short times (from seconds to hours), is to call on sources of electricity that can ramp their power production up and down quickly. Batteries are helping with load balancing, more and more as their costs fall. When a mismatch is to be avoided and available wind power is in excess, the sale of wind power can be curtailed: the wind turbine operator would be told that not all output can be placed on the grid. The unsold power is said to be "spilled." In addition, some other electricity generators can be told to produce power only at certain hours; for example, a coal plant or a hydroelectric dam would be scheduled to run during the day but not at night. Another strategy that can be pursued by the electricity generation system is to invest in transmission lines that access distant winds which are strong when local winds are weak, and vice versa, thereby smoothing out wind's contribution to the grid while at the same time creating a larger market. More generally, expanded transmission and distribution capability enables greater diversification across power generators; for example, it may foster the export of excess solar power from one region at midday to a second region where winds have subsided at the same time. The Competitive Renewable Energy Zone in Texas, which connects the state's major cities to sites in western Texas favorable to wind and solar generation, is an example. The costs of financing the transmission lines are borne by the state's electricity consumers; the benefit is greater diversification in electricity generation and a larger presence for wind and solar power. Still another enabler of high wind penetration is more flexible electricity demand ("demand-side management"). The timing of delivery of electric power to a water heater or electric car battery, for example, can be put under the control of the grid operator. Customers buy-in is fostered when there are time-variable electricity prices and smart appliances that are programmed to benefit from these prices. It becomes profitable for all parties when a washing machine is operated on a windy day rather than a calm day, for example. All these items are being encouraged vigorously during the last decade, with wind power usage has become more prevalent.[45]

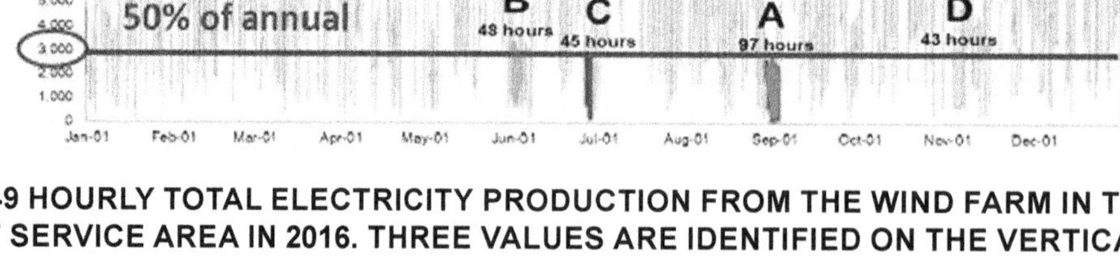

FIG. 3.49 HOURLY TOTAL ELECTRICITY PRODUCTION FROM THE WIND FARM IN THE ERCOT SERVICE AREA IN 2016. THREE VALUES ARE IDENTIFIED ON THE VERTICAL AXIS: 17,000 MEGAWATTS IS THE TOTAL CAPACITY OF THE WINDFARMS; 6000 MEGAWATTS IS ANNUAL AVERAGE WIND POWER PRODUCTION; AND 3000 MEGAWATTS IS HALF OF THE ANNUAL AVERAGE THE THRESHOLD CHOSEN FOR ILLUSTRATION PURPOSES. THE FOUR LONGEST LULLS ARE LABELED A, B, C AND D. (SOURCE: [46])

---

[44] https://acee.princeton.edu/wp-content/uploads/2019/04/AndlingerDistillate_Article8.pdf

[45] https://acee.princeton.edu/wp-content/uploads/2019/04/AndlingerDistillate_Article8.pdf

[46] Socolow, R. (2018). Lull Analysis to Characterize Windpower Intermittency, in 17th Annual Report of the Carbon Mitigation Initiative. Princeton University, April 2018. Online at https://cmi.princeton.edu/annual_reports/2017/integration/, cited in https://acee.princeton.edu/wp- content/uploads/2019/04/AndlingerDistillate_Article8.pdf.

3.16.5.3 Lull Analysis and Long Lulls

An interesting way to appreciate the variability of the wind is to use "lull analysis".[47] A lull is a period of definite duration when the wind speed is below some threshold. A lull might last from a few seconds to several days. In Figure 3.49 once again the Texas example is used to demonstrate the lull analysis. Hourly wind production is shown for an entire year, 2016. At the end of 2016, the total installed wind capacity of the Texas wind farms was about 17,000 megawatts and during that year the rate of wind-power production averaged 6,000 megawatts. Arbitrarily selected the threshold to be half the annual average value, or 3,000 megawatts, shown as a horizontal line in Figure 3.49. A lull begins when wind power output first falls below 3,000 megawatts and ends when it first ascends above that value.[48]

3.16.5.4 Grid Stability and Grid Services

Events that can disrupt the operation of the grid can occur over timescales from seconds (heavy machinery turning on or off, failure of a generator or transmission line) through hours, days, and months (power plant shutdown, routine maintenance). To minimize these vulnerabilities, the operator of a power grid takes account of specific features of each generator, such as its size, its start-up time, the maximum rates at which it can increase and decrease its output (its "ramping" rates), and its costs for electricity production. The grid operator also considers transmission constraints. Disruptions to the grid appear as frequency or voltage reductions. Frequency reductions are caused by a generator disconnecting from the grid or a new load coming online. Voltage reductions result from electrical circuit faults. Such drops create problems for those electricity consumers requiring high-quality power. In a minor frequency disruption, the grid frequency remains within its narrow "dead band" of permitted frequencies departures from the reference frequency limited to roughly 0.02 cycles per second (1 part in 3,000 for a 60 cycle-per-second grid). A larger disruption results in the grid frequency falling below the bottom of the dead band. The first wind turbines that produced grid power played little role in stabilizing the grid during a sudden and significant disruptive event. Typically, the wind turbines on a grid were immediately disconnected from it, as were other smaller, geographically dispersed (distributed") energy resources, like solar arrays. As distributed generators on the grid became more numerous, a threat to the stability of the grid emerged where all of these generators could simultaneously disconnect during a grid disruption and turn a minor event into an event with cascading impacts, where each turbine shutdown makes the grid anomaly worse. Both grid managers and the wind industry realized that wind turbines needed to be modified so that they could contribute toward minimizing the consequences of any grid disruption. Such modifications become especially important when distributed energy sources are providing a large fraction of total power, such as when winds are strong and the load is light. A modern wind turbine counters a wayward fall in frequency with strategies that are like those provided by conventional power plants. To be able to provide extra power quickly on demand (the grid may request as much as an additional ten percent of its rated power), it must not already be producing power at its maximum value for that wind speed. Instead, it must deliberately produce less power than it could, thereby creating the "headroom" to respond for a call from the grid for extra power. Such headroom can be achieved by setting the pitch of the blades in normal operation slightly away from optimal or orienting the turbine slightly away from straight into the wind.[50]

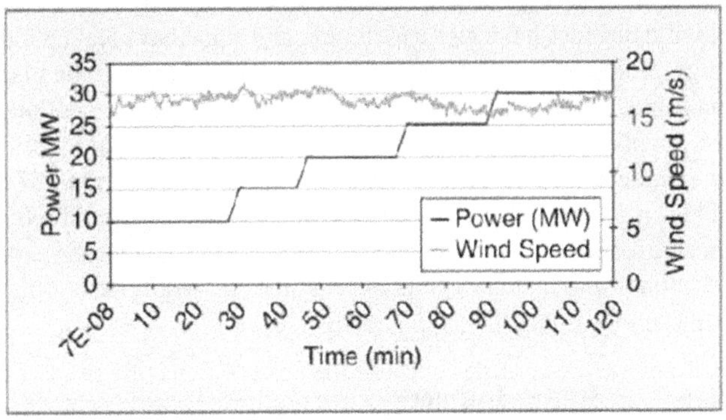

FIG. 3.50 CONTROLLED UPWARD RAMPING OF A WIND TURBINE FARM'S OUTPUT POWER, (SOURCE [49])

3.16.5.5 Farm-level Grid Services

Wind power can provide grid services at the level of the wind farm, not only at the level of the individual turbine. With the help of power electronics and advanced turbines, the operator of a wind farm can coordinate the outputs of each of the farm's turbines to keep the farm's total output within narrow limits and to control the rate at which total output ramps upward or downward. Consider the two-hour field test reported in Figure 3.50. Prior to the onset of the test, a 30-megawatt wind farm is operating at far below its rated power under nearly constant high winds (just above 15 meters per second) at only 10 megawatts; this could be the result of some strict curtailment. During the first 90 minutes, the farm output's climbs upward in the same high winds back to 30 megawatts in four five-megawatt steps, each accomplished in approximately three minutes, with output tightly controlled at each step. In short, "grid friendly" wind power is becoming the new norm.[51]

3.16.6 Environmental Issues

3.16.6.1 Wildlife Considerations

"Wind is an emissions-free source of energy" is an article issued by the Department of Energy in December 2020, repeating much of the same that was mentioned by the authors Baldwin and Seifert about the wildlife impacted by turbines used for wind energy. An article issued by National Renewable Energy Laboratory captioned "Wind turbines have some negative effects on the environment", dated December 9, 2020 mentioned that very large modern

[47] Ibid.

[48] https://acee.princeton.edu/wp-content/uploads/2019/04/AndlingerDistillate_Article8.pdf

[49] MacDowell et al. (2015). Serving the Future: Advanced Wind Generation Technology Supports Ancillary Services. IEEE Power & Energy Magazine 13(6): 22-30; cited in https://acee.princeton.edu/wpcontent/uploads/2019/04/AndlingerDistillate_Article8.pdf

[50] https://acee.princeton.edu/wp-content/uploads/2019/04/AndlingerDistillate_Article8.pdf

[51] https://acee.princeton.edu/wp-content/uploads/2019/04/AndlingerDistillate_Article8.pdf

wind turbines may visually affect the landscape. A small number of wind turbines have also caught fire, and some have leaked lubricating fluids. People do not like the sound that wind turbine blades make and cause bird and bat deaths which contribute to declines in the population of species. The wind energy industry and the U.S. government are researching ways to reduce the effect of wind turbines on birds and bats. Most wind power projects on land require service roads that add to the physical effects on the environment, in addition to producing the metals and other materials used to make wind turbine components impacting. [52]

3.16.7 Radar Impacts

Since originally reported by the authors Baldwin and Siefert the US Government (Bureau of Ocean Energy Management, BOEM with the U.S. Department of the Interior, DOI) had been studying the Outer Continental Shelf (OCS) of the US, under the National Environmental Policy Act (NEPA) for "Radar Interference Analysis for Renewable Energy Facilities on the Atlantic Outer Continental Shelf". This includes the impact of radar interference on offshore wind turbines. Presence of wind energy installations can generate a return signal to radars. Radar signals generated by objects other than the intended targets such as wind turbines are referred to as "clutter" or "interference" and represent a potential impact to be investigated under NEPA. In April of 2019, BOEM engaged Booz Allen Hamilton (Booz Allen) through Call Order 140M0119F0015 to study how proposed and hypothetical future offshore wind energy installations on the Atlantic coast may impact land-based radar systems, potential mitigations for these impacts, and how an atmospheric phenomenon, known as "ducting," influences these impacts. Investigating radar interference impacts on nine total wind farms on U.S. Atlantic coast from Massachusetts to South Carolina were investigated as shown in figure 3.51.[53]

Conclusions and Recommendations of the BOEM report include:[54]

1. Several Radar Impacts: Proposed wind farms are within Line of Sight (LOS) and Interference of several radar systems; radar interference should be of a concern for wind farm development.
2. "SeaSonde" radars present a unique challenge for offshore wind installations; however, there may be a relatively inexpensive mitigation option available for SeaSonde that involves a software upgrade.
3. Mitigation techniques available for offshore wind farms mirror those available for onshore wind farms, include a combination of operational mitigations and radar system modifications and upgrades.

3.16.8 Local Impacts

3.16.8.1 Challenges for Wind Power

A few challenges that wind turbines face are the criticisms of the noise, aesthetics, bird deaths, efficiency, and an overall "not-in-my-backyard" attitude. Noise is caused by mechanical movements of the parts inside the nacelle and from aerodynamic turbulence caused by the rotations of the blades. Common complaints are the annoyance

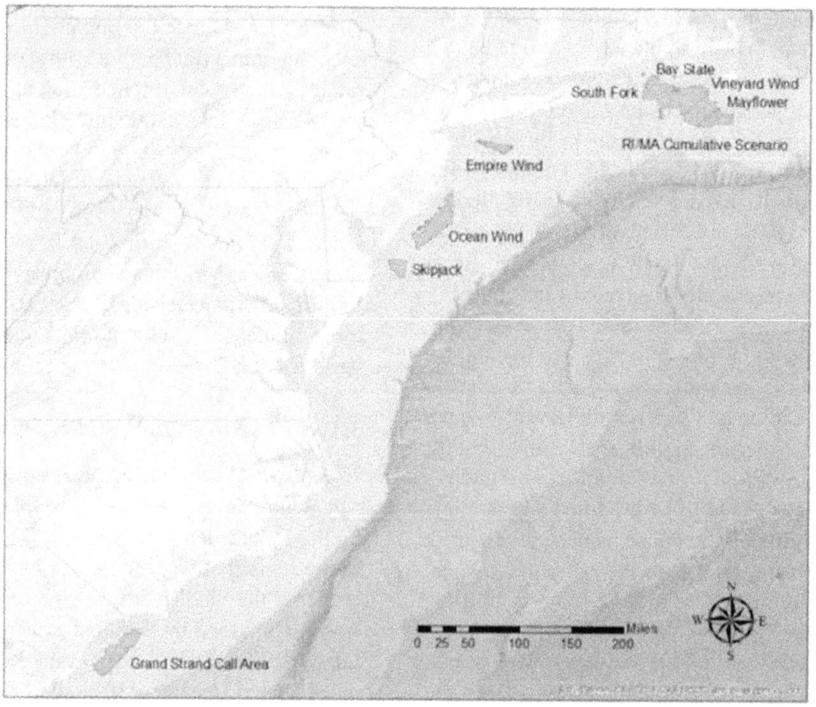

FIG. 3.51 MAP OF PROPOSED WIND FARM LOCATIONS STUDIED BY BOEM (SOURCE: [55])

[52] https://www.eia.gov/energyexplained/wind/wind-energy-and-the-environment.php, dated 12-9-2020.
[53] https://www.boem.gov/sites/default/files/documents/environment/Radar-Interferance-Atlantic-Offshore-Wind_0.pdf
[54] https://www.boem.gov/sites/default/files/documents/environment/Radar-Interferance-Atlantic-Offshore-Wind_0.pdf
[55] https://www.boem.gov/sites/default/files/documents/environment/Radar-Interferance-Atlantic-Offshore-Wind_0.pdf

of sound during the day and the deprivation of sleep during the night. The Institute of Noise Control Engineering in Europe has helped create the Wind Turbine Noise conference held every other year since 2005 where representatives from the UK, France, Italy, USA, Netherlands, Sweden, Denmark, Australia, Japan, and many others come together to discuss how to solve the issue of noise. Topics include sources of noise, designs modification, noise modeling, effects on individuals, vibrations from wind turbines, etc. Aesthetics has always been and will always be a challenging factor for wind turbines. Everyone has differing opinions on the topic and in the end, it comes to a "not in my backyard attitude." There are people who believe in clean energy and enjoy the idea of wind turbines, but it becomes a different story when they see and hear them at all hours of the day. Unfortunately, since most commercially used wind turbines are very tall and have very large rotors, they get in the way of the natural landscape. The visual effect of wind farms is a subjective issue. Another major, yet incorrect criticism is that wind turbines are the major cause of bird deaths. According to the American Bird Conservancy, in the collisions category, wind turbines come in last with a low estimate of 100,000 bird deaths in 2010 and a high estimate of 440,000 in 2009. Glass windows lead the way with a low estimate of 100,000,000 and a high estimate of one million bird deaths. Other factors of bird deaths include pesticides and toxins, fisheries, global warming, cats, invasive species, disease, and other forms of energy production such as the burning of coal. The major indirect cause of bird deaths is habitat depletion as lands are cleared for residential housing, industrial uses, and other things we use or enjoy.[56]

What are the Threats to Birds?

Many of the 1,093 species of birds protected under the "Migratory Bird Treaty" Act are experiencing population declines due to increased threats across the landscape. Of those 1,093 species, 89 bird species are listed as either threatened or endangered under the "U.S. Endangered Species Act". An additional 342 species are listed as "Birds of Conservation Concern", in one or more geographic scales (e.g., local, regional, or national). Both natural and human-caused sources of bird mortality contribute cumulative or combined effects to declining bird populations. Millions of acres of bird "habitat are lost or degraded" every year due to development, agriculture, and forestry practices. These rapidly accelerating impacts can be mitigated only through habitat restoration and protection. In addition, millions of birds are directly killed by human-caused sources such as "collisions with man-made structures". Death from natural causes is also common sources of mortality. Natural and human-caused mortality impacts are exacerbated by the landscape alterations resulting from a changing climate.[57]

How Many Birds are Killed?

True estimates of mortality are difficult to determine. However, recent studies have synthesized the best available data to estimated ranges of mortality to bird populations in North America from some of the most common, human-caused sources of bird mortality. These are listed in the table 3.8. This list addresses only human-caused sources, not natural sources. Many additional human-caused threats to birds, both direct (causing immediate injury/death) and indirect (causing delayed negative effects to health or productivity) are not on this list because the extent of their impact is either not currently well researched or easily quantified. For instance, habitat loss is thought to pose by far the greatest threat to birds, both directly and indirectly, however, its overall impact on bird populations is very difficult to directly assess. Other common human-caused and natural threats to birds that are known, but not listed include various "entanglement and entrapment threats" (e.g., open pipes and nets); predation by other animals besides cats, including humans (e.g., poaching); weather events; starvation; and "disease".[58]

TABLE 3.8: TOP THREATS TO BIRDS IN THE U.S. ACCORDING TO MEDIAN ESTIMATE OF BIRD MORTALITY ANNUALLY, 2017[59]

Hazard Type (Year)	Min Range	Max Range	Median/Avg. Estimates
Wind turbines (2010 / 2009)	100,000 (2010)	440,000 (2009)	-
Communication towers (2008 / 2012)	4 million (2008)	50 million (2008)	6.6 million (2012)
Power lines (2001 / 2014)	10 / 8 million	154 / 57.3 million	25.5 million (2014)
Roads / vehicles (2005 / 2014)	10.7 / 89 million	380 / 340 million	214.5 million (2014)
Urban Light (2009)	31.158 million	-	-
Collision Building Glass (2014)	365 million	988 million	599 .5million
Land-based Wind Turbines (2013)	140,438	327,586	234,012
Electrocutions Loss (2014)	900,000	11.6 million	5.6 million
All (2017)	549 million	5,182 million	3,324 million
All, excluding cats (2017)	542.4 million	1,477 million	2,019 million
Industry, excluding cats and vehicles (2017)	453 million	1,136.8 million	709.6 million

[56] http://www.geni.org/globalenergy/research/wind-turbine-technologies/Wind%20Turbine%20Technologies.pdf
[57] https://www.fws.gov/birds/bird-enthusiasts/threats-to-birds.php_10-17-21
[58] https://www.fws.gov/birds/bird-enthusiasts/threats-to-birds.php_10-17-21
[59] https://www.fws.gov/birds/bird-enthusiasts/threats-to-birds.php_10-17-21

3.16.8.2 Acoustic Noise

"Screech noise source structure of a supersonic rectangular jet" were investigated by Rice, E. J.(NASA Lewis Research Center Cleveland, OH, United States) and Taghavi, R. (Sverdrup Technology, Inc., Brook Park OH., United States) acquired on September 6, 2013. The near field of the screech noise source structure of a supersonic jet was studied to determine the amplitude and phase of the sound pressure near and in the high-speed flow field. The transverse structure of the unsteady pressure field was investigated so that pressure fall-off was observed. Five islands of high sound pressure level have been distinguished associated with the actual local sources of sound production. These sources of screech noise are associated with the jet shock structure with the peak region of noise level being found slightly downstream of each of the five observed shocks. All the apparent noise sources have their peak levels in the subsonic flow region. Strong cancellations in the acoustic field are observed in the downstream and sideline directions which may account for the predominant upstream propagation of the fundamental tone noise.[60]

3.16.9 Addressing Needs For Wind To Reach Its Full Potential In The U.S.

3.16.9.1 Policy

The following measures were taken with the federal production tax credit (PDC) as one of the core motivators for wind power development, see figure 3.53:[61]

- 5year extension of PTC in 2015, plus guidance allowing 4 years for project completion after the start of the construction.
- PTC phase-out with progressive reduction in the value of the credit for projects starting construction after 2016.
- PTC phases out in 20% per year increments for projects starting construction in 2017 with 80% PTC value, for projects starting construction in 2018 with 60% PTC value and for projects starting construction in 2019 with 40% PTC.

The following figure 3.52 depicts how state policies are playing a role in wind power development. State Policies Help Direct the Location and Amount of Wind Development, but Wind Growth is Outpacing State Targets, see figure 3.52.

3.16.9.2 Electric Transmission[62]

The goal of NREL's transmission integration in the US s to tackle the challenges of integrating renewables, including wind energy into the bulk-power system while maintaining safe, efficient, and cost-effective grids. NREL works with electric utilities, energy policymakers, and other industry partners to integrate large-scale renewable energy into the transmission system. Transmission and grid integration issues include data analysis and development of models to enable the electric power system to efficiently manage the variability of renewables such as solar and wind power. Wind and solar generation vary with wind speed and solar insolation. This variability affects how transmission systems with high penetrations of renewable energy sources operate. NREL researchers are identifying these effects and finding solutions to address them to enable transmission grid integration which include capabilities:[63]

- Power systems simulation – Expertise in large grid integration studies, such as production cost and dynamic modeling and simulation of large power systems

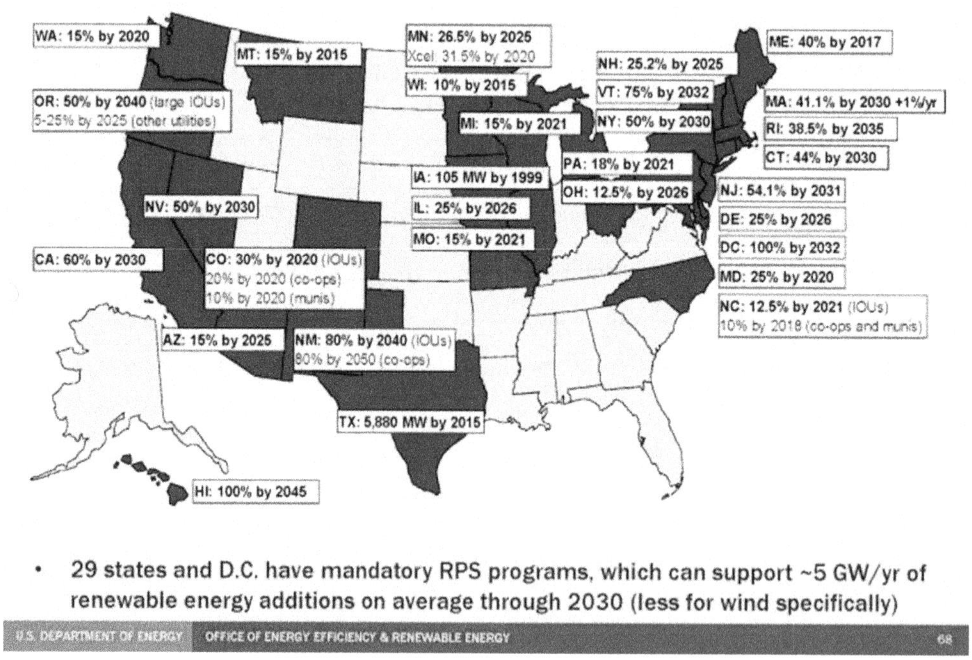

FIG. 3.52 STATE POLICIES AND WIND GROWTH (SOURCE: [64])

[60] https://ntrs.nasa.gov/citations/19920004782-10-17-21

[61] https://www.energy.gov/eere/wind/downloads/2018-wind-technologies-market-report _10-18-21

[62] https://www.nrel.gov/grid/transmission-integration.html

[63] https://www.nrel.gov/grid/transmission-integration.html

[64] "https://www.energy.gov/eere/wind/downloads/2018-wind-technologies-market-report%20_10-18-21" https://www.energy.gov/eere/wind/downloads/2018-wind-technologies-market-report _10-18-21

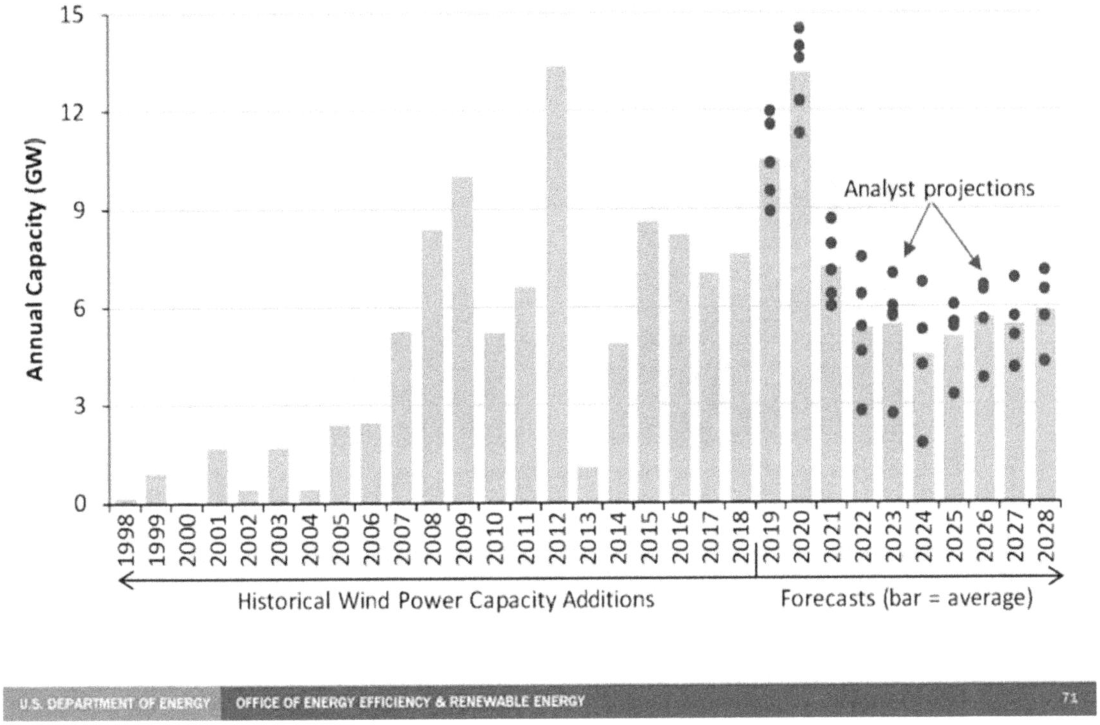

FIG. 3.53 SIZEABLE WIND ADDITIONS ANTICIPATED FOR 20192020 GIVEN FEDERAL TAX INCENTIVES; DOWNTURN AND GREATER UNCERTAINTY BEYOND 2020 (SOURCE: [65])

- Transmission system planning – Expertise in the planning of transmission-level power systems
- Transmission and distribution system co-simulation – Expertise in the co-simulation of combined transmission and distribution systems to gauge the impact of distributed energy resources on the bulk system and how bulk system operations impact distributed energy resources.
- Specific examples of the transmission studies relating to wind energy are:
- Western Wind and Solar Integration Study Exploring whether we can integrate large amounts of wind and solar into the U.S. western electric grid.
- Hawaii Solar and Wind Integration Studies Studying the effects of high penetrations of renewables on Hawaiian island grids.

FIG. 3.54 ELECTRICAL TRANSMISSION OF WIND ENERGY: (SOURCE [66])

3.16.10 Wind Farm Development

3.16.10.1 Various Wind Turbine Development

Technological advances in vertical-axis wind turbines that can generate more energy with a smaller footprint are now challenging the traditional use of horizontal wind turbines in wind farms. horizontal axis wind turbines used for commercial energy production are extremely tall so that they are able to capture the faster-moving, laminar air currents. The size of the cross-sectional area is also important as it is directly related to the power produced, which explains the large rotor on each horizontal axis turbine. When the wind blows past the large rotors, the laminar air currents become turbulent. This turbulence decreases the efficiency of any horizontal-axis turbine downwind. The effect of spacing required between horizontal axis turbines to decrease the effect of turbulence has been about 7 rotor-diameters, but the most recent study done by Charles Meneveau, a fluid mechanics and turbulence expert and mechanical engineering professor at Johns Hopkins University, suggests that they should now be placed 15 rotor-diameters apart. Unfortunately, this suggestion will further increase the already large footprint of a wind farm, which brings into question whether it is truly more efficient. One of the few horizontal-axis wind turbines that is suitable for residential and commercial areas is the Swift Wind Turbine. It can be mounted on rooftops or can be free standing, as shown in Figure 3.55. It is about 2.13 meters in diameter and has a rated power of 1.5 Kw. It captures small wind with a cut-in wind

[65] "https://www.energy.gov/eere/wind/downloads/2018-wind-%20technologies-market-report_10-28-21"

[66] https://www.nrel.gov/grid/transmission-integration.html

FIG. 3.55 SWIFT WIND TURBINE MOUNTED ON A RESIDENTIAL HOME IN HOLLAND, MICHIGAN (LEFT) AND FREE STANDING (RIGHT), SOURCE:[67]

speed of 3.58 m/s and generates about 1,200 kWh annually in 5 m/s wind speeds. The Swift Wind Turbine's specifications are comparable to vertical axis wind turbines in terms of footprint and the resulting power generated, so the major deciding factor would be aesthetics.[68]

3.16.10.2 New Wind Farms

Horizontal axis wind turbines may not be the only turbines being used in wind farms from now on. John O. Daribi, a graduate student in the department of Aeronautical Laboratories and Bioengineering at California Institute of Technology, produced a study that compared the power density of a wind farm using horizontal-axis wind turbines with a wind farm using Windspire vertical axis-wind turbines. The study compared horizontal axis wind turbines with rated powers of 2.5 and 3 MW with rotor diameters of 100 and 112 meters with the Windspire with a rated power of 1.2 kW and a rotor diameter of 1.2 meters. Figure 3.56 shows a small collection of Windspire turbines.[70]

FIG. 3.56 WINDSPIRE VERTICAL AXIS WIND TURBINES (SOURCE: [69])

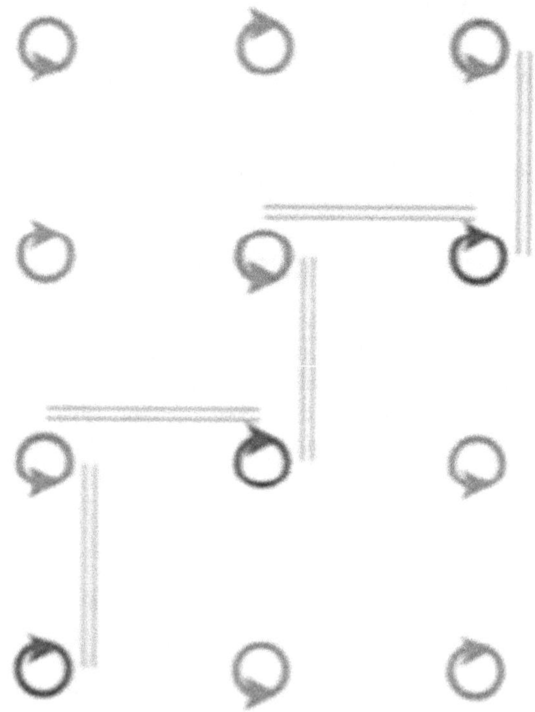

FIG. 3.57 CLOCKWISE AND COUNTERCLOCKWISE CONFIGURATIONS OF THE WINDSPIRE ARRAY: SOURCE [71]

[67] http://www.geni.org/globalenergy/research/wind-turbine-technologies/Wind%20Turbine%20Technologies.pdf, pg. 17.
[68] http://www.geni.org/globalenergy/research/wind-turbine-technologies/Wind%20Turbine%20Technologies.pdf, pg 17.
[69] http://www.geni.org/globalenergy/research/wind-turbine-technologies/Wind%20Turbine%20Technologies.pdf, pg. 19.
[70] http://www.geni.org/globalenergy/research/wind-turbine-technologies/Wind%20Turbine%20Technologies.pdf, pg. 19.
[71] http://www.geni.org/globalenergy/research/wind-turbine-technologies/Wind%20Turbine%20Technologies.pdf, pg. 20.

According to the study, the power density of wind farms using traditional horizontal-axis wind turbines today is about two or three watts produced per meter of land area. Since turbulence from upwind wind turbines will significantly decrease the efficiency of downwind turbines, they must be spaced an average of seven rotor-diameters apart. This creates a large area of unused space on the ground and therefore a much larger footprint. In the field tests done by Dabiri and his colleagues, they strategically placed three clockwise-rotating (blue) and three counter-clockwise-rotating (red) turbines to find the optimal power density as shown in Figure 3.57.

They found that a spacing of four rotor-diameters allowed the turbines downwind to recover within 5% of an isolated turbine performance. This allowed for a 48.6 m^2 footprint of a six-turbine array and an overall power density of 6 to 30 W/m^2. It is a marked improvement over the 2 to 3 W/m^2 power density of the horizontal-axis wind turbines and the results suggest that vertical axis wind turbines may be included in future wind farms.[72]

3.16.11 Wind Resource Assessment

Wind resource assessment is the process by which wind power developers estimate the future energy production of a wind farm. Accurate wind resource assessments are crucial to the successful development of wind farms. Wind prospecting begins with the use of wind resource maps, but the lack of accuracy and fine detail make them useful only for preliminary selection of sites for collecting wind speed data. With increasing numbers of ground-based measurements from specially installed anemometer stations, as well as operating data from commissioned wind farms, the accuracy of wind resource maps has improved over time in the US. In addition to the publicly available sources, maps are available as commercial products through specialist consultancies, or users of GIS software can make their own using publicly available GIS data such as the US National Renewable Energy Laboratory's High Resolution Wind Data Set. Multiple meteorological towers are usually installed on large wind farm sites. Correlations between long term weather stations and onsite meteorological towers are being established. Wind data analysis software are being augmented for the user in removing measurement errors from wind data sets and perform specialized statistical analysis. Wind flow modeling software are continuously refined to predict important characteristics of the wind resource at locations where measurements are not available. An important feature is the development of wind farm modeling software to simulate the behavior of a proposed or existing wind farm, most importantly to calculate its current energy production. Several of the current wind farm modeling software applications available include ZephyCFD, Meteodyn WT, Openwind, Windfarmer, WindPRO, WindSim, and WAsP. In recent years, a new breed of medium scale wind farm development has grown from the increased need for distributed generation of electricity from local wind resources.[73] Guidelines for wind resource assessment states the pivotal activity during the development phase of a wind project is wind resource assessment (WRA). The best practices for initiating wind development are (i) on-site wind measurement with tall towers and high-quality instruments for a period of at least 1 year; (ii) auditable data acquisition, processing, and archiving; (iii) rigorous modeling of wind flow based on a linear model for simple terrain and a computational fluid dynamics model for complex terrain; and (iv) a credible estimate of losses and uncertainties.[74] Wind resource assessment is important because it is the basis for determining initial feasibility and cash flow projections and is ultimately vital for acquiring financing.[75] NREL conducts techno-economic, resource, and infrastructure assessments by producing high-resolution maps and data at the US County level. These assessments inform their wind power industry, power system, and transportation planning the essential market designs and policies, infrastructure investments, and research programs.[76]

3.16.12 Materials Required For The Land-Based Wind Turbine Industry

Materials required for the Land-Based Wind Turbine Industry from 2010 through 2030 for Wind Energy in the United States are reported by David R. Wilburn for U.S. Department of the Interior U.S. Geological Survey. An understanding of the sources and abundance of raw materials required by the wind turbine industry and the many uses for these materials is necessary to assess the effect of this industry's growth on future demand for selected raw materials. On the assumption that wind energy will supply 20 percent of the electricity consumed in the United States by 2030. Estimates for the quantity of materials in typical "current generation" and "next generation" wind turbines were developed. The results of the study suggest that achieving the market goal of 20 percent by 2030 would require an average annual consumption of about 6.8 million metric tons of concrete, 1.5 million metric tons of steel, 310,000 metric tons of cast iron, 40,000 metric tons of copper, and 380 metric tons of the rare-earth element neodymium. With the exception of neodymium, these material requirements represent less than 3 percent of the U.S. apparent consumption for 2008. The data suggest that, except for rare-earth elements, there should not be a shortage of the principal materials required for electricity generation from wind energy. There may, however, be selective manufacturing shortages if the total demand for raw materials from all markets is greater than the available supply of these materials or the capacity of industry to manufacture components. This report provides policymakers and the public with an overview of factors currently and potentially affecting raw material requirements for land-based wind turbines. The life cycle of a wind turbine can be divided into the following five phases: construction, transportation. the equipment and supplies necessary installation, operation, decommissioning and disposition. Material Requirements of Wind Turbines typical wind turbine is reported to contain 89.1 percent steel, 5.8 percent fiberglass, 1.6 percent copper,

---

[72] http://www.geni.org/globalenergy/research/wind-turbine-technologies/Wind%20Turbine%20Technologies.pdf, pg. 20.

[73] https://en.wikipedia.org/wiki/Wind_resource_assessment

[74] https://www.adb.org/sites/default/files/publication/42032/guidelines-wind-resource-assessment.pdf

[75] https://www.bing.com/search?q=wind+resource+assessment&qs=HS&pq=wind&sk=HS2&sc=8-4&cvid=122944AE0BBA4EA7A851AE-FAE3A429D4&FORM=QBLH&sp=3

[76] https://www.nrel.gov/international/resource-and-infrastructure-assessments.html#:~:text=Energy%20Resource%20and%20Infrastructure%20Assessments%20NREL%20conducts%20techno-economic%2C,and%20infrastructure%20assessments%E2%80%94producing%20high-resolution%20maps%20and%20data%E2%80%94for%20countries

TABLE 3.9: PROJECTED USE OF MATERIALS THROUGH 2030 IN THE US FOR WIND TURBINES: (SOURCE [77])

(kg. kilogram. kg/MW. kilograms per megawatt: MW, megawatt: Nd, ncodymium: Nd-Fc-B. ncodymium. iron, and boron: Nd_20_3. ncodymium oxide; XX. not applicable]

Turbine	Steel, stainless (kg/MW)	Concrete (kg/MW)	Fiberglass (kg/MW)	Miscellaneous[1] (kg/MW)	Copper[2] (kg/MW)	Ndin magnet[3] (kg/MW)	Cast iron (kg/MW)	Total weight (kg/MW)
Current generation[4]	115.000	590,000	9.800	8.050	2.500	0	23.900	750,000
Next generation[5]	103,000	402,000	6.S00	9.300	3.000	43.2	20.000	540,000
Average change in megawatt consumption, in percent	-10	-32	-31	+ 16	+20	XX	-16	-28
	(million kg)	**(million kg)**	**(million kg)**	**(million kg)**	**(million kg)**	**(million kg)**	**(million kg)**	**(million kg)**
Average for 2010-2030[6]	1,500	6,800	no	130	40	0.38	310	8,890
Total by 2030[7]	30,000	135,000	2300	2,400	750	6	6,000	176,456

[1] Includes aluminum, plastic, epoxy resins, polymer foam, and wood.

[2] Includes copper windings, copper alloys in components, copper wire, and copper in electronics.

[3] Estimates of rare-earth clement requirements were developed based on published information on rare-earth content of wind turbine generators from Hatch (2009) and Lifton (2009). Wind turbines of the current generation typically did not use permanent magnets: therefore, the rare-earth clement requirement of these wind turbines was considered negligible. It was assumed thai 20 percent of (lie next generation wind turbines would use rare-earth permanent magnets. The remaining 80 percent would either use conventional electromagnets or ferrite permanent magnets that use no rare-earth elements. A next generation wind turbine that used a rare-earth permanent magnet would require 216 kg/MW Hd or 251 kg/MW Nd_20_3, assuming that Nd accounts for about 27 percent of die weight of a rare-earth (Nd-Fe-Bt permanent magnet. The Nd content of the average next generation wind turbine was therefore estimated as [(80 percent x 0) + (20 percent x 216 kg/MW)] = 43.2 kg/MW Nd [50 kg/MW Nd_20_3].

[4] The representative current-generation onshore wind turbine is capable of generating 1.5 MW of electricity using conventional technology that includes a three-blade rotor, steel tower, three-stage gearbox, and a wound-rotor type generator.

[5] The representative next-generation wind turbine is capable of generating 3 MW of electricity and could use more composite materials in the rotor blades, stccl-concrctc towers, and a mixed generator technology assuming 80 percent double-fed induction generator technology and 20 percent permanent magnet technology. The latter could use rare-earth elements.

[6] Thc estimate of the average amount of material that would be required annually for 2010-2030 in order to achieve the goal of wind power supplying 20 percent of the U.S. demand for electricity by 2030 based on the distribution of current and next generation wind turbines as reported in appendix 2.

[7] The estimate of the total amount of material that would be required for 2010-2030 in order to achieve the goal of having wind power supply 20 percent of the ILS. demand for electricity by 2030 based on the distribution of current and next generation wind turbines as reported in appendix 2.

1.3 percent concrete (primarily cement, water, aggregates, and steel reinforcement), 1.1 percent adhesives, 0.8 percent aluminum, and 0.4 percent core materials (primarily foam, plastic, and wood) by weight (U.S. Department of Energy, 2008). In 2008, utilities and manufacturers installed more than 5,000 utility-scale wind turbines in the United States, requiring more than 1.1 million metric tons (Mt) (1 million short tons) of iron and steel, 920,000 cubic meters (1.2 million cubic yards) of concrete, 2.4 million steel bolts, and 43,000 kilometers (km) (27,000 miles) of reinforced steel rebar (American Wind Energy Association, 2009b).[78]

3.16.13 Wind Energy Research

3.16.13.1 Grid Integration[79]

Grid Modernization is possible by "Advanced Power Electronics and Smart Inverters", NREL's advanced power electronics and smart inverter research which enables high penetrations of renewable and distributed energy resources on the U.S. electricity grid to make it cleaner, reliable, and more efficient. Integrating renewable and distributed energy resources, such as photovoltaics (PV) and energy storage devices, into the electric distribution system requires advanced power electronics, or smart inverters, that can

[77] https://pubs.usgs.gov/sir/2011/5036/sir2011-5036.pdf
[78] https://pubs.usgs.gov/sir/2011/5036/sir2011-5036.pdf
[79] https://www.nrel.gov/grid/power-electronics-inverters.html

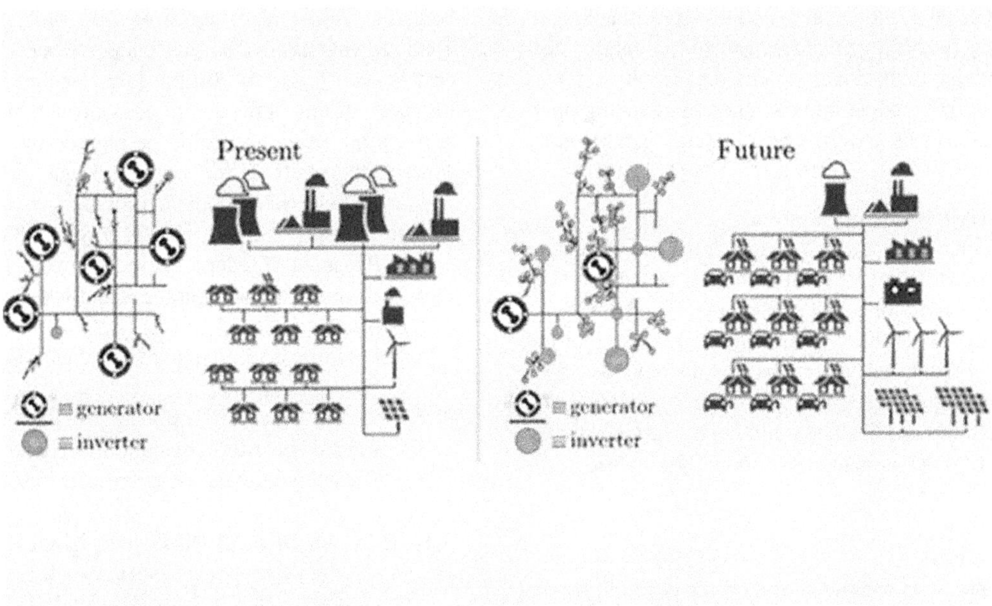

FIG. 3.58 GRID-FORMING INVERTER CONTROLS (SOURCE [80])

provide grid services such as voltage and frequency regulation, ride-through, dynamic current injection, and anti-islanding functionality. To enable this integration, NREL is designing novel wide-bandgap smart inverters, developing robust control algorithms for better inverter functionality, determining interactions between multiple smart inverters and between inverters and utility distribution systems, supporting standards development for smart inverter functionalities, and analyzing the impacts of smart inverters on distribution systems. Capabilities for grid integration include:[81]

- Modeling and simulation of smart inverters
- Design of smart inverter hardware and firmware
- Development of smart inverter control algorithms
- Development of wide-bandgap smart inverters
- Use of control and power hardware-in-the-loop techniques to determine interactions between multiple inverters at multiple points of common coupling
- Development and validation of test procedures for smart inverters to support IEEE 1547 and UL 1741
- Power system impact studies for smart inverter-interfaced distributed energy resources

Advanced Energy Industries validated its advanced PV inverter technology using NREL's power hardware-in-the-loop system and megawatt-scale grid simulators. Our utility-scale power hardware-in-the-loop capability allowed Advanced Energy to loop its inverter into a real-world simulation environment so researchers could see the impact of the inverter's advanced features on power reliability and quality. Advanced Energy's inverter will help support a smarter grid that can handle two-way flow of power and communication while reducing hardware costs. Reaching Department of Energy goals of 20% wind energy by 2030 and 35% by 2050 requires a better understanding of power system reliability at high levels of wind energy penetration. These important questions need to be addressed to enable high-penetration scenarios:

- What is the role for short-term power storage to provide power variability smoothing for wind turbine drivetrain loads?
- How can advanced wind plant controls contribute to grid reliability enhancements?
- Can system stability and reliability be maintained with high instantaneous wind penetrations?
- What is the role of energy storage to firm wind power ability to provide such controls?
- How can other enabling technologies, such as response loads, help integrate high levels of variable power?
- What testing needs to be done to enable this transformation?

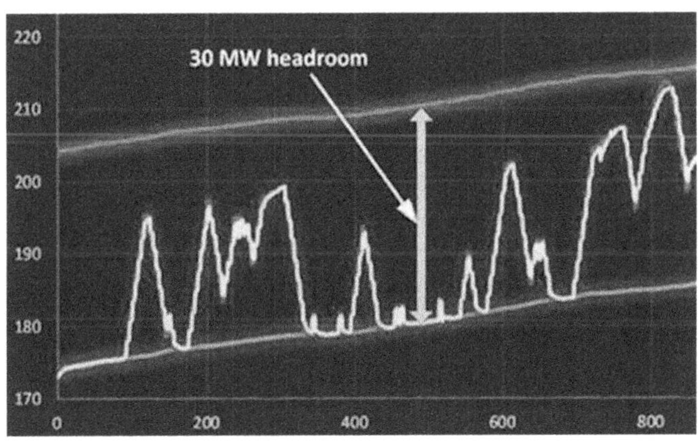

FIG. 3.59 GRID-FRIENDLY CONTROLS FOR UTILITY-SCALE PV (SOURCE[82])

[80] https://www.nrel.gov/grid/power-electronics-inverters.html
[81] https://www.nrel.gov/grid/power-electronics-inverters.html
[82] https://www.nrel.gov/grid/power-electronics-inverters.html

The goal of this project is to develop and test coordinated controls of active power by wind generation, short-term energy storage, and large industrial motor drives to provide ancillary services to the grid and minimize loading impacts, thereby reducing operation and maintenance costs and the cost of energy generated by wind power.[83]

3.16.13.2 Hybrid Energy Systems

Hybrid reinforcements have been developed. Instead of making wind turbine blade reinforcements from pure glass or pure carbon, hybrid designs trade weight for cost. For example, for an 8m blade, a full replacement by carbon fiber would save 80% of weight but increase costs by 150%, while a 30% replacement would save 50% of weight and increase costs by 90%. Hybrid reinforcement materials include E-glass/carbon, E-glass/aramid. The current longest blade by LM Wind Power is made of carbon/glass hybrid composites.[84]

3.16.14 Conclusions

Wind capacity additions continued at a robust pace during the past decade. Wind power has been a significant source of new electric generation capacity in the US in recent years. Supply chain is diverse and multifaceted, with strong domestic content for nacelle assembly, towers and blades. Turbine scaling is significantly boosting wind project performances, while the installed cost of wind projects has declined. Wind power sales prices and levelized cost of energy are at all-time lows, enabling economic competitiveness despite low gas prices. Growth beyond current cycle remains uncertain and could be blunted by declining federal tax support, expectations for low natural gas prices and solar costs, and modest electricity demand growth.[85]

The outlook for future wind energy growth will depend upon[86]:

- Phaseout of federal tax incentives.
- Continued low natural gas and wholesale electricity prices.
- Potential decline in market value as wind penetration increases.
- Modest electricity demand growth.
- Limited near-term demand from based on state policies.
- Growing competition from solar in some regions.

3.16.15 Additional Bibliographical References:

1) Energy and Power Generation Handbook Established and Emerging Technologies, edited by K. R. Rao, ASME Press, New York, 2011.
2) Wind Energy for Power Generation Meeting the Challenge of Practical Implementation by K. R. Rao, Volumes 1 & 2, This Springer book is published by Springer International Publishing AG part of Springer Nature, Gewerbestrasse 11, 6330 Cham, Switzerland.

[83] https://www.nrel.gov/grid/power-electronics-inverters.html
[84] https://en.m.wikipedia.org/wiki/Wind_turbine_10-24-21
[85] https://www.energy.gov/eere/wind/downloads/2018-wind-technologies-market-report.pdf
[86] https://www.energy.gov/eere/wind/downloads/2018-wind-technologies-market-report.pdf

CHAPTER

4

WIND ENERGY RESEARCH IN THE NETHERLANDS

Peter Eecen and K. R. Rao[1]

4.1 INTRODUCTION

Interest in the application of modern wind energy grew in the Netherlands in the 1970s when the limit of fossil fuels became clear. Wind energy has been an important source of energy in The Netherlands for centuries and the country was known for the many wooden wind mills especially in the coastal regions. At the same time, research activities in the modern wind energy were started, which have led to the relatively large wind energy research community in the Netherlands today.

Wind energy research activities in the Netherlands are predominantly performed at the Energy research Centre of the Netherlands (ECN) and the Delft University of Technology (DUT). Both institutes are involved in wind energy research since the start of the modern wind turbines, the 1970s. These institutes match their research programs with each other as close as possible. ECN Wind Energy has a research staff of 55 scientists and DUT has a research staff of 15 permanent researchers and more than 35 PhD students. Another institute dedicated to wind energy research is the foundation Knowledge Centre WMC that has been founded by the DUT and ECN in 2003 with an additional research staff of 25 scientists. Although the major part of the wind energy research is concentrated in these institutes, many other universities and scientific institutes contribute to the research with dedicated and specialized research.

The current wind energy research and associated industrial activities are taking place in an international context, mostly the European context, therefore the research activities not only take account of the long-term energy research program of the Dutch government, such as the long-term energy research program EOS [1], but also of the R&D priorities defined in the international context, such as the Strategic Research Agenda (SRA) of the wind energy sector [2]. The three wind energy research organizations are well represented in international bodies such as European Wind Energy Association (EWEA), European Academy of Wind Energy (EAWE), International Energy Agency (IEA), International Electrotechnical Commission (IEC), International Network for Harmonised and Recognised Measurements in Wind Energy (MEASNET), European Wind Energy Technology Platform (TPWind) and the European Energy Research Alliance (EERA).

4.2 WIND ENERGY IN THE NETHERLANDS

In the last 20 years, the amount of wind energy has grown at a rate of approximately 30% per year. In Europe, at the end of 2008 a total amount of 64.5 GW wind energy has been installed onshore and 1.5 GW offshore. In the Netherlands, the installed wind energy capacity amounts 1921 MW onshore and 228 MW offshore. During 2009, the Netherlands decommissioned 106 turbines (total capacity 34.8 MW) and installed new turbines totaling 101.4 MW to have a total of 2216 MW installed wind power which generated 4% of the national electricity demand. Wind energy has an important role in the extension and replacement of electricity production capacity.

The official target for the Netherlands for the implementation of renewable energy, set in a European context, is 14% in 2020. However, the Netherlands has the ambition to reach the share of renewable energy of 20% in 2020. In order to reach that ambition, the share of renewable electricity generation in the total electricity production should reach 35%. Based on this ambition, the Netherlands has set the target for installed wind energy capacity in 2020 at 4000 MW onshore and 6000 MW offshore.

Currently, the cost of electricity production using onshore wind turbines reaches the cost of fossil Electricity production (6-8 ct/kWh). Offshore, the cost of wind energy electricity production is strongly depending on the complexity of the wind farm, such as distance to the coast, water depth, soil, etc. The cost offshore is in the order of 12 to 18 ct/kWh. However, it is expected that the relatively young offshore wind technology will experience a

[1] Peter Eecen was the original author of this chapter that appeared as Chapter 9 in the "Energy and Power Generation Handbook - Established and Emerging Technologies", published in ASME in 2011. This chapter has been updated by Editor K. R. Rao in Section 4.10. K.R. Rao has provided relevant write-ups from on-line sources in the public domain covering "Wind Energy Research in The Netherlands since 2009 to date".

strong learning curve, leading to significant cost reductions. This is one of the reasons that the Dutch government balances its funding between supporting the implementation and funding further innovation by supporting wind energy research with a focus on the further development of offshore wind energy.

Geographically, the Netherlands has a central position with regards to the North Sea. In the next decades, offshore wind energy will be a significant part of the total wind energy capacity in the countries surrounding the North Sea. The conditions are relatively favorable: the average wind speed is high, the water depths are moderate (20 to 50 m), and there are many harbors to access the farms. In order to realize the ambitious targets, substantial R&D effort is required. The wind energy research program in the Netherlands has therefore been focused on offshore applications since the beginning of the century. Not only because most capacity must be installed offshore, but in addition, the offshore application has the highest challenge in knowledge, technology, reliability, installation and maintenance that is also applicable to onshore applications. The knowledge institutes in the Netherlands active in the field of wind energy are collaborating intensively. The long-term collaboration between the Energy research Centre of the Netherlands ECN and Delft University of Technology DUWIND has led to among others the common foundation Wind turbine Materials and Constructions WMC where blade tests are performed.

In the 1990s, the Netherlands had several wind turbine manufacturers, among which Lagerwey, Nedwind en Windmaster. Although these manufacturers did not develop to large global players, the Netherlands still has significant knowledge in the field of development and manufacturing of wind turbines. Currently, in the Netherlands wind turbines are being developed by for instance XEMC Darwind, 2B-energy, VWEC, Lagerwey Wind, STXHarakosan and EWT. In addition, the Netherlands has a strongly developed industrial sector in the field of offshore technology, which is heavily involved in the installation of offshore wind energy. The activities range from offshore engineering to the development of installation and maintenance vessels and foundations. Large Dutch players are among others Ballast Nedam, Mamoet van Oord, Heerema, IHC Merwede, Fugro, and MSC Gusto. Because of the strong Dutch knowledge position in offshore wind energy by the collaborative applied research activities, the Dutch industry is involved in the Construction of offshore wind farms in The Netherlands, Denmark, Germany and the United Kingdom.

4.3 HISTORIC VIEW TO 1990

The first proposals for producing electricity from wind in The Netherlands date from the 1920s. After some experiments in the 1930s, there were more experiments until the 1970s without too much success. Especially in the 1960s, the fossil fuels were cheap and the general idea was that nuclear energy would provide cheap energy for a long time. The Energy research Centre of the Netherlands in that time was called Reactor Centre of the Netherlands (RCN). In 1975, a national committee reported that wind energy is the preferred option for large scale renewable energy, and remarkably, these large numbers would not be installed only onshore, and specifically offshore wind energy in the North Sea should be investigated. A committee guided by RCN had the assignment to define the first Dutch national research program on wind energy.

The first funding of Dutch wind energy research was organized by the National Wind Energy Research Program (NOW) phase in the period 1976 to 1985 which was followed by the 1986 to 1990 NOW phase to stimulate technological, economical and environmental research related to wind energy. The intention of these programs was to give The Netherlands a leading role in the development of wind energy given the rich wind history. And indeed, these programs facilitated the development of wind energy as a realistic energy option for the next decades. The 1986 to 1990 Integral Wind Energy Program (IWP) was intended to initiate the large-scale application of wind energy in the Netherlands and to provide an incentive for the further development of cost-effective wind turbines. In 1990, these programs led to a large involvement of Dutch industry in wind energy research and contributed to the accumulation of wind technology know-how in the Netherlands. Examples are the know-how developed in the field of noise emission reduction, material expertise and fatigue properties that led to significant improvements of wind turbines. However, while The Netherlands significantly invested in wind energy research, the development of large scale wind energy was not realized. The national realization of wind turbines always was far below the targets and also the national industry was largely overtaken by international competition.

In the first phase of the NOW research program, the comparison was made between horizontal and vertical axis wind turbines. Technical options were investigated and no choice could be made then. Both options were considered for the exploratory phase and experimental turbines of both types were realized. Although there were no solid arguments to not further develop the vertical axis wind turbine and given that the horizontal axis turbine was the dominant design in the international market, a decision was made to stop the further development of vertical axis wind turbines. In the mean time Polymarin built a floating 15 m Darrieus turbine in the Gaasperplas in Amsterdam in the early 1980s.

From 1977 to1978, a larger turbine with horizontal axis was designed, the HAT25. The name indicated the rotor diameter of 25 m of this turbine. ECN, the National Aerospace Laboratory NLR and Eindhoven university of Technology were responsible for parts of the research program. The 25-m HAT (horizontal axis turbine) experimental wind turbine was erected at the site of ECN and the turbine had the typical half concrete, half steel tower. It has been operated by ECN since August 1981 and extensive measurement campaigns have been performed at this turbine. The experiments have been used for validation of codes, testing of control strategies, analyses of load cases, etc. Alternative rotors have been installed, such as the FLEXHAT rotor and many more. Some prominent measurements on the 25-m HAT turbine are the axial force measurements to verify the computer program PHATAS, which calculates static and dynamic loads in the large wind turbine components and dynamic loads in the large turbine components. These were used to determine excitation and response frequency spectrums for different modes of operation and to determine the fatigue characteristics of the machines by means of the rain flow counting method.

In 1985 a prototype of the advanced 1 MW NEWECS-45 was built and in 1986 a start was made with the experimental wind farm Sexbierum, near the city of Sexbierum in the Netherlands. This wind farm consisted of 18 turbines with rotor diameter of 18 m rated at 300 kW totaling 5.4 MW. Seven towers were installed for meteorological measurements. Especially the wake measurements in this wind farm were unique for that time. Also the effect of the wind farm on the environment, and especially birds, was part of the research, where it was concluded that wind farms do not pose an excessive threat to birds.

In the Netherlands, the research activities on wind energy always did have a large international focus and was based on international cooperation. Already in the IEA R&D WECS [3] program that was initiated in 1977, the Netherlands participated with the objective to perform cooperative research, development and demonstration, and exchange of information in the field of wind energy utilization. From the early 1970s, the Netherlands, just as many other countries were exploring the use of wind energy. After the oil crisis, an urge to collaborate between R&D institutes in wind energy led to the start of the International Energy Agency Implementing Agreement for Co-operation in the Research and Development of Wind Turbine Systems [4]. A manual for structural safety analysis of wind turbines was prepared and technical feasibility studies were carried out on multi-unit offshore wind farms. Already in 1981, it was concluded that it was technical feasible to install a large, multi-unit offshore wind farm with 50 to 200 turbines on individual support structures, with rotor diameters between 50 and 110 m in water depths of 10 to 45 m. The study group on recommended practices completed a study of power performance testing and studied recommended practices for wind turbine costing, evaluation of fatigue, acoustics, safety and reliability, quality of power, etc. As early as 1980, a new task was initiated by ECN on wake effects of wind turbines. The objective to better estimate the performance of arrays of wind turbines led to the theoretical and experimental modeling of turbine clusters, among others experiments in the Dutch TNO wind tunnels [5, 6]. The measured data still are used for analyses and validations today. In addition, generated turbulence levels and associated dynamic effects were part of the studies. The Dutch TNO wake interaction model MILLY was tested against wind tunnel experiments. In 1983 the Netherlands joined the IEA agreement for cooperation in the development of Large-Scale Wind Energy Systems [3] with ECN as the contracting party. One of the activities was to prepare the specifications for the 3 MW 80 m GROHAT turbine with an industrial consortium.

4.4 HISTORIC VIEW 1990 TO 2000

The wind energy research activity in The Netherlands in the period from 1990 to 2000 period was concentrated at ECN Wind Energy and the Delft University of Technology. Additional activities were carried out by the Dutch organization for Applied Scientific Research TNO and research and development by the industry. ECN focused on aerodynamics, durability and fatigue, electrical conversion and regulation, criteria for certification and standards, and offshore applications. The primary focus of the ECN program is the improvement of the institute's research potential in order to be able to react adequately to possible assignments from industry and therefore the program is oriented towards the development of measuring methods, simulation models, test procedures, etc. The development of measuring methods is facilitated by the availability of the HAT25 turbine at the Petten site. In the same period, the Delft University of Technology plays an important role in research and researcher training, with predominant focus on fundamental aspects, such as aerodynamics and loads. The 1991 to 1995 TWIN program administered by NOVEM was followed by the 1996 to 2000 program. The main focus in these programs are further research in rotor aerodynamics, especially dynamic and 3D effects and development of engineering rules; design tools for rotor development, o.a. buckling and optimization through cost functions; inventory of extreme wind conditions; reduction of emission noise, o.a. through design and field testing of rotor blades with serrated edges and empirical research in which serrated edges and tip shapes are tested in wind tunnels; standards and certification; and design and construction of light weight turbine concepts such as the FlexHat research [7]. The FlexHat program aimed at diminishing internal peak loads, by employing the dynamics of the rotor instead of withstanding them. Flexibility is introduced in the four main degrees of freedom of the rotor, resulting in a variable speed conversion system with soft characteristics, rotor control by means of passive activation of the tip in combination with variable speed, an elastomeric teeter with elastomeric teeter limiters (bumpers), and a flex beam which is moderately flexible in flap wise direction and stiff in torsional direction. The components were tested on the HAT25 research turbine, located at ECN Petten.

In 1997, NOVEM carried out a feasibility study of a demonstration project of a near shore wind farm. The wind farm is meant to gain experience and knowledge of offshore installation, construction and operations. The idea to have a demonstration offshore wind farm was steadily further developed and the wind farm became operational in 2007. The realization of the farm included an extensive monitoring and evaluation program. The farm is highlighted further in this Chapter.

Around the same time, engineering design codes came available, like the wind field generator for aeroelastic codes SWIFT, an aerodynamic correction method for 3D effects TIDIS, an optimization tool for blade design BLADOPT, and a tool for calculations on noise emissions from blades SILANT. In 1999 the DUT developed the method NewGust to quickly generate extreme turbulence gusts for load calculations. These codes have been used by industry and most of these codes are still in use today.

The objective of the project 3D effects in stall was to improve methods that are used in order to predict the effect of rotation on sectional aerodynamic coefficients, particularly the lift coefficients. The project was a cooperation between ECN, DUT and the National Aerospace Laboratory NLR. In the first stage, following the ideas of Herman Snel, boundary layer equations in a rotating frame of reference were developed by NLR and subsequently put in integral form. In parallel, the aerodynamic analysis program XFOIL was purchased and improved by ECN and DUT resulting in a more accurate prediction of lift coefficients near stall. In the second stage, the integral form of boundary layers in the rotating frame of reference was implemented in the improved XFOIL by NLR. The resulting program called RFOIL was validated using the experimental aerodynamic field data from ECN's HAT25 experimental wind turbine. In addition, power curve predictions were validated by experimental data. In the third stage, an engineering method was developed by ECN, allowing the effect of rotation on the lift coefficient to be obtained without having to perform a calculation with RFOIL. The main result of the project was the code RFOIL, which is more accurate than its predecessor in rotating cases like wind turbines. The RFOIL code is to date still freely distributed by ECN and is in use with most wind turbine manufacturers and wind energy researchers.

ECN coordinated the JOULE II project Dynamic Inflow [8], research on the validation of 3D effects in stall. These effects describe the wake induced unsteadiness and non-uniformity of the flow in the rotor plane. Aim of the project was the definition and implementation of engineering models within computer codes for dynamic load calculations. These models have been qualified by means of comparison with both existing sophisticated models (dynamic vortex wake calculations) and experiments on turbines with a large range of rotor diameters. Important dynamic inflow effects were found at fast pitching transients and at yawed flow conditions. The effects were predicted well with the newly developed engineering methods. The projects provided many insights in the behavior of the induced velocities in the rotor plane under several conditions. Measurements and calculations showed important effects of dynamic inflow on the mechanical loads at fast pitching transients and at yawed conditions. At fast pitching transients large overshoots in the loads were apparent. At yawed conditions, the influence of the skewed wake on the induced velocities effected the phase and amplitude of the azimuthally binned averaged flat wise moments. For partial span pitch conditions and for wind gusts, the dynamic inflow effects were much less. Direct evidence of dynamic inflow was found in the wake flow measurements in the wind tunnel at pitching transients and yawed conditions. An interesting result was that under yawed conditions, the skewed wake does not only effect the axial induction (which was expected from helicopter aerodynamics), but also the in plane velocities. The dynamic inflow effects were predicted well with the engineering methods, developed in the project.

Delft University of Technology coordinated the Joule project "Structural and Economic Optimization of Bottom Mounted Offshore Wind Energy Converters" (Opti-OWECS) with the aim to reduce cost of electricity by extending the technology and demonstrating practical solutions for offshore wind turbines. An innovative integral design methodology was developed, the OWECS design approach, considering all components of an offshore wind farm. The design solution is then considering criteria like levelized production costs, adaptation to local site conditions, dynamics of the system, installation effort as well as availability of the turbines. This also required a novel offshore wind farm cost model which was developed at DUT.

During this period, the Dutch wind energy research was strongly involved in international cooperation, like numerous European projects and participation in IEA Annexes. The Dutch researchers have been heavily involved in the development of design guidelines and IEC International Standards.

For instance, TASK XVIII — Enhanced Field Rotor Aerodynamics Database has been operated by ECN [9]. In 1998, the ExCo approved Task XVIII to extend the database developed in Task XIV and to disseminate the results. The objective of the ANNEX-XIV was the coordination of full-scale aerodynamic test programs on wind turbines, in order to acquire the maximum of experimental data at minimum costs. In these full-scale aerodynamic test programs local aerodynamic quantities (forces, inflow velocities, inflow angles) are measured at several radial positions along the blade. The supply of local aerodynamic data, is a major step forward in understanding the very complicated aerodynamic behavior of a wind turbine. In conventional test programs only blade (or rotor) quantities are measured. Usually these quantities are integrated over the rotor blade and they are not only influenced by aerodynamic effects, but also by mass effects. Then the local aerodynamic properties of the blade can only be derived indirectly, introducing an uncertainty. Since aerodynamic field experiments are typically very time consuming, expensive and complicated because of the large volumes of data and the extensive data reduction it is very advantageous to cooperate. And because specific turbine configurations that are investigated experimentally may exhibit a very different aerodynamic response characteristics, the combination of measurement data on very different facilities will provide much more insight about the general validity of aerodynamic phenomena.

In 1999 a new Dutch wind energy research strategy was formulated by the government, and with input from Dutch wind turbine and blade manufacturers, engineering firms and end users such as utilities, insurance companies and certifying institutes the following priority subjects were set:

1. New developments: offshore, innovative materials and recycling.
2. Testing and measuring: condition monitoring systems, wind turbine test facilities.
3. Databases: failure statistics of wind turbines and components.
4. Design tools: reliability, wind turbine control and aerodynamics.

The research programs at ECN and DUT were shaped after these priorities and the research projects were defined accordingly.

In 1999 several projects were started under the Ecology, Economy and Technology funding scheme: One of the projects researches the possibilities of producing large blades from ecologically friendly materials. The other is the DOWEC project.

The Dutch Offshore Wind Energy Converter project (DOWEC) included NEG Micon, LM Glassfiber, Van Oord ACZ, DUT and ECN and was started as a concept study for offshore wind turbines [10]. The goal of the DOWEC Concept Study was to make an inventory of all wind turbine concepts in order to select the most optimal concept for a 5 to 6 MW offshore wind turbine. In the first phase the DOWEC concept study aimed at the choice of the optimal wind turbine concept. These turbines should be able to withstand the severe wind and wave conditions at the Dutch North Sea. In the design process, the wind turbine was not treated as an isolated system, but the designs of different wind turbine concepts were evaluated as an integral part of a complete large-scale offshore wind farm. All significant properties like the structural loads, the power performance, the system reliability, the costs of the electric infrastructure, maintenance costs and installation costs were determined for the optimized designs. The concept study resulted in five feasible concepts for far offshore wind turbines of 5 to 6 MW. These concepts were quantitative ranked based on the cost of generated energy. Furthermore, qualitative criteria like development risk and market potential were taken into consideration when finalizing the choice of concept. From these concepts, one was selected by industry for further development. The industrial tasks resulted in the engineering, purchase, and construction of a 2.75 MW wind turbine. The turbine has been erected and tested at the ECN Wind turbine Test site Wieringermeer. The final result of the DOWEC project should have been the realization of a 6 MW offshore wind turbine which is optimized for exploitation in large wind farms in the North Sea addressing all design and operational aspects in a cost-effective way. The realization of this turbine never happened, but the design of the 6 MW turbine was delivered in 2003. The design is publicly available and the information has been used extensively. For instance the 5 MW reference turbine defined by NREL is largely derived from it and many aspects of the DOWEC 6 MW design can be found in modern offshore wind turbines installed today (Fig. 4.1).

At the end of the decade, ECN developed and patented a novel technique to visualize the stall behavior of rotor blades. The power performance of stall regulated turbines in some cases varied considerably and to quantify the cause, a measurement technique was required, which could be applied on large commercial wind turbines, was fast enough to monitor the dynamic changes of the stall pattern, and was not influenced by the centrifugal force and most importantly did not disturb the flow. The technique developed and patented at ECN was called Stall Flag Method [11] and was composed by small reflecting sheets covering with a hinged non-reflecting sheet. When glued to the blade the hinged non-reflecting blade will open upon stall and the reflecting part becomes visible. During night time, the reflecting parts can be recorded by camera if the turbine is lit by a light source. With this technique problems can be searched for, optimum locations for stall strips and vortex generators can be found and insight can be obtained into the problem of multiple stall. The statistical stall behavior is characterized from the sequences of thousands of analyzed images. This leads to the visualization of the stall pattern, the blade azimuth angles and the rotor speed. It also measures the yaw error and the wind speed from the optical signals of other sensors, which are recorded simultaneously. For example, the delay in the stall angle by vortex generators can be measured with an accuracy of one degree from the stall flag signals. Many experiments have confirmed the independence of stall flags from the centrifugal force and that stall flags respond quickly to changes in the flow.

FIG. 4.1: THE NM90 2.75 MW WIND TURBINE (ALSO KNOWN AS DOWEC TURBINE) AT THE ECN TEST SITE IN THE WIERINGERMEER AS SEEN FROM THE MEASUREMENT MAST (109 M) (*SOURCE:* [ECN, TAKEN BY R. NIJDAM])

4.5 HISTORIC VIEW 2000 TO 2010

In 2000, almost 40 MW of wind capacity was installed in the Netherlands. This is a continuation of the slow but steady installation rate of 40 to 50 MW in the last 4 years. At the time, the national target was 2750 MW installed wind power in 2020 and therefore it was necessary to go offshore. In 2000, preparations for the necessary step to offshore were taken at the technical and administrative level. The research in wind energy became focused on offshore application and a demonstration offshore wind farm was tendered. The consortium which was called Noordzeewind and consisted of Shell Renewables, energy company Nuon International, bank ING Bank, engineering firm Jacobs Compimo, and NUON owned project developer WEOM won the tender for the Near Shore Wind farm. According to the requirements in the environmental effect report for the Near Shore Wind farm, a monitoring and evaluation program had to be carried out. The outlines of this program, which cover environmental, economical and technical aspects were drafted early 2000. According to these outlines, this program started with an evaluation of the undisturbed situation. The entire program aimed at learning as much as possible from the near shore demonstration wind farm in order to better realize the future larger offshore wind farms that are necessary to reach the national targets.

As a follow-up to the JOULE project "Database on Wind Characteristics," Sweden, Norway, United States, The Netherlands, Japan, and Denmark formulated IEA Task XVII. There was an urgency to initiate an international effort to collect, describe and store high-quality wind data measured by several wind energy related projects. The unique database on wind characteristics consists of quality controlled and documented wind field time series supplemented with tools that enable easy access and a simple analysis through an internet connection. The statistics database contains a list of all time series, the derived statistics together with other parameters. The raw time series (74.6 GB) consist of files of 10-minute periods in a common format. The accessibility of the database has been developed during the operational years of the database and is quite easy to use and it is comprehensive. From numerous areas over the world, controlled and high quality data have been added to the database. Important data consists of wake measurements, three component offshore measurements, extreme complex terrain measurements and long-term data sets covering time spans of at least 20 years. Reports [12, 13] are available on description of data that have been added by ECN to the international wind database [14].

An interesting spin off of the research at DUT is the development of the Ampelmann [15]: a motion compensated platform to safely access offshore wind turbines. The initial feasibility study, preliminary design, and proof of concept of an offshore access system named Ampelmann was supported by SenterNovem. The Ampelmann is a ship-based self stabilizing platform that actively compensates all vessel motions to make offshore access safe, easy and fast. The base of the Ampelmann is mounted on the deck of the vessel or barge and takes care of the motion compensation, keeping the top side of the Ampelmann stationary. The gangway, which is mounted on the top side, can now be easily deployed onto the offshore structure enabling safe offshore crew transfer. To achieve this, the vessel is equipped with a set of motion sensors and a Stewart platform. The motions of the ship are continuously registered in six degrees of freedom by the motion sensors on the deck of the vessel and instantaneously fed into a control system in order to keep the top plate of the Stewart platform motionless compared to the fixed world. On this top plate, a transfer deck is installed. By extending a gangway between the transfer deck and the offshore structure, the structure can be accessed in an easy and safe way, even in high waves. The Demonstrator project was sponsored by We@Sea, Delft University of Technology, Shell, Smit, Heerema fabrication, SMST, and Ecofys with the common goal to further develop and test its design, compensation performance, operational procedures and safety systems in the harsh offshore conditions of the North Sea. Currently, various Ampelmann systems are in use for daily delivery of personnel to fixed platforms in the North Sea.

In 2005, ECN completed the wind resource atlas of the Dutch part of the North Sea for several heights. ECN developed the atlas by combining data from two sources: the numerical weather prediction model Hirlam and the meteorological stations at the North Sea, thus combining a source with dense spatial sampling and a source with accurate (measured) values. At the ECN website, maps with the mean wind speed at 60, 90, 120, and 150 m above sea level are published. ECN can for every location at the North Sea and for each height produce a time series of 10 years based on the Hirlam data which is validated by using the measurement stations. The wind resource in such a location is expressed in terms of the wind speeds and wind directions, the turbulence intensities, and the stability classes (Fig. 4.2).

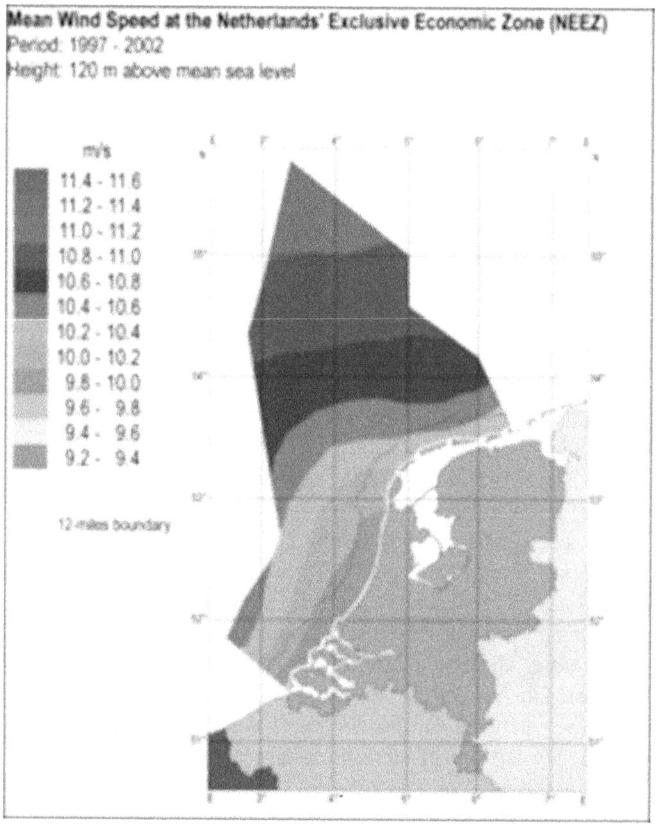

FIG. 4.2: MEAN WIND SPEED AT THE DUTCH PART OF THE NORTH SEA AT 120 M IN HEIGHT (*SOURCE*: [ECN, ARNO BRAND])

4.5.1 UpWind

UpWind is a large European project funded under the EU's Sixth Framework Programme (FP6). The project looks towards the wind power of tomorrow, more precisely towards the design of very large wind turbines (8 to 10 MW), both onshore and offshore.

Furthermore, the research also focuses on the requirements to the wind energy technology of 20 MW wind turbines. The challenges inherent to the creation of wind farms of several hundreds MW request the highest possible standards in design, complete understanding of external design conditions, the design of materials with extreme strength to mass ratios and advanced control and measuring systems geared towards the highest degree of reliability, and critically, reduced overall turbine mass. The aim of the project is to develop the accurate, verified tools and component concepts the industry needs to design and manufacture new, cost effective type of turbines. UpWind focuses on design tools for the complete range of turbine components addressing the aerodynamic, aero-elastic, structural and material design of rotors. The UpWind consortium, composed of 40 partners, brings together the most advanced European specialists of the wind industry. In most work packages, researchers from the Netherlands are involved and the following work packages have coordinators from the Netherlands: Metrology, Upscaling and Rotor structure and materials.

4.5.2 Smart Rotors

For many years DUT is carrying out research work on smart rotors in the context of the Dutch INNWIND and the EU UPWIND projects. The objective of smart structures for rotor blades is to alleviate significant blade loads by applying spanwise-distributed load control devices without incurring lower reliability or higher maintenance. DUT concentrates on the investigation of concepts, feasibility, and integrated design; aerodynamics; structural integration; control, identification, and experimental investigation of developed models. DUT proved the concept in their wind tunnel. Researchers observed load reductions between 70% and 90% and concluded a significant amount of fatigue blade load alleviation is possible. This was confirmed by research at ECN that is concentrating on the application of synthetic jets for this purpose. DUT has carried out experiments in the Open Jet Facility with a rotating 2 m diameter wind tunnel model.

4.5.3 WE@SEA

The 4-year WE@SEA program on offshore wind energy is funded from the Dutch national natural gas fund. This program is run by a consortium consists of companies in offshore technology, wind energy technology, offshore wind farm development, logistics, investors, energy consultants, environment, and other stakeholders and was operated from 2004 to 2008. The program of 26 million concentrated on medium-term research in the research lines: offshore wind power generation; spatial planning and environmental aspects; energy transport and distribution; energy market and finance; installation, operation and maintenance, and dismantling; education, training, and knowledge dissemination; and the PhD@ Sea project at DUT. The PhD@Sea research subjects were divided over the research lines of the overall program. Its topics are: large blades; wind turbine concepts; morphology of the North Sea bed; grid stability of large-scale integration of wind energy in electrical power systems; park-grid interaction; offshore access through the Ampelmann; reliability, availability, maintainability, and serviceability analyses and scenarios.

4.5.4 Controlling Wind and Heat & Flux

Over the years 1999–2003, ECN invented and patented two wind farm control techniques, the one called Heat & Flux, the other Controlling Wind. The concept Heat & Flux aims at maximizing the power output of a wind farm by adjusting the axial induction of the windward turbines below their individual optimum for power production, which means making them more transparent for the wind than usual by realizing an axial induction factor below the Lanchester–Betz optimum of 1/3. This will reduce the velocity deficit in the wake and increase the output of the downwind turbines. Other benefits are decreased average loading of the upwind turbines and decreased fatigue loading of the turbines in the wake. ECN quantified the effects of "Heat and Flux" farm control. For this purpose models have been developed for calculating the power and energy production with "Heat and Flux" operation. The models have been developed in close interaction with wind tunnel experiments on model turbines and farms and have been tested in full-scale experiments on a row of five 2.5-MW turbines. Qualitatively, the wind tunnel tests and the field experiments revealed positive effects of simple "Heat and Flux" control settings on several occasions. Quantitatively, the accuracy and reliability of the wind tunnel measurements are questionable because of the large scatter and the conditions that deviate from full scale in various aspects. The full-scale experiments point at comparable optima for "Heat and Flux" control settings as model predictions of the model developed by ECN. This method has been implemented in ECN's Control Tool for designing wind turbine control algorithms.

FIG. 4.3: THE MEXICO WIND TUNNEL EXPERIMENT IN THE DNW WIND TUNNEL (*SOURCE*: [ECN PHOTO TOON WESTRA])

4.5.5 Wind Tunnel Experiment Mexico and Mexnext

In the past, the accuracy of wind turbine design models has been assessed in several validation projects [16]. They all showed that the modeling of a wind turbine response which can be either the power or the loads is subject to large uncertainties. These uncertainties mainly find their origin in the aerodynamic modeling where several phenomena such as 3-D geometric and rotational effects, instationary effects, yaw effects, stall, and tower effects, amongst others, contribute to unknown responses, particularly at offdesign conditions. The availability of high quality measurements is considered to be the most important pre-requisite to gain insight into these uncertainties and to validate and improve aerodynamic wind turbine models. For this reason, the European Union project "MEXICO: Measurements and EXperiments In COntrolled conditions has been carried out [17]. The project was coordinated by ECN and was carried out with ten institutes from six countries. They co-operated in doing experiments on an instrumented, 3-bladed, 4.5-m diameter wind turbine placed in the 9.5 m^2 open section of the Large Low-speed Facility (LLF) of German Dutch Wind Tunnel (DNW) in the Netherlands (see Fig. 4.3). The LLF is described later in this Chapter. The measurements were performed in December 2006 and resulted in a database of combined blade pressure distributions, loads, and flow field measurements that can be used for aerodynamic model validation and improvement.

A similar previous experiment was performed by the National Renewable Energy Laboratory (NREL) in the National Aeronautics and Space Administration (NASA) Ames wind tunnel [18] on a 10-m diameter turbine. An obvious difference between the two types of experiments lies in the larger size of turbine diameter the latter experiment. But in addition to the NASA-Ames experiment, the Mexico experiment also included flow field measurements. Especially the flow field of inflow and wake are interesting, where the flow field is important for the understanding of discrepancies between calculated and measured blade loads. This is mainly because the load calculations are done in two steps. First, the flow field around the blade, the induction, is calculated, and second from that the loads are derived. Where in conventional experimental programs, only blade loads are measured, it is not possible to distinguish between these two sources of discrepancies. The addition of flow field measurements should open up this possibility.

At the end of the MEXICO project, due to budget reductions, the database was still in a rather rudimentary form and only limited analyses had been carried out. Especially since it is a huge amount of data which requires a lot of effort to analyze. Therefore it is beneficial to organize the analysis of the MEXICO data in a joint project under IEA Wind in which various countries share this task. Given the platform for discussion and interpretation of the results the outcome of the data analysis shall be better than the summed result from individual projects.

ECN is the Operating Agent of the IEA Wind Task 29 called MEXNEX(T) which has the objective to improve aerodynamic modes used for wind turbine design. In this task the access of the

Mexico data is provided and a shared thorough analysis of the data takes place. This includes an assessment of the measurement uncertainties and a validation of different categories of aerodynamic models. The insights are compared with the insights that were gained within IEA Wind Task 20 on the NASA-Ames experiment and other wind tunnel experiments.

Special attention is paid to yawed flow, instationary aerodynamics, 3-D effects, tip effects, non-uniformity of the flow between the blades, near wake aerodynamics, turbulent wake, standstill, tunnel effects, etc. These effects are be analyzed by means of different categories of models like CFD, free wake methods, engineering methods, etc. As such, the Task will provide insight on the accuracy of different types of models and the following activities are performed.

1. Processing and presentation of the measured data and the associated uncertainties. High quality measurement data are provided to the partners to facilitate and compare calculations.
2. Analysis of tunnel effects. Since the 4.5-m diameter wind turbine model was placed in the open jet section of the LLF facility (9.5 m x 9.5 m) the ratio of turbine diameter over tunnel size may make the wind tunnel situation not fully representative of the free stream situation. The tunnel effects have been studied with advanced CFD models. Supporting information on tunnel effects will also be obtained from eight additional pressure measurements, which were measured with taps in the collector entrance. These pressures measure the speedup in the outer flow (outside the wake) needed for the mass conservation of the tunnel flow.
3. Comparison of calculational results from different types of codes of the various partners with the MEXICO measurement data. This is proving to be a thorough validation of different codes and it provides insights into the phenomena that need further investigation.
4. Deeper investigation into the observed phenomena. A deeper investigation of different phenomena using isolated sub-models, simple analytical tools, or by physical rules. These phenomena that are investigated include 3-D effects, instationary effects, yawed flow, non-uniformity of the flow between the blades (i.e., tip corrections), and the wake flow at different conditions.
5. Comparison with results from other (mainly NASA-Ames) measurements. It is investigated whether these findings are consistent with results from other aerodynamic experiments, particularly the data provided within IEA Wind Task 20 by NREL from the NASA-Ames experiment.

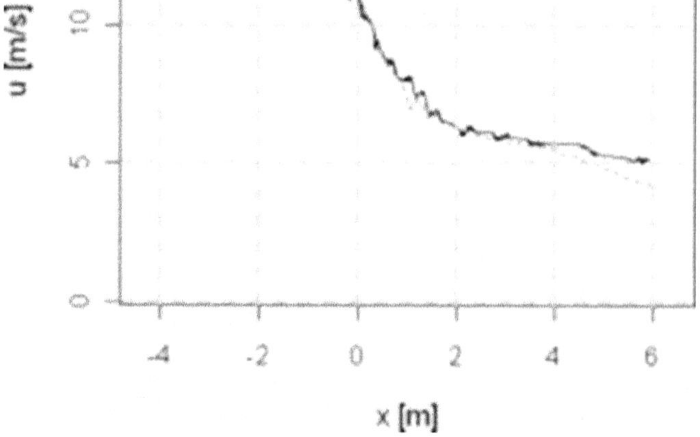

FIG. 4.4: MEXNEXT RESULT: AXIAL TRAVERSE FROM PIV MEASUREMENTS COMPARED TO AWSM PREDICTIONS (*SOURCE:* [ECN])

A large variety of results have been obtained, the data are analyzed on quality and tunnel effects. PIV measurements in the wake have been analyzed, and a comparison has been made between the observations from the NASA-Ames and MEXICO experiments. The effects from airfoil imperfections have been estimated which is important for the comparison with models. A typical example of the comparison between calculated and measured results are shown in Fig.4.4 where the results are compared to predictions by the ECN code AWSM, a numerical code based on the Generalized Prandtl's Lifting Vortex Line Method.

4.5.6 DU Airfoil Data

In the last 15 years DUT has developed many airfoils for wind turbine application [19]. These airfoils are called the DU airfoils and are applied world-wide in wind turbine blades ranging from 6 to 60 m. The development of the DU airfoils has been supported by the European Commission, the Dutch Ministry of Economic affairs and the DUT in the framework of the educational and research program of the section Wind Energy of the Faculty of Civil Engineering and Geosciences, which is at present the Wind Energy section of the Faculty of Aerospace Engineering. The twodimensional aerodynamic characteristics of the DU airfoils have been extensively verified in the Delft University low-speed lowturbulence wind tunnel for various Reynolds numbers. In addition to the clean configuration for Reynolds numbers ranging from 1 million to 3 millions, the characteristics of several airfoils with vortex generators, zigzag tape, trip wires, trailing edge wedges or Gurney flaps of 1% and 2% chord heights have been established. In some cases a stethoscope was used to determine the transition location. The DU airfoils are included in the ATG software package of ECN.

4.6 RESEARCH PROGRAMS

The research programs described here concern the programs of ECN Wind Energy, Knowledge Centre WMC and the Wind Energy group at DUT, DUWind.

4.6.1 Energy Research Centre of the Netherlands ECN and Knowledge Centre WMC

The Energy Research Centre of the Netherlands is dedicated to the research in efficient use of energy and infrastructure, deployment of renewable energy sources, clean conversion of fossil fuels and development of energy analyses and policies. While targets for the near and medium term have been set in Europe, substantial further acceleration of new technologies is required to meet the long-term goals of reducing the dependence on fossil fuels and the CO_2 emissions. ECN aims to carry out groundbreaking research that will have a major influence on energy transition and brings technologies to every stage of development. The strength of ECN lies in its portfolio, which enables the development of new generations of technologies. Many technologies developed by ECN have reached maturity in recent years, increasingly resulting in third party economic activity. ECN focuses on the needs of government and industry. In addition, it conducts contract research for companies and governmental institutions and to a large extent for the European Union. ECN collaborates intensively with other knowledge institutes and universities.

ECN Wind Energy has the mission to develop high-quality knowledge and technology for large-scale cost effective application of wind energy and transfer these to the market. ECN Wind Energy has the ambition to contribute substantially to lowering the production costs of offshore wind energy to a level at which it is competitive with fossil-fuel generation in 2020.

The long-term research program of ECN Wind Energy is combined with the long-term research program of Knowledge Centre WMC because of the close co-operation between the two organizations and the fact that the programs are complementary and dependent on each other. ECN Wind Energy and WMC organize their research in four priority areas:

1. Rotor and Farm Aerodynamics
2. Integrated Wind Turbine Design
3. Operation and Maintenance
4. Materials and Structures

The research is supported by extensive experimental facilities and an Experiments & Measurements group that is MEASNET and ISO17025 accredited.

In the field of **Aerodynamic** research, ECN Wind Energy aims at optimizing the aerodynamic performance of the wind turbine rotor and of the wind farm as a whole. By developing new knowledge, methods and design software, the results of the long-term research can directly be transferred to the industry. An additional result of the research is that Intellectual Property is generated. The tools and Intellectual Property are used for research purposes of turbine and blade manufacturers and project developers or are directly applied. A unique feature of its research is the combination of theoretical and numerical research with the various experimental test facilities: wind tunnels, scale wind farm and full-scale test wind farm in the Wieringermeer. The aerodynamic research is divided in rotor aerodynamics and wind farm aerodynamics.

Research in the field of rotor aerodynamics aims at the development of advanced aerodynamic design tools. ECN has a tradition in developing BEM, but also developed the aerodynamic free wake model based on lifting line panel method AWSM. A new code is being developed where also the boundary layer is being modeled. This method where the external pressure field of the rotating panel method that includes the wake is combined with a three dimensional unsteady boundary layer model will lead to an accurate method that is significantly faster than full Navier–Stokes field solvers. It is called ROTORFLOW. In a Dutch consortium ECN has developed the RFOIL 3D code, a modified version of the XFOIL software. The progressively advanced aerodynamic design tools are used to design and model future large wind turbine rotors with increased accuracies.

Research in the field of wind farm aerodynamics aims at the understanding of the flow field in and around wind farms, which is especially relevant for the construction and operation of large offshore wind farms. Insight in wake effects in wind farms is essential. Computational fluid dynamics (CFD) are applied for the further understanding of flow phenomena in wind farms. The research comprises the development of validated wake models and CFD wind farm models using commercial solvers as well as the development of dedicated wind farm aerodynamics CFD solvers. Interesting model challenges are combined, such as the modeling of the atmospheric boundary layer, the flow inside the wind farm, the modeling of the turbine including its control, the wind farm control as well as the optimization of this system.

The FarmFlow program is a validated tool developed by ECN to calculate the wake effects of large offshore wind farms. It is unique that it calculates both wake losses and added turbulence levels with an accuracy currently unmatched. The Farm Flow tool allows the user to accurately optimize the lay-out of the wind farm with respect to power output and, in conjunction with aero-elastic codes, the possibility to perform design calculations on wind turbines that are placed in wind farms (Fig.4.5).

In the field of **Integrated Wind Turbine Design** research, ECN Wind Energy and WMC aim at the improvement of the design process in industry by developing new, improved models, concepts and tools for the integrated system design. ECN focuses on aero-elasticity, control and concept studies. WMC is dedicated to the development of the integrated design tool and the blade design module. The research in the field of integrated design is among others.

Aero-elasticity: rotor aerodynamic models are coupled to structural dynamic models and the resulting aero-elastic models are verified and validated against measurements and used for integrated design. Improvements are needed for simulations concerning oblique inflow, individual pitch, large deformations and complex drive train and support structure geometries. Coupling the improved aerodynamic models to structural dynamic models (multi-body or FEM) improves the results obtained from simulations. To enable a straightforward coupling the ECN develops the Aero-module. This module couples the aerodynamic codes to aeroelastic codes, such as TURBU and PHATAS which have been developed by ECN.

FIG. 4.5: ECN FARMFLOW WIND FARM CALCULATION MODEL (*SOURCE*: [ECN])

Research in wind turbine control aims at the improvement of the present control algorithms, and the development of new ones, mainly targeting reduction of the cost of energy. To this end, the following objectives are pursued:

- optimal power and rotor speed control,
- load reduction in the mechanical components of the wind turbine,
- high availability and reliability of the wind turbine,
- easy and robust commissioning of the controller,
- coping with all grid requirements.

ECN developed the user-friendly ECN Control Design Tool which provides the wind turbine control engineer with an excellent set of tools in an open-source MATLAB environment, covering the complete process of turbine modeling, control design, stability analysis and compilation of the final controller into executable code for linking to advanced simulation software for load calculations or to dedicated hardware. Furthermore, the following four major control components of the integral approach are investigated

- Optimized feedback control (OFC), aiming at removing limitations for up-scaling of wind turbines, such as high turbine loads and stability problems,
- Extreme event control (EEC), dealing with prevention of unnecessary standstill and/or increased loads due to extreme events,
- Fault-tolerant control (FTC), having the task of detection of minor (sensor) faults followed by recovery actions,
- Optimal shutdown control (OSC), for preventing the accumulation of damage in cases of a turbine shut-down caused by a severe failure.

ECN develops design tools for large wind turbines and these tools are constantly improved in order to address aeroelastic stability issues and to minimize cost of energy. The advanced tools are developed to reduce uncertainties in design load calculations, through an improved modeling of the physics involved. ECN upholds the quality and validity of its computer programs through participation in European benchmark studies and through feedback from field experiences. Aeroelastic and hydromechanical behavior of a wind turbine are modeled using sophisticated tools like, e.g., AeroModule (including BEM and AWSM for rotor aerodynamics) and PHATAS and TURBU for dynamic load modeling of the turbine and support structure in time domain and frequency domain respectively. Wind and wave fields for use in the time domain code are generated with the SWIFT and ROWS code respectively. For quick assessments of aeroelastic turbine behavior the code BLADMODE is used and development and testing of custom made commercial control algorithms for wind turbines is done with the Control Design Tool. The BLADOPT: program optimizes rotor blade geometry for lowest cost of energy. WMC has developed the FOCUS code which includes several of these design tools.

Research in the field of Electrical Systems and Components aims at the design and optimization of electrical systems in a wind turbine that is connected to the wind farm grid; statically and dynamically. ECN developed, in collaboration with TUD Electrical Power Processing, two types of models: EeFarm and DynFarm. These programs are used with the developed Farm Flow code that calculates the wake losses in wind farms to optimize offshore wind farm designs.

For more than 15 years ECN Wind Energy is developing methods and tools to optimize the **Operations and Maintenance** (O&M) of offshore wind farms and in fact, ECN is one of the few institutes carrying out long-term R&D on this topic. The scope of work comprises the development of models and software (Decision Support Tools) to analyze the O&M aspects, to determine costs and downtime, identify cost drivers, and to optimize the O&M strategy and the development of diagnostics and measurement techniques for those situations where condition based maintenance is preferred instead of preventive or corrective maintenance. ECN's leading role in the field of O&M for offshore wind farms has been recognized for many years. Major project developers make use of ECN's knowledge and models to develop O&M strategies for their offshore wind farms. Turbine designers apply the knowledge in the design process of offshore wind turbines.

WMC aims with the research on **Materials and Structures** at increasing its knowledge of the structural behavior of the blade and its materials under the complex loading and the environmental conditions and to make this knowledge applicable for wind turbine and blade manufacturers. With this knowledge current designs and the design of the next generation of blades are further optimized.

4.6.2 Delft University of Technology DUT

Research on wind energy at the Delft University of Technology began 30 years ago, starting with the tip vane project, an aerodynamic research project at the faculty of Aerospace Engineering. Nowadays DUWIND is the wind energy research organization of the Delft University of Technology. Since an increasing number of research questions require a multi-disciplinary approach, DUWIND was established in August 1999 as an interfaculty research organization, specifically for wind energy. World-wide, DUWIND now is one of the largest academic research groups in wind energy. Its research program covers almost all aspects of modern wind turbine technology, and is undertaken across 5 faculties in 13 groups. The five faculties are 1. Faculty of Aerospace Engineering, 2 Faculty of Civil Engineering and Geosciences 3.Faculty of Electrical Engineering, Mathematics and Computer Science, 4 Faculty of Mechanical and Materials Engineering and 5. Faculty of Technology, Policy and Management. Each of the research groups at these faculties has its own specific expertise therefore the DUWIND research group covers a wide area of wind energy research.

The research program of DUWIND has been reformulated and is published in October 2008. The program encompasses approximately 60 PhD positions, which implies a doubling in size of DUWIND compared to 2007. From the start the core of the research program is the technology development of wind turbines and wind farms, including the underlying basic research. The program focuses on longterm research efforts, mainly performed by PhD researchers. Long term is to be understood as leading to useful results in a 5to 10-year time frame after starting the PhD research. The program is made up and executed by the research groups at the five TU-Delft faculties, and is "bottom-up": a new topic is included only when it can be embedded in an existing expertise group, and when funding is found. The PhD students will have activities in five sub-programs:

- Unsteady aerodynamics
- Smart structure rotors

- Design methods
- Offshore components and design
- Dutch wind energy in Europe.

The focus of DUWIND program is on the development of turbine and wind farm technology, ranging from basic research through technology development to design support for the industry. Being a university, DUWIND provides courses for students and for professionals in the wind energy industry. As is indicated before, DUWIND closely co-operates with the wind energy group of ECN. Many projects are performed jointly, and mutual use is made of the research facilities. DUWIND and ECN together form the Dutch node of the European Academy of Wind Energy EAWE. DUWIND is part of the international wind energy research community. It is one of the founding members of the European Academy for Wind Energy and has started the Academy conference series 'The Science of making Torque from Wind'. DUWIND contributes to the Wind Energy Technology Platform of the European Commission, the European Energy Research Alliance (EERA) and to the wind energy programs of the International Energy Agency. It has participated in all Framework Programs of the European Commission.

4.7 EXPERIMENTAL RESEARCH INFRASTRUCTURE IN THE NETHERLANDS

4.7.1 Blade Testing and Material Research at WMC

Knowledge Centre WMC is a research institute for materials, components and structures. The major activities are fundamental and applied research on fiber reinforced plastics and wind turbine structures. WMC is active in research projects for the European and Dutch governments and industry and performs blade and material tests on contract basis for the international industry. WMC develops design tools, the FOCUS software, that are being used worldwide by many of the largest wind turbine manufacturers. Results of the research work are published and presented internationally.

The Knowledge Centre WMC has a history from 2000, when it was started as "WMC-Group" in the Delft University of Technology doing quite a lot of full-scale testing work on rotor blades. However, the test facility at Delft University of Technology became too small for the next generation of large rotor blades. Therefore, in order to keep track with the market's needs another, larger location needed to be found to continue the activities. A larger test facility was built in Wieringerwerf, in the northern part of the Netherlands where a new industrial site and yacht-basin "Waterpark Wieringermeer" was developed.

For the new laboratory a location was chosen close to open water, on the borders of the IJsselmeer. Now the sometimes very large structures that are brought for testing can also be transported by water. Starting 2003, as the centre moved from Delft to Wieringerwerf, the former "WMC-Group" of the DUT continued its work as a new foundation with a new name: Knowledge Centre WMC, established by the Delft University of Technology and the Energy research Centre of the Netherlands ECN. With its links to both organizations the new Knowledge Centre WMC can continue to combine fundamental and applied research on wind turbine and fiber reinforced plastics structures.

With the background of continuous testing and development of tests, WMC is actively involved in international standardization committees. For the experimental research WMC has a laboratory for material research and a large test area with a dedicated strong floor enabling testing of structures of over 60 m in length, such as rotor blades for large wind turbines. Testing machines up to 300 tons are available for material and component testing, both static and fatigue. The facility is one of the largest of its kind and has its own mechanical, electro technical and hydraulic workshops for development and maintenance of equipment and test rigs.

The research topics for WMC in national and European research projects are amongst others:

- Material behavior under complex and fatigue loading
- New materials
- Condition monitoring
- Reliability of design codes
- Reliability of full-scale testing
- Connection methods
- Design recommendations and standards on full-scale testing of wind turbine rotor blades

On the other hand, WMC is performing more fundamental research activities in the context of many international projects. WMC initiated the EU project OPTIMAT BLADES to investigate the behavior of Fiber Reinforced Plastics and WMC is work package leader for materials research in the UPWIND project. The projects executed for industrial research purposes include:

- Testing of materials
- Numerical analyses
- Design evaluation
- Verification tests on mechanisms
- Strength and fatigue behavior of components and connections
- Development of design software: the integral wind turbine design software FOCUS
- Full-scale verification tests on (sub)components such as:
  —Wind turbine rotor blades (or parts thereof)
  —Wind turbine pitch bearings and hubs
  —Connections

For larger wind turbines, the potential power yields scales with the square of the rotor diameter, but the blade mass scales to the third power of rotor diameter (square-cube law). With the gravity load induced by the dead weight of the blades, this increase of blade mass can even prevent successful and economical deployment of very large wind turbines. In order to meet this challenge and allow for the next generation of larger wind turbines, higher demands are placed on materials and structures. This requires more thorough knowledge of materials and safety factors, as well as further investigation into new materials with a higher strength to mass ratio. Furthermore, a change in the whole concept of structural safety of the blade is required. WMC is performing research activities in this field in order to improve both the empirical and fundamental understanding of materials, extend the material database, study effective blade details, establish of tolerant design concepts and probabilistic strength analysis and establish material testing procedures and design recommendations.

4.7.2 ECN Wind Turbine Test Site Wieringermeer

At the Petten site, ECN has operated a wind turbine test site for many years. The size of the turbines allowed at the site is limited and therefore it is not in use anymore since the year 2000.

From 2003, ECN has expanded its wind turbine facilities with a wind turbine test site in the municipality of Wieringermeer. The site is located just south of the village Kreileroord, at about 30 km distance from ECN's main offices in Petten (Fig.4.6). This unique facility is a combination of:

Four locations for prototype wind turbines up to 6 MW, recently expanded with a fifth location for prototype wind turbines up to 10 MW.
Five Nordex N80/2500 research wind turbines; these wind turbines are equipped for experimental research.
Three meteorological masts, 108 m, 108 m, and 100 m high equipped with atmospheric measurement equipment.
A scaled wind farm consisting of ten turbines rated 10 kW and 14 meteo masts up to 19 m height.
A measurement pavilion with offices, a computer centre and advanced glass fibre based data acquisition system.

The research wind farm — consisting of five Nordex N80 wind turbines — enables ECN to perform wind farm specific research and development programs. The site also comprises five prototype locations. These locations enable manufacturers to test, optimize and certify prototypes together with ECN. Supporting facilities are three meteo towers, a 36 MVA grid connection, data collection equipment and a test site control center. A fourth meteo mast of 100 m height is expected early 2011. The test site also accommodates a unique facility, which allows ECN to perform accurate wind field measurements in a scale wind farm. This Scaled Wind Farm — consisting of ten Aircon P10 turbines plus fifteen measurement towers — allows for the development and testing of wind farm specific control strategies. The site has a favorable wind climate: the average wind speed at 100 m height is 8.3 m/s. With this wind climate, not only do the five Nordex N80 turbines produce about 30,000,000 kWh per year from which the income of the green electricity gives a solid base for the financial exploitation of the site, the wind conditions allow for fast fulfillment of the experimental capture matrices of the test turbines.

FIG. 4.6: ECN WIND TURBINE TEST SITE PETTEN (*SOURCE*: [ECN, PICTURE BY ARIS HOMAN])

Many research programs have been carried out at the experimental facility of ECN, also many European programs. The fact that the research turbines are owned by ECN and a proper research agreement has been reached with the manufacturer, allows ECN to optimally use these research turbines for its R&D programs. What was obvious from many defined programs is that it is very difficult to execute these programs in commercial wind farms since wind farm operators are reluctant to facilitate research programs that might conflict with maximum energy production.

ECN investigates aspects like operation and maintenance strategies; wake effects, noise effects, effect on birds and condition monitoring. Other important activities are the development of improved wind farm control strategies and new advanced measurement techniques.

Currently, in 2010, plans are made to extend the research wind farm. Some fourteen turbines in the range from 5 MW to 8 MW are to be set up over the coming years, at least seven will be dedicated offshore turbines. ECN intends to purchase seven turbines for further research activities and a further seven proto-type test locations will be made available. A fifth proto-type location has already been created and a XEMC-Darwind turbine is being built and will be tested in 2011.

4.7.3 ECN Scale Wind Farm

A recent addition to the ECN experimental research infrastructure is the ECN scale wind farm. It has been designed to further advance the knowledge in wind fields in and around wind farms, including the understanding of wakes and turbine-turbine interaction. The development of large-scale offshore wind power implies the construction of even larger wind farms. At the moment, the large uncertainties connected to the wind field in the wind farm leads to financial risks when investing in these large wind farms. Furthermore, the cost of wind energy should be reduced even further. Therefore, improved models are required describing the flow within and around wind farms so that optimized wind farm control strategies can be developed. The high-quality data of the ECN scale wind farm are used for the development and validation of wind farm aerodynamic models and wind farm control strategies (Fig.4.7).

The understanding of unsteady wind fields within and around wind farms could be greatly increased when more detailed models or measurements of the wind field in a wind farm would be available. The same applies to understanding the response of many turbines within a wind farm on the mutual wakes. One reason for this is that adequate measurements are lacking and the models are not (yet) accurate enough. In full-scale wind farms meteorological masts are very expensive and the number of masts is thus limited. On the other hand, the value of modeling wind farms in wind tunnels is limited due to scaling effects. Therefore, ECN has overcome this problem by building the scaled wind farm facility which consists of relatively small wind turbines together with many measurement masts that measure the wind conditions in the wind farm and above the wind farm. The scale of this wind farm is not too small to alleviate the dominant scaling effects and the scale is not too large to permit the building of sufficient meteorological masts.

FIG. 4.7: THE SCALE WIND FARM IN BETWEEN TWO PRO-TOTYPE TURBINES (*SOURCE*: [ECN])

FIG. 4.8: THE SCALE WIND FARM (*SOURCE*: [ECN])

The ECN scaled farm consists of ten permanent magnet, direct drive, pitch-controlled wind turbines. The turbines have 10 kW rated power, a rotor diameter of 7.6 m and a hub height of 7.5 m. It is essential that the researchers have full access to the hardware and software of the wind turbines. The scaled farm has been designed in a way that allows ECN performing experiments without any risks for the environment as well as the turbines themselves. As a result, ECN is able to adapt the controllers as well as the turbines for the dedicated experiments. A dedicated wind farm controller has been installed. Inside and around the wind farm a network of fourteen measurement masts has been installed, which measure the wind velocity field from 3.6 m to 19 m height. This covers the rotor area and up to one rotor diameter above the rotor. The large number of meteorological masts within the wind farm permits to measure at the same time single, double, triple and quadruple wakes while simultaneously measuring the external conditions with three nearby 108 m meteorological masts. The unusually densely spaced wind measurements gives the unique possibility to capture the complete wind field, which gives valuable additional information compared to the usual measurement of the wind speed at a single location. Furthermore, most of the wind measurements will be performed using 3D sonic anemometers thus capturing the three wind velocity vectors of the wind field.

The scaled farm is located in ECN's large wind turbine test field in Wieringermeer in between the prototype turbines. The scaled wind farm and its surroundings are characterized by flat terrain, consisting of mainly agricultural area, with single farmhouses and rows of trees. The lake IJsselmeer is located at a distance of 1 km East of the scaled wind farm. Great care has been taken to ensure undisturbed inflow of the wind in the scaled wind farm.

This worldwide unique research facility shall give further insights in the field of wind farm aerodynamics, wake interaction and wind farm control. The high quality data are used for the development and validation of wind farm aerodynamic models and wind farm control strategies. This will allow operating a wind farm at maximum efficiency while guaranteeing at the same time a maximum in reliability and a minimum in mechanical and electrical loads. A unique feature installed in the wind farm controller is that the controller can adjust the yaw of the turbines to exact positions, as well as the pitch angle. Since it has been measured that the yaw angles of the ECN research turbines are seldom equal when the wind is along the row, this feature will make comparison with models more easy and the experimental results more accurate. The side by side comparison of the two rows allows the demonstration of small differences due to changes in turbine or wind farm control (Fig.4.8).

4.7.4 Experimental Facilities at Delft

DUWIND has access to all experimental facilities of TU-Delft. The most important ones that have an important role in the wind energy research are:

- The wave tank of the faculty of Civil Engineering and Geosciences, where the scale model of the motion-compensating offshore access system, the Ampelmann, has been tested.
- The Low Speed Low Turbulence Tunnel of the faculty of Aerospace Engineering, where all of the Delft University airfoils have been measured. The test section is 1.25*1.80 m, with a max wind speed of 120 m/s. The LST has an very low turbulence level: <0.1%.
- Structures and Materials Laboratory of the faculty of Aero-space Engineering, where innovative composite materials, such as thermoplastic composites and smart materials, are produced and tested.
- The Smart Structures laboratory at the Delft Centre for Systems and Control, faculty of Mechanical, Maritime and Materials Engineering is set up to run real time control on structures and to test the performance of active laminates produced at the materials laboratory.

Especially for wind energy research, the wind tunnels are important and are elaborated in the next section.

4.7.5 Wind Tunnels at TU Delft

The Delft University of Technology has a large selection of wind tunnels available with a wide range in wind speed from low speed to large speed. The most recent addition, the Open Jet Facility, is very relevant for wind energy application. The tunnels are applied for fundamental research as well as for applied research and many students use the tunnels for their practical. The wind tunnels available at DUT are indicated in Table 4.1.

TABLE 4.1 LIST OF WIND TUNNELS AT DELFT UNIVERSITY OF TECHNOLOGY

Low-speed windtunnels					
Name	**Type**	**#**	**Dimensions test section ($W \times H$)**	**Cross-section**	**V_{max} (m/s)**
BXF	oj	4	5 × 10 cm	rectangular	25
M-tunnel	oj/cc	1	40 × 40 cm	rectangular	35
W-tunnel	oj	1	40 × 40 cm	rectangular	35
BLT	cc	1	125 × 25 cm (× 540 cm)	rectangular	38
LTT	cc	1	180 × 125 cm	octagonal	120
OJF	cc	1	285 × 285 cm	octagonal	35
High-speed windtunnels					
Name	**Type**	**#**	**Dimensions test section ($W \times H$)**	**Cross-section**	**M-Range**
TST-27	bd	1	280 × 270 mm	rectangular	0.5–4.2
ST-15	bd	1	150 × 150 mm	rectangular	0.7–3.0
ST-3	co	1	30 × 30 mm	rectangular	1.5–3.5
HTFD	bd	1	350 mm	circular	6.0–11.0

4.7.6 The Delft Open Jet Facility

In 2009, at the Delft University of Technology, a new octagonal 500 kW 30 m/s open jet facility began operation that can test model rotors up to 1.8 m for concepts such as flexible smart dynamic rotors and controls. For this kind of experiments the large advantage of the OJF over the other wind tunnels at DUT is that it has an open jet and an outlet diameter of almost three meters and can handle very large models that may obstruct the airflow quite considerably. The new wind tunnel offers more possibilities than ever before for teaching, such as laboratory courses involving model rotors, and research; parameter studies can be executed effectively and efficiently. The initial plans for the OJF were made in the 1980s in response to growing interest in wind energy. The development was delayed because of several reasons, one of them the large budget that was needed for the construction. When the wind group at DUT became part of the Faculty of Aerospace Engineering in 2003 it was possible to seriously proceed with the construction of the facility. The requirement was that the OJF should be versatile and also be used for other research than research into wind turbines. Construction of the OJF started in November 2006 and the opening of the tunnel was in 2009.

The dimensions of the OJF are very impressive. A large fan powered by a 500 kW electric motor enables it to achieve a maximum speed of around 120 km per hour. Air is rotated 180 degrees through a long diffuser and two rows of corner vanes. It then passes through a short diffuser before entering the "settling chamber." Here, five fine-mesh screens reduce the turbulence and velocity deviations in the airflow. Via a contraction the air is then blown into the test section as an even jet stream and cooled at the end by an enormous cooling radiator and guided back to the fan. Some of the many unique technical features in this project are the high power motor, the modular adjustable frequency drive, the thermal sensors, and the large cooling fan (Fig.4.9).

The OJF is mainly used by PhD students, graduates and members of the permanent academic staff. It fulfils an important role in research into the aerodynamic effects that wind can have on buildings and ships as well as in the field of sports. The vast majority of the models are made in the faculty's own workshop. To accommodate these research activities, the OJF is constructed to be extremely versatile. The wind energy group at DUT uses the OJF extensively,

FIG. 4.9: THE ROTOR OF THE WIND RACE CAR OF ECN IN THE OJF (*SOURCE:* ECN, PICTURE BY P. EECEN)

for instance for the verification and validation of the various calculation models. In the context of the large scale European research project Upwind experiments in the OJF contribute to the definition of new concepts for wind turbine blades, including smart rotors.

4.7.7 German–Dutch Wind Tunnels at DNW

DNW, the German–Dutch Wind Tunnels, is a non-profit foundation and was established by the German Aerospace Center DLR and the Dutch National Aerospace Laboratory NLR. Its headquarters are in the Noordoostpolder in the Netherlands and it has wind tunnels situated in a number of locations in the Netherlands and Germany. DNW operates its own large, low-speed facility and the aeronautical wind tunnels of DLR and NLR. DNW provides solutions for the experimental simulation requirements of aerodynamic research and development projects. These projects can originate in the research community (universities, research establishments or research consortia) or in the course of industrial development of new products. Most of the industrial development projects originate in the aeronautical industry, but the automotive, civil engineering, shipbuilding and sports industries have also benefited from DNW's capabilities.

The wind tunnel of DNW used for the experiments in the European project Mexico is the large low-speed facility (LLF). The DNW LLF wind-tunnel facility allows testing over a wide range of conditions and flight regimes. Due to the modular design of the wind tunnel a number of test sections can be used. These include open jet, 9·5 m × 9·5 m ($0 < V < 62$ m/s), 8 m × 6 m ($0 < V < 116$ m/s) and 6 m × 6 m ($0 < V < 152$ m/s) closed test sections. The open jet windtunnel configuration allows large model heights above the ground at zero to low speeds. The closed jet test sections enable testing to be conducted over a higher range of airspeeds. The Mexico experiment, described earlier has been performed in this wind tunnel.

4.7.8 The Offshore Wind Farm Egmond aan Zee (OWEZ)

In the beginning of the century, the government formulated a target for wind energy that required the development of offshore wind farms due to the restricted possibilities onshore. Besides the focus of the wind energy research on offshore developments, an offshore demonstration wind farm should be built relatively close to shore in order to demonstrate the feasibility, but more importantly acquire knowledge and subsequently decrease the cost of offshore wind energy. The government decided on the final location for the demonstration project 100-MW Near Shore Wind farm. The location is situated in Dutch territorial waters of the North Sea between 10 and 18 km from the coast near the village of Egmond aan Zee. After a tender procedure, the Egmond Building Combination (EBC), a joint venture of Ballast Nedam and Vestas, built the 108 MW wind farm on the order of Shell Renewables and Nuon. The Offshore Wind farm Egmond aan Zee (OWEZ) comprises 36 Vestas V90 wind turbines of 3 MW each and associated support systems. Each wind turbine is connected by a transition piece to a steel monopole foundation, piled to a penetration depth of about 30 m. The power generated is transmitted through three 34 kV cables to shore, which land north of IJmuiden harbour. A substation, located near Wijk aan Zee, transforms the voltage from 34 kV to 150 kV and transmits the power into the national grid. Investment costs are around 200 million and financed on balance by Nuon and Shell. The wind farm initially has been operated by EBC under a 5-year warranty, operations, and maintenance contract. The wind farm produced approximately 350 GWh per year.

In the context of realizing the wind farm, an extensive Monitoring and Evaluation Program (MEP) was carried out. In that context, a meteorological mast was installed and data from the turbines were collected. The meteorological mast has a height of 106 m above sea level and measures the wind conditions at three levels. Data collected in the MEP includes contractual issues, project organisation; permits; technical description of the design, support structure, wind turbines, and electrical design; assembly and installation; planning versus execution; budget; health, safety, security, and environment management; risk management; financing, insurance, and power purchase agreements; quality assurance management; requirements and qualifications; monitoring and evaluation program; and lessons learned [20]. This information has been made available [21]. Commercially sensitive data that has been collected during the execution of the MEP is on the subjects: corrosion and lightning; dynamics of turbines; aero-elastic stability; scour protection; electricity production, disruptions, failure data, availability, maintenance, and reliability; power quality, grid stability, and power forecasts; and wind turbine P–V curve and wake effects. ECN and DUT have executed several projects with selections of this data under an NDA agreement with NoordzeeWind.

4.7.9 The Offshore Wind Farm 'Prinses Amalia Windpark'

The second offshore wind farm in Dutch National waters of the North Sea is Prinses Amaliawindpark. This wind farm is situated some 23 km offshore from IJmuiden, in block Q7 of the Dutch continental shelf. This is just outside the 12-mi zone south-west of OWEZ. The wind farm consists of 60 Vestas V80 wind turbines of 2 MW each with a total capacity of 120 MW. The water depth at the site is between 19 m and 24 m. The project is owned and developed by a group of companies of ENECO Holding NV, Econcern BV, and Energy Investment Holdings. It was built by Vestas Wind Systems A/S, Van Oord Dredging, and Marine Contractors BV under separate construction contracts. Initially, the windfarm is operated by Vestas Offshore, an affiliate of Vestas, under a 5-year warranty, operations, and maintenance contract. During its operation, more and more research is carried out at the 'Prinses Amaliawindpark', also in the context of the FLOW program.

4.8 SUMMARY

This Chapter is an attempt to provide insight in the wind energy research of the Netherlands. The result is not complete and must be regarded as a personal selection of the activities that have been performed in all the years of wind energy research in the Netherlands. A more extensive overview can be found in Verbong [22], written in Dutch. The research in wind energy is concentrated at the wind energy department of the Energy research Centre of the Netherlands ECN and the interfaculty wind energy department DUWIND at Delft University of Technology. Both institutes have been involved in wind energy research from the start in the 1970s and closely match their research programs. ECN Wind Energy has a research staff of 55 scientists and DUT has a research staff of 15 permanent researchers and more than 35 PhD students. Together

ECN and DUT belong to the top 5 of the international wind energy research groups.

In the Netherlands, the wind energy research is supported by an extensive experimental infrastructure. The Knowledge Centre WMC that has been founded by the DUT and ECN has a research staff of 25 scientists and is a research institute for materials, components and structures. WMC is performing blade tests for large wind turbines to 60 m in length. ECN has a research wind farm where proto-type wind turbines are tested, where a research farm of 5 full-scale turbines are used for research activities and where a scale wind farm is located for research on farm control and wind farm aerodynamic research. At DUT a large selection of experimental facilities are being used for wind energy applications. The most prominent facilities are the wind tunnels, of which the Open Jet Facility is the most recent addition.

4.9 ACRONYMS AND INTERNET

2B-energy	www.2-benergy.com
AgencyNL	www.agentschapnl.nl, formerly known as Sen terNovem www.senternovem.nl
DUT	Delft University of Technology — www.tudelft.nl
EAWE	European Academy of Wind Energy — www. eawe.eu
ECN	Energy research Centre of the Netherlands — www.ecn.nl
EERA	European Energy Research Alliance — www. eera-set.eu
EWEA	European Wind Energy Association — www.ewea.org
EWT	Emergya Wind Technologies — www.ewt international.com
IEA	International Energy Agency — www.iea.org
IEC	International Electrotechnical Commission — www.iec.ch Lagerwey Wind www.lagerwey-wind.nl
LSEO	Landelijke Stuurgroep Energie Onderzoek
MEASNET	International Network for Harmonised and Recognised Measurements in Wind Energy — www.measnet.org
MEP	Monitoring and Evaluation Program executed at the Offshore Wind farm Egmond aan Zee
NLR	Nationaal Luchten Ruimtevaartlaboratorium — www.nlr.nl
OWEZ	Offshore Wind farm Egmond aan Zee — www.noordzeewind.nl
RCN	Reactor Centre of the Netherlands STX-Harakosan www.stxwind.com
TNO	Dutch organization for Applied Scientific Research
TPWind	European Wind Energy Technology Platform — www.windplatform.eu
VWEC	VWEC Wind Energy Consult
WMC	The Knowledge Centre Wind turbine Materials and Constructions — www.wmc.eu
XEMC	Darwind www.xemc-darwind.com

4.10 REFERENCES

1) Energie Onderzoek Subsidie (EOS) carried out by Agentschap NL (www.agentschapnl.nl, www.senternovem.nl/eos).
2) Strategic Research Agenda, EWEA 2005.
3) IEA Large-Scale WECS Annual Reports, 1978 to 1990.
4) IEA R&D WECS Annual Reports, 1978 to 2010.
5) P.E.J. Vermeulen, Report of the IEA Technical Meeting on Wakes and Clusters,, February 2-3 1981, MT-TNO 81-04676.
6) P.E.J. Vermeulen, P.J.H. Builtjes, J.B.A. Vijge, Mathematical modeling of wake interaction in wind turbine arrays, MT-TNO 81-01473.
7) G. A. M. van Kuik, J. W. M. Dekker, The FLEXHAT program, technology development and testing of flexible rotor systems with fast passive pitch control, Journal of Wind Engineering and Industrial Aerodynamics, Volume 39, 1992, Pages 435-448.
8) J.G. Schepers, H. Snel, Final results of the EU JOULE projects 'DYNAMIC INFLOW', presented at the ASME '96, ECN-RX-95-062.
9) J.G. Schepers et al., Final report of IEA ANNEX XIV. Field Rotor Aerodynamics, ECN-C-97-027.
10) H.B. Hendriks and M. Zaaijer, DOWEC Dutch Offshore Wind Energy Converter 1997 – 2003. Executive summary of the public research activities.
11) G.P. Corten and H.F. Veldkamp, Insects can halve wind-turbine power, NATURE, VOL 412, 42-43, 5 July 2001.
12) P.J. Eecen, S.A.M. Barhorst, 'Data added to International Wind Database, IEA Annex XVII', ECN-C-03-128.
13) P.J. Eecen, S.A.M. Barhorst, 'Extreme wind conditions, Measurements at 50m Meteorological Mast at ECN, Petten', ECN-C-04-019.
14) G.C. Larsen and K.S. and Hansen, Structure and Philosophy, RISOR-1299.
15) www.ampelmann.nl.
16) J.G. Schepers et al. Verification of European Wind Turbine Design Codes, VEWTDC. European Wind Energy Conference, EWEC, Copenhagen, July 2001.
17) H. Snel, J.G. Schepers and B. Montgomerie. The Mexico project the database and first results of data processing and interpretation In 'The Science of Making Torque from the Wind', August 2007.
18) S. Schreck. IEA Wind Annex XX: HAWT Aerodynamics and Models from Wind Tunnel Measurements. NREL/TP-500-43508, NREL, December 2008.
19) W.A. Timmer and R.P.J.O.M. van Rooij, Summary of the Delft University wind turbine dedicated airfoils, AIAA-2003-0352.
20) Offshore Windfarm Egmond aan Zee, General report OWEZ_R_141_20080215.
21) www.senternovem.nl/offshorewindenergy/technology_nsw/www.monitoring_mep-nsw/index.aspwww.noordzeewind.nl.
22) G. Verbong, Kwestie van lange adem, de geschiedenis van duurzame energie in Nederland (history of renewable energy in the Netherlands).

4.11 WIND ENERGY IN THE NETHERLANDS SINCE 2011-INTRODUCTION

"Wind Energy in Netherlands" documented in Chapter 9 by Eecen in the ASME's 2011 classic publication "Energy and Power Generation Handbook Established and Emerging Technologies" edited by K. R. Rao [1], has been updated by Editor-in-Chief of the current Renewable Energy Series, K. R. Rao. K. R. Rao is also the author of the monumental publication "Wind Energy for Power Generation Meeting the Challenge of Practical

Implementation", in two volumes and nearly 2000 pages published by Springer in 2019 [2], is extremely competent and knowledgeable to update this chapter by Eecen due to the non-availability of Eecen to perform the task.

This update has been provided in section 4.11 and follows the same topical outline initially contributed by Eecen in the 2011 publication. Eecen with his professional and scholastic expertise provided information which is applicable even to-day, however the topical discussions covering what happened since 2011 to date need to be up-dated. Even though the topics are retained as much as possible, a continuation of the same is not appropriate. Thus, discussions are going to be as applicable for the topics listed in this update and cannot be attributed do the original write-up of Eecen. All of the material in this update is from information available in the on line public domain and is documented in the footnotes pertinent to the discussions. Material in this update can be easily researched from the footnotes provided and readers have an opportunity to even look up to the latest positions of the topic at the time of browsing the manuscript.

As of October 2020, wind power in the Netherlands has an installed capacity of 4,990 MW, 19% of which is based offshore. In 2019, the wind turbines provided the country with 12% of its electricity demand during the year.[2] Netherlands reached an Energy Agreement for Sustainable Growth. Offshore wind turbines will then supply 3.3% of the Netherlands' total energy needs. The Coalition Agreement and the Climate Agreement (2019) include a commitment to continue the successful offshore wind energy policy. This will result in a total offshore wind power capacity of 11 GW by 2030, which is enough to supply 8.5% of all the energy in the Netherlands. The total installed capacity of offshore wind power in the Netherlands is around 2.5 gigawatts (GW) in 2021 and should increase to at least 4.5 GW by 2023. This commitment is enshrined in the Energy Agreement for Sustainable Growth. Offshore wind turbines will then supply 3.3% of the Netherlands' total energy needs.[3]

The Netherlands is working on a transition to a sustainable, reliable, and affordable energy supply. The Netherlands wind energy market is expected to grow at more than 9.5% during the forecast period 2020-2025. Wind energy in the Netherlands powers about 5.7 million homes and cuts the CO_2 footprint of electricity by 12%. The Netherlands is expected to build the world's largest offshore wind farm by 2027, along with a 2.3-square-mile artificial island to support it. The wind farm is expected to be capable of producing 30 GW of power.[4]

Onshore farms in The Netherland's wind farms made up 80% of the new installations with 11.8 GW. This was 22% lower than the government's pre-COVID forecast. The Netherlands installed the most wind power capacity in 2020 (1.98 GW), 75% of that was offshore wind. The Netherlands installed the most wind capacity use due to strong offshore additions. Offshore wind made up 20% of new installations in Europe with 2.9 GW of new capacity connected to the grid in 2020 of which The Netherlands installed half of that capacity. Offshore installations were driven by the completion of the Borssele Wind Farm zones, the three wind farms were awarded to different consortia in 2016 and 2017. 220 GW of wind power capacity are now installed in Europe, 11% of this figure is offshore, and Netherlands has 5 GW. In The Netherlands, installed offshore wind capacity more than doubled but most of the new capacity came online at the end of the year and so did not impact the annual figures in 2021. In the Netherlands an increase of 1.6 GW of wind energy installations are expected in 2021. In 2021-2025 new onshore wind installations in The Netherlands was 20 GW. According to Realistic Expectations Scenario, between 2021 and 2025, offshore installations in The Netherlands were 4.4 GW. In cumulative terms of total capacity, Europe would reach 318 GW of installed capacity by the end of 2025 of which The Netherlands will be above the 10 GW.[5]

As of May 15, 2021 there are over 1200 Dutch windmills all over the country some of the most impressive mills can be found at the *Kinderdijk* and the *Zaanse Schans*. Blades of windmills can be adapted to the weather. Most of the Dutch windmills have very similar blades. They mostly consist of big frames. Back when these windmills were designed, this design was chosen because of the Dutch weather. Winds can blow quite hard in the Netherlands with the frame-like structure makes sure the blades have some flexibility and cannot be blown off by a strong wind. The Netherlands also has a lot of modern windmills. They are placed on land, but there are also a lot of windmills in the North Sea. A large part of Dutch renewable energy comes from windmills. From the 21.8 billion kW renewable energy produced in the Netherlands, 49 percent is from windmills! Countrywide, 111 billion kW is used yearly. The highest windmill in the world is built in the Netherlands, although this is a modern windmill, it is massive. This machine is almost 245 meters *(803 ft)* high! This huge windmill was built by the American multinational company General Electric in the harbor of Rotterdam. Each of the blades is 100 meters *(328 ft)* long, which is as long as a football field. The tower is higher than the *Big Ben*, the *Tower of Pisa*, and the *Arc de Triomphe* stacked on each other. The turbine produces 12 mw, which is enough to power 16,000 European households! The testing period started in 2019 and would last for 5 years, meaning the machine will be removed in 2024. If these tests end up being successful, we might see wind farms consisting of these giants being built, which could provide energy for up to one million households. Many Dutch windmills are on the UNESCO world heritage list. The famous *"Kinderdijk"* is an example of this. This is a dyke filled with picturesque windmills, and most certainly worth a visit. The *Beemster* region is also a world heritage site. There are roughly 1000 historic windmills in the Netherlands. There are a lot of windmills in urban areas. Windmills were usually built in the countryside, but as cities grew, they started to get swallowed by the cities. Nowadays there are many beautiful windmills in the middle of city centers. Some of the historic windmills are

[2] https://en.wikipedia.org/wiki/Wind_power_in_the_Netherlands

[3] https://www.government.nl/topics/renewable-energy/offshore-wind-energy

[4] https://www.mordorintelligence.com/industry-reports/netherlands-wind-energy-market

[5] https://s1.eestatic.com/2021/02/24/actualidad/210224_windeurope_combined_2020_stats.pdf#:~:text=Wind%20energy%20in%20Europe%20-%20 2020%20Statistics%20and,restrictions%20to%20the%20moment%20of%20people%20and%20goods.

TABLE 4.2: NEW ADDITIONS, TOTAL WIND CAPACITY AND THE SHARE OF WIND IN THE ELECTRICITY DEMAND IN 2020 IN NEDERLAND'S[6]

New Installations MW (2020)			Cumulative Capacity MW			%age Share of Wind of EU, 2020		
On Shore	Offshore	Total	On Shore	Offshore	Total	On Shore	Offshore	Total
486	1,493	1,979	4,174	2,611	6m784	9%	3%	12%.

still being used for drainage. Most Dutch windmills can turn the directions of the blades. The weather often changes in the Netherlands, and so does the direction of the wind. To make the windmills more efficient, the Dutch designed their windmills in such a way, that the top of the mill can turn. In this way, the blades can catch wind no matter the direction the wind is coming from.[7]

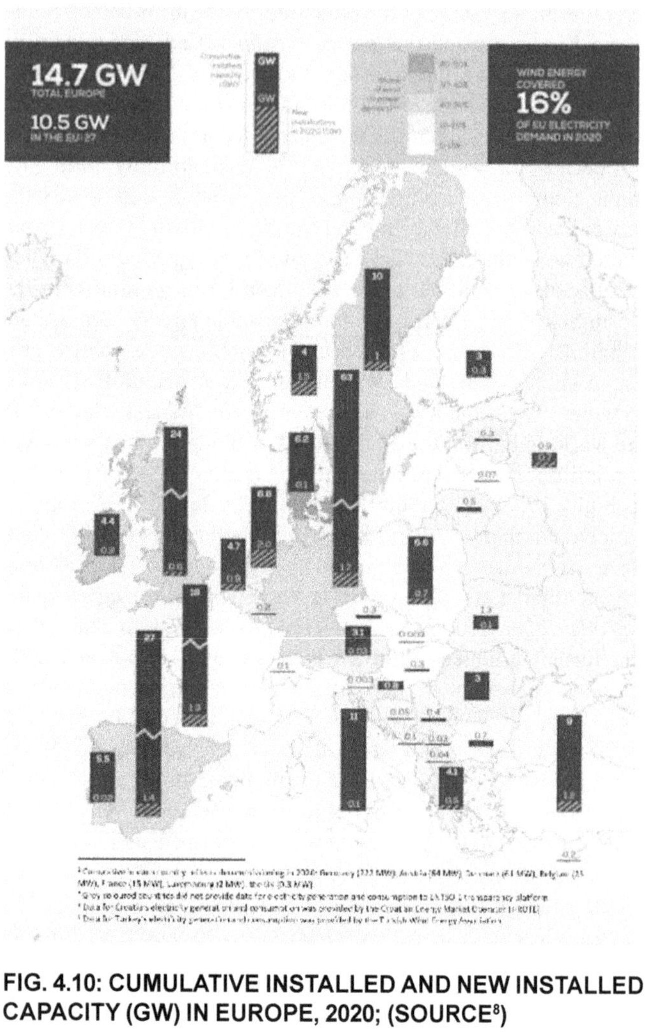

FIG. 4.10: CUMULATIVE INSTALLED AND NEW INSTALLED CAPACITY (GW) IN EUROPE, 2020; (SOURCE[8])

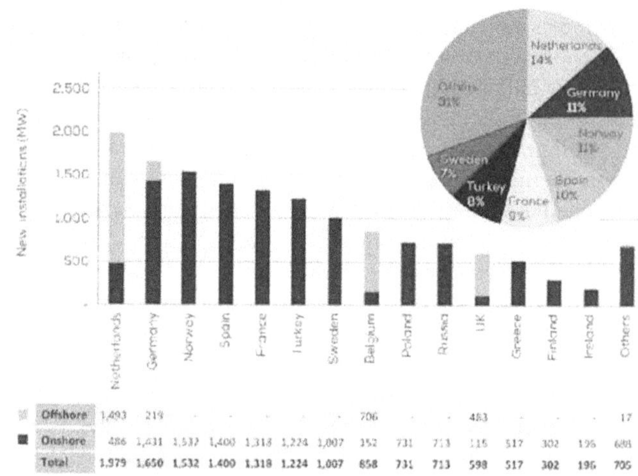

FIG. 4.11: NEW WIND INSTALLATIONS IN 2020 IN EUROPE PER COUNTRY (SOURCE:[9])

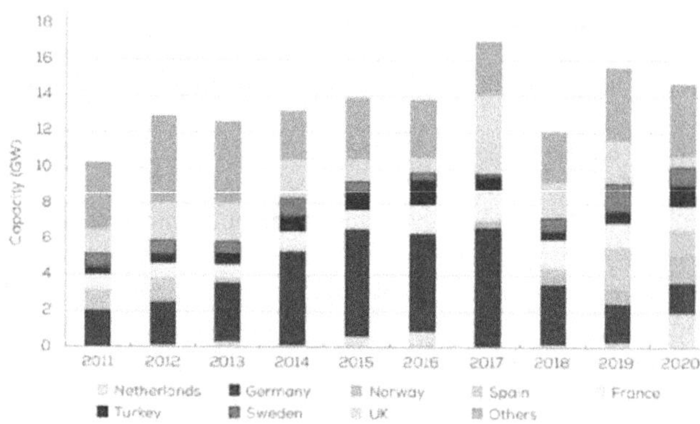

FIG. 4.12: WIND INSTALLATIONS IN EUROPE, INCLUDING NETHERLANDS, 2020 (SOURCE:[10])

---

[6] Ibid.

[7] https://aboutthenetherlands.com/dutch-windmill-facts/

[8] https://s1.eestatic.com/2021/02/24/actualidad/210224_windeurope_combined_2020_stats.pdf#:~:text=Wind%20energy%20in%20Europe%20-%20 2020%20Statistics% 20 and,restrictions%20to%20the% 20moment%20of%20people%20and%20goods.

[9] Ibid.

[10] Ibid.

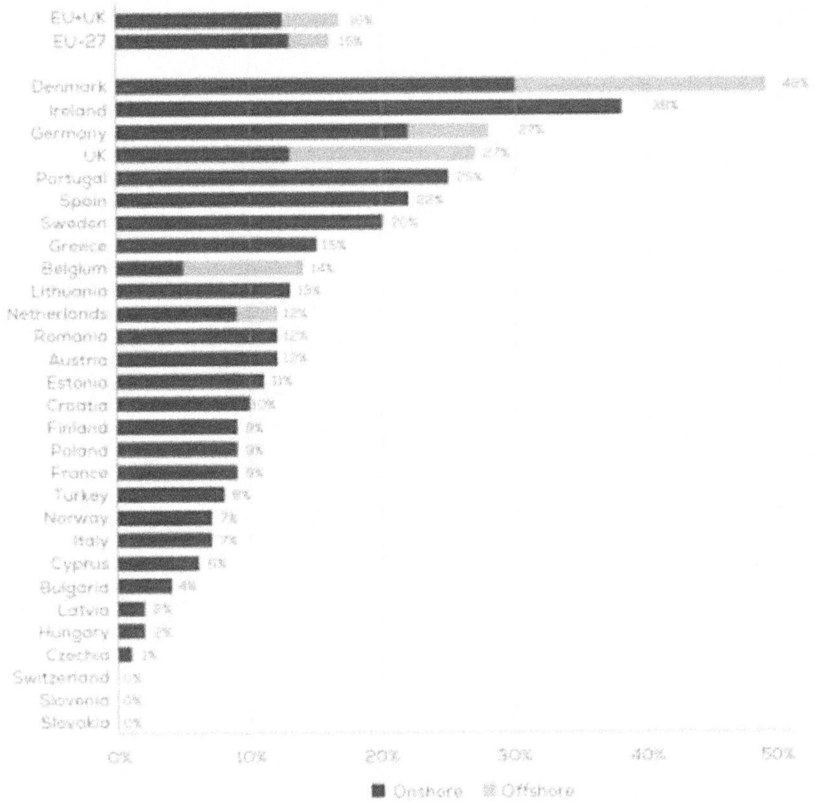

FIG. 4.13: PERCENTAGE OF ANNUAL DEMAND BY COUNTRIES, INCLUDING NETHERLANDS IN 2020. (SOURCE:[11])

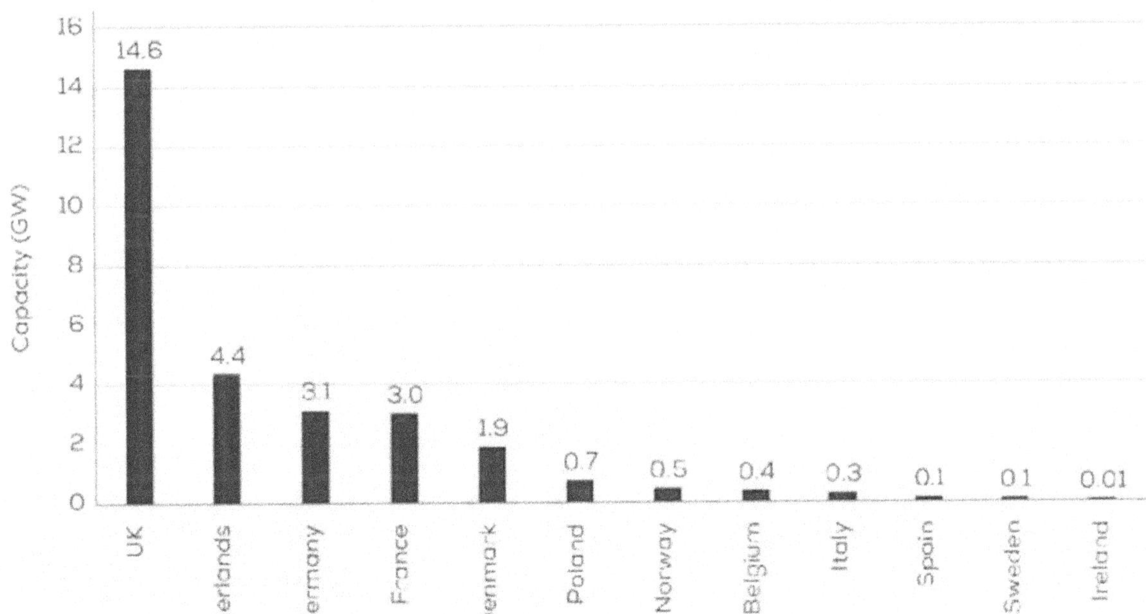

FIG. 4.14: A SCENARIO OF NEW OFFSHORE INSTALLATIONS IN EU, 2020 (SOURCE:[12])

---

[11] Ibid.

[12] Ibid.

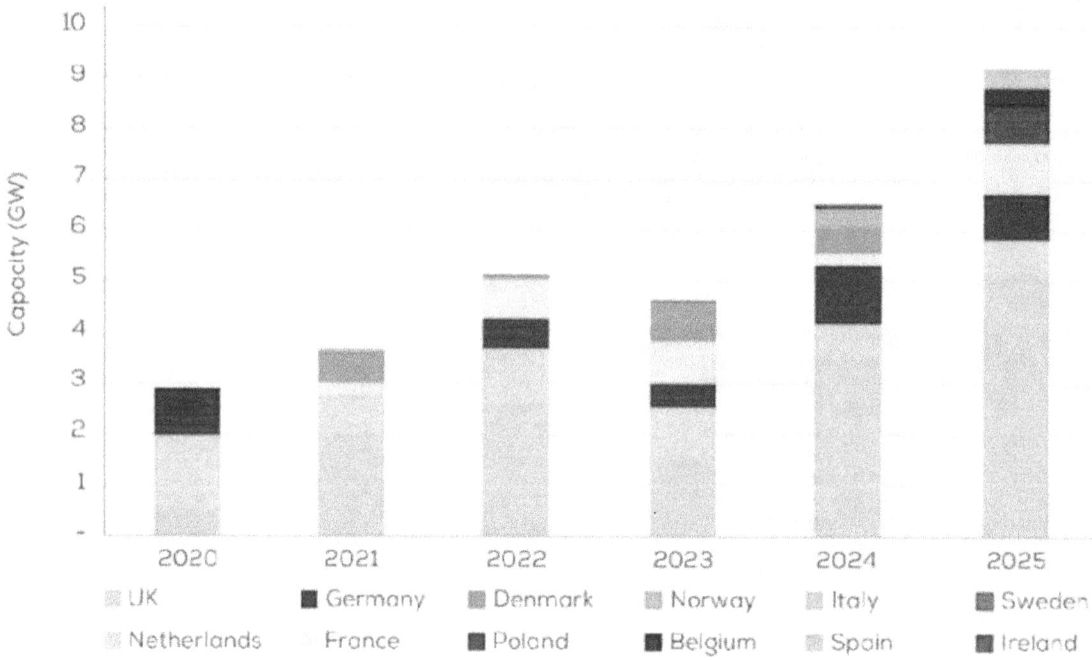

FIG. 4.15: A SCENARIO OF REALISTIC OFFSHORE EXPECTATIONS 2020 THROUGH 2025 (SOURCE:[13])

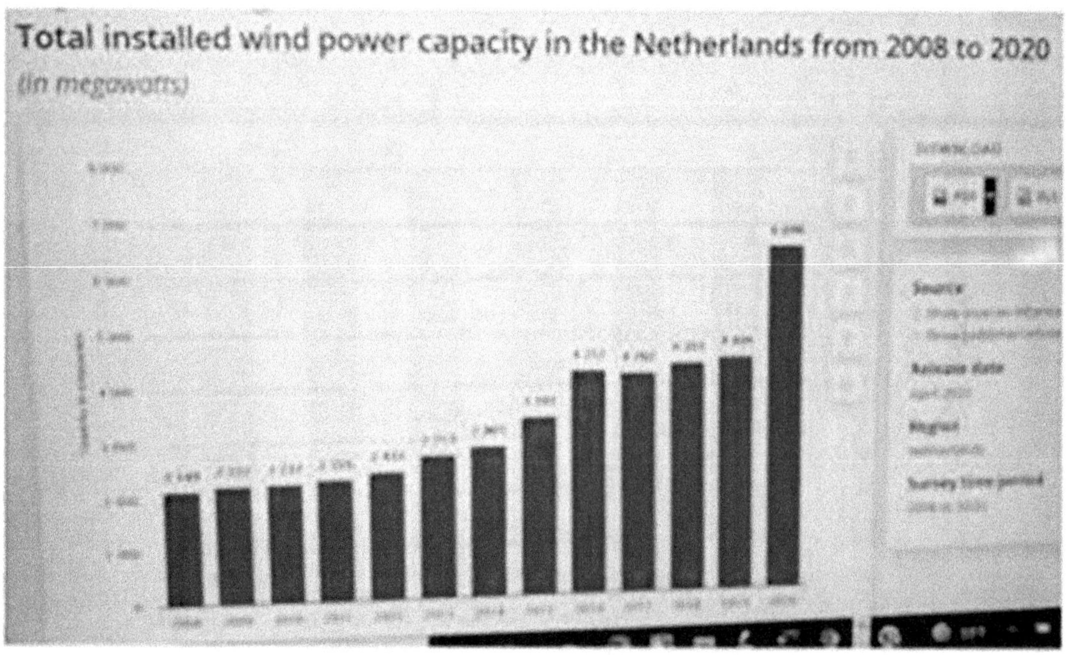

FIG. 4.16: TOTAL INSTALLED WIND POWER CAPACITY (MW) IN THE NETHERLANDS FROM 2008 TO 2020 (SOURCE:[14])

---

[13] Ibid.

[14] Ibid.

The Netherlands triplicated its total installed wind power capacity from 2,149 megawatts in 2008 to around 6,600 megawatts in 2020 as shown in figure 4.16.[15] Europe is one of the regions with the largest installed wind power capacity worldwide.

4.11.1 Wind Energy In The Netherlands Since 2011

The wind energy in Netherlands has been addressed in this update in two sections, in the first one introducing the subject in 4.11 and the second one in this section addressing the items addressed by Eecen in the initial write up [2] as appropriate for the current update. As such the items considered with developments since the 2011 write-up of Eecen for "Upwind", "Smart Rotors", "Role of Wind with Heat and Flux", "Wind Tunnel Experiments" and "DU Airfoil". The only exception is "WE@SEA" which continues unabated since the initial write-up of Eecen that does not require any add-on information [2].

4.11.1.1 UpWind[16]

As enunciated by Eecen in Upwind [1], similar to it is Energy DTU[17] which is the Technical University of Denmark,[18] a globally leading center for wind energy with technical-scientific competences in the international front and with a unique integration of research, education, and public/private sector consulting.[19] The various objectives of Upwind includes "Blade Defect Forecasting"[20], the aim of which is to develop blade defect forecast tool for seasonal repair prediction based on environment parameters at high temporal and spatial scales for 10-15 years ahead. Key environmental parameter will be used in the forecasting for operational costs in the coming repair season for blades in operation and for site specific design requirement. The "expected outcome" is a tool that precisely and robustly can predict erosion at turbine blades. The erosion prediction will be based on environmental parameters like rain, wind, and lightning. This project is supported by Innovation Fund Denmark, Grand Solutions and the project began on December 1, 2019, expected to run through November 30, 2021.

Another important testing conducted as a facet of Upwind is "Blatigue", fast and efficient fatigue test of large wind turbine blades. The objective of "Blatigue" is to develop fast and efficient fatigue test methods for large wind turbine blades and to develop equipment to excite the blades under such tests. The standard certification tests, that blades are tested against today, are not representing the real world very well. In reality blades are exposed to torsion and bending in different directions at the same time. The industry consequently needs fatigue test methods that better match the loads to which the blades are exposed to in real operational conditions. A multi-axis fatigue test methods can do exactly that. As blades become larger the time needed to perform the necessary certification tests becomes longer. This is converting into a challenge for the industry as 100m blades are expected to take more than a year to test. There is a need to shorten the test time and multi-axis test methods can do that as flap and edgewise loads can be tested at the same time. Testing within the project will take place in the new Large Scale Facility at DTU Wind Energy Risø Campus. The project started on November 1. 2016 and will run until May 31. 2020. Total project budget was 30.35 mio. DKK. This was updated by Charlotte Hede Linde on 31 May 2021.

Another significant objective of Upwind is to study "Wind Turbine Blade Erosion for reducing the largest uncertainties of wind turbine blade performance". The objective of erosion is to enable longer lifetime of wind turbine blades at multi-MW machines. Wind turbines at sea suffer unexpected fast erosion on the leading edge of blades such that blades needs to be replaced long before expected. It is speculated that the cause for fast erosion is the rough weather at sea and high tip speeds. The overall aim is to investigate the combined environmental conditions in the marine atmospheric boundary layer with regard to winds, precipitation and atmospheric composition of aerosols in the Northern European seas where offshore wind farms are located based on novel remote sensing observations on the ground and from space and atmospheric modelling. The expected outcome is to help the industry with in-depth meteorology and for

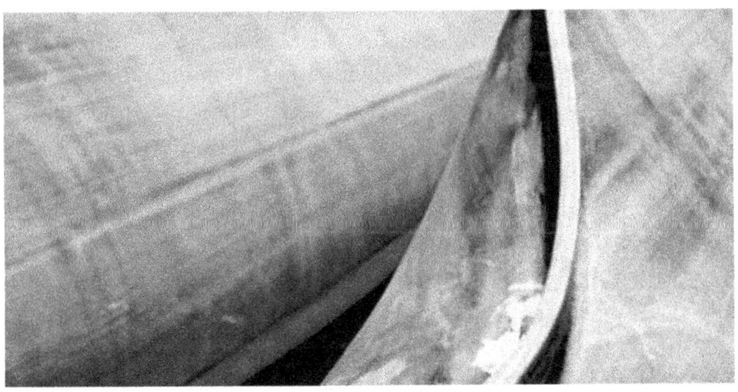

FIG. 4.17: BLADE DEFECT FORECASTING (SOURCE:[21])

FIG. 4.18: BLATIGUE: FAST AND EFFICIENT FATIGUE TEST OF LARGE WIND TURBINE BLADES. (SOURCE[22])

[15] https://www.statista.com/statistics/421525/total-wind-power-in-thenetherlands/#:~:text=The%20Netherlands%20triplicated%20its%20total%20installed%20wind%20power,the%20Netherl ands%20from%202008%20to%202020%20%28in%20megawatts%29, Published by Madhumitha Jaganmohan, Apr 26, 2021

[16] https://windenergy.dtu.dk/english/research/research-projects/completed-projects/upwind

[17] https://windenergy.dtu.dk/english/research/research-projects

[18] https://www.dtu.dk/english/About/profile

[19] https://windenergy.dtu.dk/english

[20] https://windenergy.dtu.dk/english/research/research-projects/blade-defect-forecasting

[21] Ibid.

[22] https://windenergy.dtu.dk/english/research/research-projects/blatigue

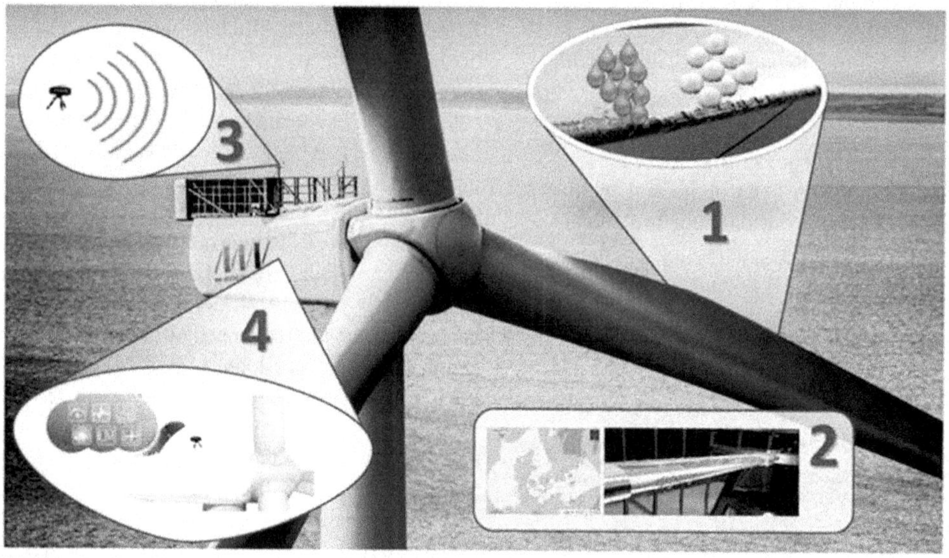

1. **Research hypothesis:** Erosion damage is mainly generated during heavy precipitation (big drops of rain or hail), which occurs in a very little fraction of the turbines operation time. By reducing the tip speed of the blades in these few hours a significant extension of the leading edge lifetime can be obtained with negligible loss of production.

2. **Methodology:** Define rain and hail erosion classes to quantify leading edge blade in-field and in lab testing. Correlations between rain intensity, droplet size, impact speed, materials properties, etc. will be established.

3. **Measurement Device:** Low-cost prototype for precipitation measurement on site and real time warning device enabling modern control of wind turbines.

4. **Erosion safe mode:** A safe mode control based on the erosion classes to control the wind turbine, reducing the tip speed under severe conditions – preventing aerodynamic degradation and reducing maintenance costs.

FIG. 4.19: EROSION: WIND TURBINE BLADE EROSION REDUCING THE LARGEST UNCERTAINTIES OF BLADE EROSION. (SOURCE:[23])

avoiding blade erosion at leading edges. The project which began on April 1. 2017 is expected to run until December 31. 2021.[24]

Yet another of Upwind activities is "IEA International Energy Agency (IEA, www.iea.org) Wind TCP Task 41: Distributed Wind",[25] the objective of this to identify and explore studies of particular Danish interest of Distributed Wind (DW) for cost effective technology development and integration into a continuously evolving energy system. This is done by collaborating and contributing to the IEA Wind TCP Task 41 international activities on DW turbine technology development and assessment in a series of dedicated work packages (WPs). IEA Wind TCP Task 41 is an international network centered on international collaboration and coordination in the field of DW. The purpose is to accelerate the development and deployment of DW technology as one of the leading generation source in global renewable markets, the facilitation of easier and faster DW integration into electrical grids, increasing thus the competitiveness of wind and accelerating the replacement of fossils fuels.

This project is organized into five work-packages (WPs), which closely follow the first five IEA Wind TCP Task 41 planned work-packages WP0: Management, coordination and dissemination; WP1: DW technology design standards for small and mid-sized wind turbines; WP2: Data information catalog for DW research; WP3: Integration of DW into evolving electricity systems; WP4: Outreach and expand collaboration of ongoing R&D DW activities. The expected outcome of the the project aims at building up a stakeholder network of relevant Danish players within the area of DW technology and organize and strengthen the Danish influence and participation in IEA collaborations, both bringing the long experience of Danish actors into play and to learn from others around the world. By supporting the work of the IEA Wind TCP Task through various publications, data sets collections and reports, this project will achieve and consolidate the Danish knowledge and experience within this area, increasing thus furthermore the competitiveness of wind and accelerating the replacement of fossil-based fuels. DTU Wind Energy will contribute to the IEA Wind TCP Task 41 through communication, exchanging information, sharing results and carrying out concrete analyzes and investigations in the shape of reports and publications. Overall expected results are enhanced international collaboration and coordination in the field of DW to accelerate the development and deployment of DW technology. Recommendations for potential modifications on standards will be used to reduce the cost and development time of DW projects globally and help the

[23] Ibid.

[24] https://windenergy.dtu.dk/english/research/research-projects/erosion

[25] https://windenergy.dtu.dk/english/research/research-projects/iea-wind-tcp-task-41

FIG. 4.20: DECOWIND (SOURCE:[26])

DW manufacturers to enter and operate in global market. Compendium of standards improvement recommendations. Specification and development of data sharing catalog and portal allowing DW stakeholders to collaborate and communicate easily, globally. Assessment of different levels of DW system control, including a classification for DW control strategies for high contribution distributed grids, including different capabilities of grid support and likely data needs for future wind systems. Support to provide flexibility and grid services will increase the value of DW and demand for DW in the grid. Development of a content management system including data, software, associated publications. Data sharing in both process and practice, providing researchers and the wider industry access to global information that can be used to address wider challenges facing DW. Expanded analyses and collaboration on the integration of DW technologies, focusing on new and advancing markets such as off-grid, microgrids, and advanced distribution networks. Collaboration with ongoing R&D activities that address specific challenges associated with DW technologies (IEA Wind TCP Task 19, IEA Wind TCP Task 25, IEA Wind TCP Task 26, IEA Wind TCP Task 28, IEA Wind TCP Task 36). This EUDP-project will gather and disseminate information about DW to the Danish stakeholders and provide useful recommendations and guidelines to IEA deliverables that can be used by both Danish industry, researchers and society at large. This EUDP-project will also make it possible to promote the results of IEA Wind TCP Task 41 in national and international forums. This project started on January 1. 2020 and runs until 31.12.2022.[27]

Upwind and Energy DTU projects interlace targeting several wind energy projects with international collaboration. One of them is "DecoWind"[28] with the objective of targeting "Noise from Wind Turbines" which is one of the main barriers to public acceptance and further onshore deployment of wind energy. The DecoWind project is to address a fundamental gap in knowledge about noise generating from wind turbines and propagating to receivers/neighbors living nearby. The expected outcome with the the main aim of this project is to maximize wind farm annual energy production (AEP) without increasing noise annoyance. The project target is to increase AEP by 3-5% within existing noise regulations and pave the way for new scientific regulations. This aim is realized by (1) developing high-fidelity noise-chain models that can accurately predict atmospheric noise propagation from source to receiver; (2) integrating the developed noise-chain models into EMD's WindPRO as well as Siemens Gamesa Renewable Energy (SGRE)'s siting tool; (3) updating recommendations for noise regulation; (4) developing wind farm operational strategy (WFOS) for optimal AEP operations of wind turbines in wind farm and demonstration, and (5) commercialization. This project started on December 1. 2018 and runs for three years, with the total project budget of 18.23 mio. DKK. This report was updated by Charlotte Hede Linde on 31 May 2021.[29]

Hybrid Energy Storage[30] objective was a project is led by CEE (DTU Elektro) with DTU Wind Energy, Vestas and DEIF[31] as partners. The project will develop, test and optimize hybrid energy storage systems, i.e. combinations of different storage technologies with complementary properties. Initially, three storage technologies will be characterized in the laboratory facility SYSLAB at DTU Risø Campus. The performance of the individual technologies will be compared, and a sizing tool will be developed to determine the best combination of storage units for a given application. The expected outcome is that DTU Wind Energy and DEIF will then develop such an application a hybrid storage system integrated into a wind turbine and demonstrate it in our V27 wind turbine at DTU Risø Campus. Results from this demonstration will be used to further refine the sizing and optimization tool. This project began on September 2020 and runs for two years, with a total project budget of 3.97 mio. DKK. This has been updated by Charlotte Hede Linde on 31 May 2021.[32]

4.11.1.2 Smart Rotors

During the last decade research in the field of smart rotor has advanced significantly. Fundamental aerodynamics, structural and control concepts have been established and simulators created for distributed flaps on wind turbine blades, which are considered the most promising option. A proof of concept has created under laboratory conditions. However, the results obtained under these conditions can only be partially transferred to the real application as the control authority of smart rotors is limited compared to full pitch control. The steps that need to be taken before smart rotors can be successfully exploited are in the design of reliable systems that can operate under environmental conditions without inspections. Other potential advantages of distributed control need to be established, such as the effect on other components of a wind turbine, for example the gear box or the power system. Finally, it is necessary to investigate what benefits can be achieved if blades are designed with distributed control right from the start instead of applying control schemes to already existing turbines. The up-scaling of wind turbines and especially offshore turbines to achieve more cost effective designs, poses a great challenge to the designers. Traditional design approaches get closer to their limits. New methods in the design and control of the blades are required to facilitate a further

[26] https://windenergy.dtu.dk/english/research/research-projects/decowind
[27] https://windenergy.dtu.dk/english/research/research-projects/iea-wind-tcp-task-41
[28] https://windenergy.dtu.dk/english/research/research-projects/decowind
[29] https://windenergy.dtu.dk/english/research/research-projects/decowind
[30] https://windenergy.dtu.dk/english/research/research-projects/hybrid-energy-storage
[31] https://stateofgreen.com/en/partners/deif-wind-power-technology/
[32] https://windenergy.dtu.dk/english/research/research-projects/hybrid-energy-storage

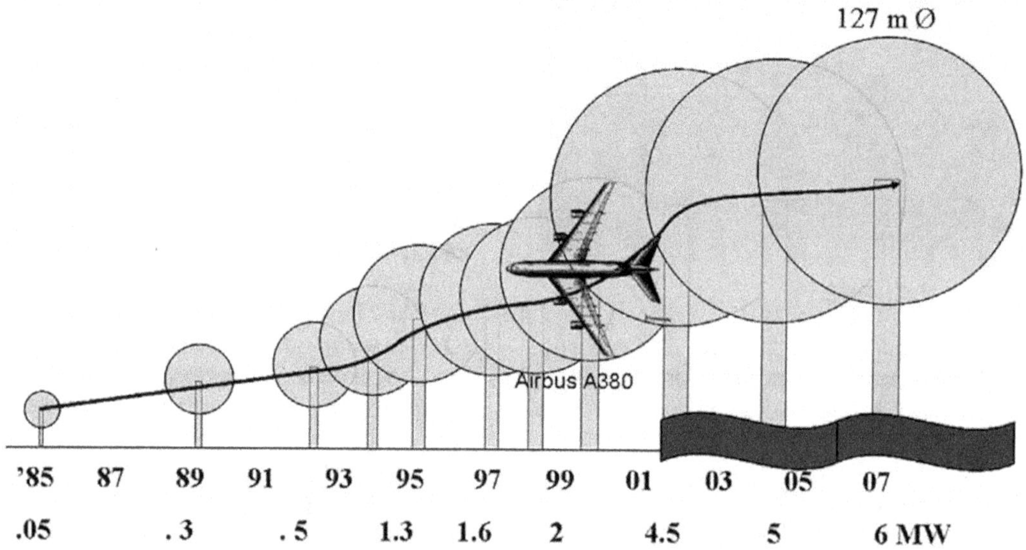

FIG. 4.21: THE INCREASE OF TURBINE STATE OF THE ART SIZE IN TIME. (SOURCE[33])

increase in turbine size. A method is the application of sophisticated load control techniques, the so-called 'smart rotor'. In this method, the blades are equipped with distributed aerodynamic control surfaces. During the last decade, the idea of the smart rotor has matured from an abstract concept to a proof of concept stage. It lies in the nature of the problem that a multidisciplinary approach needs to be chosen, including research in aerodynamics, control theory, material science and wind turbine structures. Various research institutes, among which DTU RISØ, TU Delft, University Stuttgart and Sandia National Laboratories, have been investigating the smart rotor concept and its subdomains, leading to advances in the design and analysis of such wind turbines. During the last decade 5 PhD theses have been devoted to this topic at TU Delft and at the Danish Technical University. This paper will analyze the progress made and indicate how the next steps towards the implementation of aerodynamic control surfaces in blades can be taken.[34]

Fatigue is one of the biggest issues in wind turbine design. Most of the sources for fatigue loads are related to changes in the airflow around the blades due to, for instance, turbulence, tower shadow or jaw misalignment. Therefore, systems are proposed that counteract these load fluctuations, which are both deterministic and stochastic in nature. The systems aim at influencing the lift and drag at different stations along the blade's length. This way, the aerodynamics along the blade can be controlled and the dynamic loads and modes can be dampened. This is called the 'smart' rotor concept. Such systems are already used in aerospace, both with airplanes and helicopters, although in the latter case mostly experimental. They aim at reconfiguration of the wing or rotor, flight control or vibration reduction. The concepts are often implemented using adaptive materials such as piezoelectric materials and shape memory alloys (SMAs). Piezoelectric materials are mostly implemented for high frequency actuation, in which low strains but high actuation forces are required. SMAs can exert very high forces and can recover very large strains, but within a limited bandwidth. At the Delft University of Technology, a series of experiments is conducted in which the feasibility of the concept is proven and in which several control issues are being addressed. For instance, it has been shown that the control concept can be based on the structural response of the blade to the flow disturbance and that the presence of natural modes has a large influence on the performance of the system. Currently, horizontal axis wind turbine (HAWT) manufacturers are facing several challenges related to both a fast increase in turbine size (see fig. 4.21) and market growth. upscale existing blade designs. However, the boundaries of current technologies are being appraised with gravitational loading becoming a serious design problem. This fluctuating load comes on top of the already high fatigue loads that a turbine has to suffer because of fluctuating aerodynamic loading. These are caused by changes in the wind field and by the rotation of the blade in this wind field. Examples of these are wind shear, yaw misalignment, tower shadow and fluctuations in inflow. These loadings will also increase with increasing size, for instance by the more severe wind shear experienced by larger machines. In fact, fatigue is already one of the biggest design drives and current blades are dimensioned for at least 10^8 cycles. Mitigating the amplitude of the fluctuations could lead to not only a longer service life of blades, but also to lighter blades and a reduced down time due to fatigue load induced component failure. In general, mitigating fatigue loads will lead to a large reduction in the price per kilowatt hour.[35]

4.11.1.3 Delft University of Technology (DU) Airfoil Concepts

The accuracy of airfoil polar predictions is limited by the usage of imperfect turbulence models. Can machine-learning improve this situation? Will airfoil polars teach the effect of turbulence on skin-friction? These questions are answered by

[33] https://www.witpress.com/Secure/elibrary/papers/9781845642051/9781845642051014FU1.pdf

[34] iopscience.iop.org/article/10.1088/1742-6596/555/1/012008/pdf

[35] "Implementation of the 'smart' rotor concept" by Anton W. Hulskamp & Harald E. N. Bersee, Faculty of Aerospace Engineering, Delft University of Technology, The Netherlands cited in https://www.witpress.com/Secure/elibrary/papers/9781845642051/9781845642051014FU1.pdf

refining turbulence treatment in the Rfoil code[36], boundary layer closure relations are learned from airfoil polar data. Two turbulent closure relations, for skin friction and energy shape factor, are parametrized with a class-shape transformation. An experimental database is then used to define code inaccuracy measures that are minimized with an interior point gradient algorithm. Results show that airfoil polars contain exploitable information about turbulent phenomena. Inferred closures agree with direct numerical simulation results of skin friction and the new code predicts drag more accurately. Maximum lift remains under-predicted, but Rfoil maintains its robustness and suitability for optimization of wind energy airfoils. Wind turbine airfoils operate in high Reynolds flows with intricate eddies that cannot be resolved in practical simulations. Flow solvers model the effect of unresolved turbulent phenomena by combining mechanistic insight with closure relations. Closures inject empirical knowledge into simulations and dominate errors in airfoil predictions.[37]

In order to investigate the effect of aerodynamic characteristics of S series airfoils on the straight-bladed vertical axis wind turbine (SB-VAWT), numerical simulations and wind tunnel experiments were carried out using a small SB-VAWT model with three kinds of blade airfoils, which are asymmetric airfoil S809, symmetric airfoil S1046, and NACA0018 used for performance comparison among S series. The aerodynamics characteristics researched in this study included static torque coefficient, out power coefficient, and rotational speed performance. The flow fields of these three kinds of blade under static and dynamic conditions were also simulated and analyzed to explain the mechanism effect of aerodynamic performance. According to the results, the SB-VAWT with airfoil S1046 has better dynamic aerodynamic characteristics than other two airfoils, while the SBVAWT with airfoil S809 is better in terms of the static characteristics. As the most suitable airfoil for SB-VAWT, the S series airfoil is worth researching deeply. In recent years, as one of vertical axis wind turbines (VAWT), the straight-bladed vertical axis wind turbine (SB-VAWT) has developed rapidly and attracted attention of scientists due to its advantages such as wind direction independence, simple structure, and unique shape[38] In 2012, Mohamed analyzed the performance investigation of VAWT using new airfoil shapes.[39] Xu et al. using nonsymmetrical airfoils DU06-W-200 as the research object investigated the influence of installation on the performance of SB-VAWT.[40] Xu et al. analyzed the aerodynamic performance of thickened DU series airfoils[41]. Yang and Li studied an improved VAWT airfoil on the basis of airfoil 4418 of NACA aeries[42]. In 2016, Jia et al. studied three kinds of DU series airfoils including DU25, DU30, and DU35 and found the influence of relative thickness on aerodynamic performance[43]. Based on the research mentioned above, it can be found that the research on SB-VAWT blade airfoils mainly focus on NACA series, DU series, FFA series, and FX series. However, there is a lack of research on S series airfoils which is widely used in HAWT. Therefore, in this study, two typical kinds of S series including S1046 and S809 were selected to investigate whether this kind of blade airfoil is suitable for SB-VAWT or not. NACA0018 airfoil was also selected for the comparison study. The aerodynamics characteristics researched in this study included static starting torque coefficient, out power coefficient, and rotational speed performance. The flow fields of three kinds of blade under both static and dynamic conditions were also simulated and analyzed. This study can provide a good reference for the blade airfoil research on SBVAWT.[44]

4.11.2 Research Centers And Programs In The Netherlands Since 2011

4.11.2.1 ECN (Energy Research Center, The Netherlands)

The Energy Research Centre of the Netherlands (ECN) is one of the largest energy research institute in Europe and holds a strong international position. ECN develops knowledge and technology that enable a transition to a sustainable energy system. The main office is located in Petten a town in the Dutch province of North Holland and has offices in Amsterdam, Netherlands. As of April 1, 2018, the Energy Research Centre of the Netherlands (ECN) has joined forces with the Netherlands Organization for Applied Scientific Research (TNO) and is now 'ECN part of TNO'. From January 2020, the name 'ECN part of TNO' will be replaced by 'TNO Energy Transition'. ECN founded the European Energy Research Alliance (EERA) and participates in the Intergovernmental Panel on Climate Change and the International Energy Agency (IEA).[45]

ECN plays a leading role for sustainable energy for a clean future in which ECN has the technology and expertise needed to rise to this multidisciplinary challenge and play a leading role in

[36] https://odonata-research.com/references/wind-turbine/AIAA-2003-0352.pdf

[37] http://pure.tudelft.nl/ws/portalfiles/portal/45735896/pdf

[38] Fang Feng "Research on Aerodynamic Characteristics of Straight-Bladed Vertical Axis Wind Turbine with S Series Airfoils", Cited in https://www.hindawi.com/journals/ijrm/2018/8350243/#conflicts-of-interest

[39] M. H. Mohamed, "Performance investigation of H-rotor Darrieus turbine with new airfoil shapes," Energy, vol. 47, no. 1, pp. 522–530, 2012.

[40] Z.-H. Xu, Y. Zhang, B. Yang et al., "Effects of installation method of asymmetric airfoils on performance of H-type wind turbine," Acta Energy Solaris Sinica, vol. 34, no. 6, pp. 933–937, 2013.

[41] H. Xu, H. Yang, and C. Liu, "Numerical value analysis on aerodynamic performance of DU series airfoils with thickened trailing edge," Transactions of the Chinese Society of Agricultural Engineering, vol. 17, 2014.

[42] C.-X. Yang and S.-T. Li, "Study of post stalled airfoil of a H-type vertical axis wind turbine," Journal of Lanzhou University of Technology, vol. 41, no. 1, pp. 51–54, 2015.

[43] Y.-L. Jia, P. Peng, Q.-J. Li et al., "Effects of relative thickness of airfoil on aerodynamics of DU airfoil," Machinery Design & Manufacture, vol. 3, 2016.

[44] Y. Li, Y. Zheng, S. Zhao et al., "A review on aerodynamic characteristics of straight-bladed vertical axis wind turbine," Acta Aerodynamica Sinica, vol. 35, no. 6, pp. 368–382, 2018 cited in https://www.hindawi.com/journals/ijrm/2018/8350243/#conflicts-of-interest

[45] https://en.wikipedia.org/wiki/Energy_Research_Centre_of_the_Netherlands.

realizing a clean and sustainable future. ECN energy innovation helps to ensure a clean and sustainable future in the following ways:

- Wide range of scientific disciplines to supply the government and the business sector.
- Carry out extensive research in the fields of sustainable energy, energy efficiency and clean ways of using fossil fuels and provide independent advice to the Dutch government concerning its energy policies.
- Unique mix of know-how and research infrastructure builds a bridge between scientific knowledge and its practical application in the business sector. Organizations are invited to use our extensive test facilities to test technologies and innovations.
- Energy is an area in which the Netherlands excels, which is why the Dutch government has assigned the energy sector as one of nine 'Top Sectors' which offers opportunities to raise the profile of the Netherlands internationally and to strengthen its global competitiveness.
- Organized research, energy innovation, technological development and collaboration between sector participants in seven public-private partnerships known as Top Consortia for Knowledge and Innovation (TKIs). ECN is an active participant in five of these TKIs: Solar Energy, Offshore Wind, Biobased Economy, Gas and Energy Savings in process technology (ISPT).
- The transition to a reliable, sustainable and competitive energy system is one of the Societal Challenges in the European Union (EU) Horizon 2020 research program.[46]

FIG. 4.22: WMC TECHNOLOGY CENTER, WIERINGERWERF, THE NETHERLANDS (SOURCE:[47])

4.11.2.2 WMC (Knowledge Center)

Knowledge Centre WMC Technology Center Netherlands is now part of Lunderskov Møbelfabrik (LM) Wind Power, a GE Renewable Energy business. In 2003 the Group continued in Wieringerwerf under the name "Knowledge Centre WMC" WMC Technology Center Netherlands/LM Wind Power. On Jul 18, 2018, WMC Technology Center Netherlands started as "WMC-Group" at Delft University of Technology, where it performed full-scale testing on rotor blades for wind turbines since 1984 as part of the department of Civil Engineering.[48] Organization of Knowledge Center, WMC (Wind Turbine, Materialsand Constructions) is part of Delft University of Technology, Kluisgat 5, 1771 MV Wieringerwerf, (http://www.wmc.eu), with the mission to carry out research on the materials, components and structures for wind turbines. The WMC-Group of the Delft University of Technology continues its work as the new foundation of Knowledge Centre WMC that has been founded by the Delft University of Technology (TUD) and the Energy Research Centre of the Netherlands (ECN). Since September 2003, the Knowledge Centre WMC is located at a new test facility in the northern part of The Netherlands. In the various research projects experimental and numerical work is combined. Fundamental research projects (e.g., on new materials) as well as contract work for the industry (e.g., full scale blade testing or component verification tests) is carried out. WMC contributes to national and international committees.[49]

Since August 2018 WMC offered services under the umbrella of LM Wind Power, providing the wind industry and other industries with mechanical testing of Rotor Systems, Full-scale Components, Structures and Sub-components, and Materials. In addition, WMC continued to develop the FOCUS6 integral wind turbine design software, to meet the dynamic needs of the global wind turbine design community. The test hall has a dedicated strong floor enabling testing of large structures such as rotor blades, subcomponents, hub/bearing assemblies, yawing mechanisms, nacelles and other structural components for wind turbines. WMC designs test set-ups use versatile testing and monitoring equipment, which can be configured to accommodate large and heavily loaded structures from civil engineering (bridges, lock doors), support structures, crane components, offshore equipment and more. WMC Technology Center, home to one of the most innovative wind turbine rotor test facilities of its kind is situated along the border of the Ijsselmeer, which enables the offshore transport of large structures to the facility.[50]

By bringing WMC into the global network of LM Wind Power Technology Centers enabled rotor hub testing for new GE turbines, blade and other testing, digital tools, research and similar services to the wider wind industry in Netherlands and elsewhere. Focus was on customers providing testing and validation expertise to customers in the wind industry and beyond. Apart from a solid basis in the wind industry, achieved ample experience in testing and

[46] https://www.ecn.nl/energy-research/index.html

[47] https://www.lmwindpower.com/en/services/technology-centers/wmc-the-netherlands

[48] https://search.yahoo.com/search?fr=mcafee&type=E211US0G0&p=WMC+KNOWLEDGE+CENTER+MATERIAL+%26+CONSUTRUCTION+WIKIPEDIA

[49] https://www.narcis.nl/organisation/RecordID/ORG1240788/Language/en.

[50] https://www.lmwindpower.com/en/services/technology-centers/wmc-the-netherlands

FIG. 4.23: WMC SERVICES IN CLUDEFULL SCALE BLADE TESTS (SOURCE[51])

validation of structures for (a) tidal turbines, (b) offshore oil & gas, (c) marine sector, and (d) civil engineering, and (e) automotive. With leadership in technology and professionalism, customers from all sorts of test solutions and wind turbine design tools come to this facility. The importance was on customer focus starting with smart, cost-efficient solutions, not wasting time during the process and delivering excellent results that enable customers to further develop their products.[52,53]

Full-scale blade tests program for wind turbine rotor blades is in accordance with IEC 61400-23, actively involved in the development of the standards for blade tests. WMC services include structures and sub-components tests with the extensive expertise and testing capacity for various sectors. Most of the software for the measurement and control systems is developed inhouse, giving flexibility in carrying out very complex tests on all kinds of structures and components.[54]

Each year, WMC performs thousands of tests on material coupons and small structures. WMC performs standardized material qualification tests for material selection or design data, advising purpose-built custom test programs, e.g., for developing and validating specialized fatigue life prediction models.[55]

FOCUS6 is an integrated modular tool to design wind turbines and components like rotor blades. For nearly two decades FOCUS6 has been used by global wind turbine original equipment manufacturers (OEMs), engineering companies for blade design, certification bodies, research institutes and universities. FOCUS6 integrates onand offshore wind turbine design with blade and support structure design. Head Quartered at LM Wind Power Jupitervej 6, 6000 Kolding, Denmark.[56]

4.11.2.3 Blade Testing and Material Research at WMC

On July 18, 2018, the acquisition of the WMC test facilities in Wieringerwerf, Netherlands helped the industry to test rotor hubs for GE blades and offer a range of testing services to other industry

FIG. 4.24: WMC MATERIAL TESTS' SERVICES (SOURCE: [57])

partners. The facility provided rotor hub testing for new GE turbines, advanced wind turbine rotor blades, composites and testing for digital tools, research and similar services to the wider wind industry in Netherlands and elsewhere. Thus, these test facilities contributed to LM Wind Power's research and development capabilities for the design and manufacture of advanced wind turbine blades and other components, in addition to significant educational contribution to graduates and postgraduates from the best Dutch Universities including Delft, Twente, Eindhoven and Amsterdam. LM Wind Power and GE's growing presence plays a significant role in the strategic green industry in the Netherlands in addition to providing employment for valuable senior, technical jobs with a global management office at World Trade Center, Schiphol and a research facility in Heerhugowaard. Following this acquisition by GE, a world leader in renewable energy, it is now the first company in the wind industry to become fully carbon neutral. [58]

Researchers at TU Delft recently teamed up with Dutch architecture firm Mecanoo to design a bird-friendly wind turbine that can transform wind energy into electricity without any mechanical

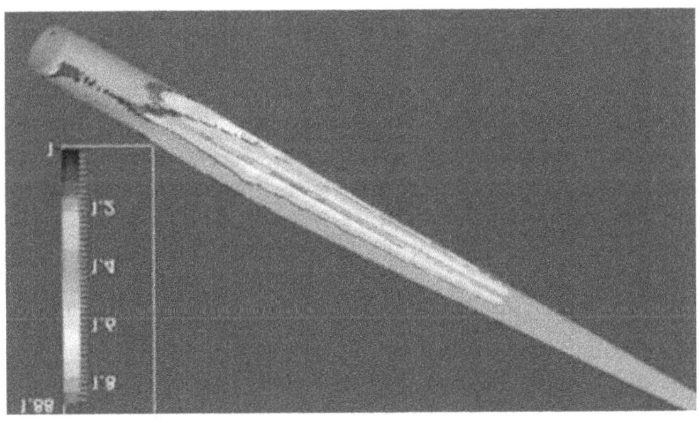

FIG. 4.25: WMC'S FOCUS6 PROGRAM (SOURCE:[59])

---

[51] info.wmc@lmwindpower.com.
[52] Ibid.
[53] https://www.lmwindpower.com/en/stories-and-press/stories/news-from-lm-places/wind-turbine-blade-test-center-wmc- acquired-in-the-netherland
[54] https://www.lmwindpower.com/en/services/technology-centers/wmc-the-netherlands
[55] Ibid.
[56] Ibid.
[57] Ibid.
[58] https://www.lmwindpower.com/en/stories-and-press/stories/news-from-lm-places/wind-turbine-blade-test-center-wmc- acquired-in-the-netherlands.
[59] Ibid.

FIG. 4.26: PICTURE OF WMC FACILITY (SOURCE: [60])

FIG. 4.27: TU DELFT AND MECANOO UNVEIL NEW BLADELESS BIRD-FRIENDLY WIND TURBINE DESIGN CLEAN ENERGY WIND BUILDING (SOURCE[63])

moving parts. The turbine, dubbed the EWICON, was installed at the Delft University of Technology in March. 04/02/2013. The Electrostatic Windenergy Convertor (EWICON) is a wind turbine that has no moving parts. Designed by Mecanoo, the prototype uses the technology developed by the EWI faculty in collaboration with Wageningen University and several companies involved in a government ecology project. The bladeless wind turbine uses the movement of electrically charged water droplets to generate power. It can be installed both onshore and offshore, or mounted on a roof. The EWICON's steel frame supports a framework of horizontal steel tubes. Within the framework, electrically charged droplets are created and blown away by the wind. Their movement creates an electric current that is then passed on to the grid. The remarkable wind turbine produces zero noise pollution, and it doesn't cast shadows, and because there are no moving parts, it has much lower maintenance costs than conventional wind turbines.[61]

4.11.2.4 An Advanced Method for Wind Turbine Wake Modeling [62]

An advanced wake modeling for wind turbine wakes is based on the wake direction since there is no information traveling upstream. Various components of this model including near wake, root vortex and atmospheric boundary layer stability models are analyzed at Energy Research Center of the Netherlands (ECN), Petten, The Netherlands. the Active Wake Control concept that modifies the pitch and yaw angles of upstream wind turbines in order to increase the overall power production and reduce the loading of the turbines. A validation test case is presented where the results obtained from the wake model are compared to the lidar data obtained from an offshore wind farm. Wind turbines extract energy from the flow field which results in wakes that contain less energy and more turbulence than the undisturbed flow and this leads to less energy extraction and higher fatigue loads for the downstream turbines. In large wind farms, most turbines are located in the wake of one or more turbines. The overall loss of a wind farm is very much dependent on the conditions and the lay-out of the farm, but it can be in the order of 5-20%. Apart from the loss in energy production an additional wake effect is formed by the increase in turbulence intensity, which leads to higher fatigue loads. In this sense it becomes important to understand the details of the wake behavior to improve and/or optimize a wind farm layout. Improvements are presented for the existing ECN wake model which constructs the fundamental basis of ECN's Farm Flow wind farm wake simulation tool. The ECN wake model is called WakeFarm1–4 which is based on the original UPMWAKE model proposed by Crespo et al., that simulates the wind turbine wakes by solving the steady parabolized Navier-Stokes equations in perturbation form in three-dimensions. The basic background flow is modeled by an atmospheric wind profile model based on Monin-Obukhov similarity theory. Because the axial-pressure gradient term is neglected in the parabolized Navier-Stokes equations, the perturbation variables are initialized by a near wake model. Schepers pointed out the problem in the near wake and used an empirical velocity-deficit profile as a boundary condition for the far wake. This approach depends on a data-fit with experimental data and the physics of the flow are not modeled explicitly. Schepers and Van der Pijl proposed a model for the near wake based on the free-wake vortex method where the wind turbine is modeled by an actuator disc model and the wake is represented by discrete constant strength vortex rings. They obtained the solution with a panel method. A new near wake model is based on a free wake-vortex method, where the radius of the wake and vorticity strength of discrete vortex rings are varied as suggested by Øye. The induced velocities are obtained by a semi-analytical solution of the Biot-Savart law. A need for a model that extends to the entire boundary layer height is obvious considering the sizes of modern wind turbines. The ECN wake model is extended further based on the numerical solution obtained by the ECN wake model using Gryning model compared with the solution obtained by surface layer model and with the available data obtained by measurements on ECN's Wind turbine Test field Wieringermeer (EWTW). Active Wake Control (AWC) is an approach of operating

[60] Ibid.
[61] Ibid.
[62] https://www.researchgate.net/ publication/322309791_An_advanced_method_for_wind_turbine_wake_modeling
[63] https://inhabitat.com/the-dutch-windwheel-is-not-only-an-incredible-circular-apartment-building-its-also-a-silent-wind- turbine/

wind farms with the aim at improving the overall wind farm performance in terms of power production at below rated wind conditions. It consists of a pitch-based approach (also known as Heat & Flux), and a yaw-based approach (called Controlling Wind). In this study the AWC concept, all the governing equations of the ECN wake farm model are presented. Then the new near wake modeling is discussed and the results compared with the original near wake modeling and EWTW data as well as the results obtained for various near wake implementation cases are shown. The details of the atmospheric stability model are given and the comparison with the solution obtained for the original surface layer model and with the available data obtained by EWTW measurements are presented. Details of AWC concept and quasi-dynamic implementation is given. Some recent results of a validation test case are presented and finally the conclusions are summarized.[64]

4.11.2.5 DUWIND, DELFT NL.[65]

TU Delft Wind Energy Institute, Delft University of Technology, is located at Kluyverweg, NL (duwind@tudelft.nl) instituted in 2015 with TU-Delft experts playing an important and active role in it. R&D programs are the ambition of DUWIND research applications. The research based scientific disciplines involved a cross section of the research subjects, in collaboration with the European Academy for Wind Energy (EAWE) emphasizing on the long term research challenges in wind energy. EAWE-vision along with DUWIND-vision developed the R&D program 2015-2020, with education in wind energy (related) topics was the principal focus of the core subjects in which DUWIND-members were involved. TU-Delft wind energy research goes back to the late seventies. Currently DUWIND spans across six faculties of the TU Delft: Aerospace Engineering (AE), Electrical Engineering, Mathematics and Computer Sciences (EEMCS), Civil Engineering and Geosciences (CEG), Mechanical, Maritime and Material Engineering (3ME), Technology Policy and Management (TPM), and most recently Applied Sciences (AS). The main objectives were 1: Wind in sustainable power supply systems: (a) System Integration; (b) Socially Responsible Innovation, and (c) Regulated Markets and Entrepreneurship; 2: Wind Power Station: Wind Farm Design, and Asset Management; and 3: Wind Energy Converters: Rotor Design, Generator Design, Support Structure Design, Vertical Axis Wind Turbines, and Airborne Wind Energy.[66]

DUWIND R&D program 2015-2020 has three levels of analysis and design: (a) Wind in sustainable power supply systems with interaction of large scale wind power with other energy sources and consumers in the electricity system. This level of the Duwind research program is new and offers cooperation with other sustainable energy sources, under the umbrella of the Delft Energy Initiative. (b) Wind power station: here the main topics of designing and operating an offshore wind farm as power station are included. Integration of all design aspects for lowest cost of energy as well as advanced asset management are the two research lines. (c) Wind energy converters: in this level the turbine itself, its components and required background knowledge, as well as alternative types on wind energy converters are addressed. Optimization of the design for lower cost of energy and longer lifetime are the purposes of this research program.[67]

The DUWIND R&D program 2015-2020 is a strong and wide set of multidisciplinary ambitions which show the strength of the interaction of the different groups within TU Delft and the devotion to improving wind energy research and industry at large:[68]

1. System Integration: Objectives are providing an interdisciplinary and cross sectorial system perspective on the development of offshore wind; Integrated technological and socio-economic modeling and analysis of hybrid power systems; and development of innovative concepts of energy networks, storage and demand side management in support of offshore wind.
2. Socially Responsible Innovation: Objectives are analyzing the specificities of stakeholders' acceptance for offshore wind, its technological and institutional consequences and possibilities for resolution; Development of a value sensitive design for offshore energy systems.
3. Entrepreneurship: Objectives are analysis of innovative regulatory and financial arrangements; Positioning offshore wind as a competitive means of power production; Development of entrepreneurial strategies and business models.
4. Wind Farm Design: Objectives are Understanding and dealing with the mutual influence of involved disciplines; Understanding and dealing with uncertainty and a dynamic design frame work; Guidelines, methods and instruments for effective optimization of wind farms.
5. Next Generation of Asset Management: Objectives are Detailed integrated understanding of wind energy systems that considers information from the component level all the way up to the decisions taken in the control room Optimize the lifetime of ageing assets; Optimal design life of wind farm components (support structures, blades); Long term cost saving through the implementation of condition based maintenance practice.
6. Rotor Design: Objectives are Improve efficiency of multi-megawatt innovative wind turbines by more refined, faster, and integrated multidisciplinary design methodologies; Reduction of development next generation wind turbine rotor risk and time; Improve wind turbine blade manufacturing techniques.
7. Drive Train and Electrical Conversion: Objectives are Improved understanding of the operating principles of the drive train and their limitations; Improved understanding of the reliability issues in drive trains; Improved understanding of the control issues in drive trains and the consequences for the wind turbine design.
8. Support Structure Design: Objectives are Detailed understanding of soil-structure interaction; Reduction of safety factors; Noise reduction during installation and in service; Development of novel support structure concepts; and Integral modeling.

[64] https://www.researchgate.net/ publication/322309791_An_advanced_method_for_wind_turbine_wake_modeling

[65] https://d2k0ddhflgrk1i.cloudfront.net/TUDelft/Onderzoek/Thematische_samenwerking/Onderzoeksinstituten/TU%20Delft%20 Wind%20Energy%20Institute%20%28DUWIND%29/DUWIND_brochure 10LQ_.pdf

[66] Ibid.

[67] Ibid.

[68] Ibid.

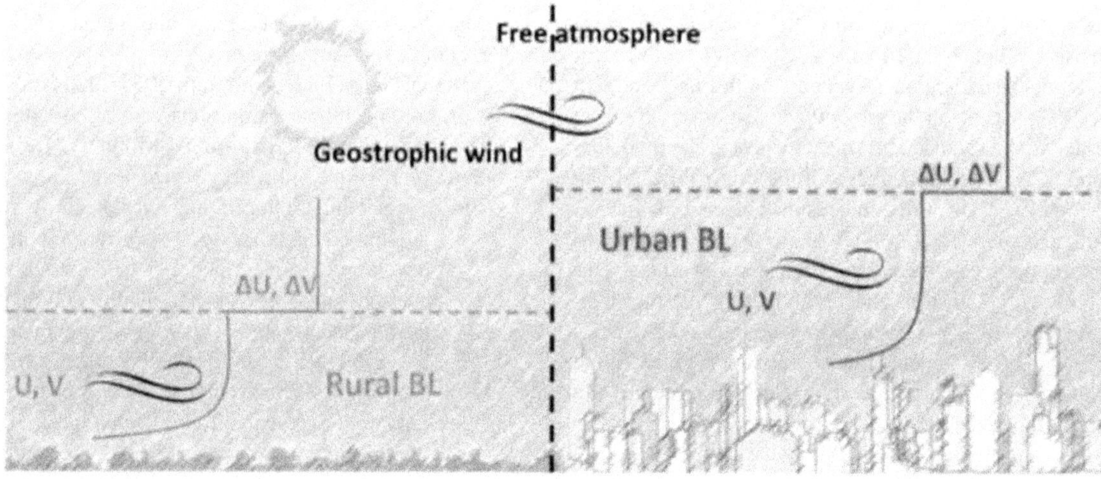

FIG. 4.28: SCHEMATIC OVERVIEW OF THE MODEL DEPICTING THE RURAL REGION INDICATED IN THE GREEN, THE URBAN COLUMN IN THE RED. THE DASH LINE REPRESENTS THE BOUNDARY LAYER DEPTH; U, V AND ΔU, ΔV REFER TO THE ZONAL AND MERIDIONAL WIND AND THEIR RESPECTIVE JUMPS TO THE GEOSTROPHIC WIND. (SOURCE:[69])

9. Vertical Axis Wind Turbines: The Vertical Axis Wind Turbine concept may be the best solution to meet the challenges and opportunities of up-scaling and offshore wind energy, especially of floating offshore wind energy. The research on VAWTs at TU Delft builds upon research in rotor design. The research at TU Delft focuses on rotor and wind farm design, with a focus on aerodynamics and aeroelastic design, including optimization for cost of energy. The TU Delft research develops knowledge, tools and technology to support the development of a 20 MW floating offshore rotor that would allow for a cost of energy below 100€/MWh. Objectives are to Develop the knowledge, tools to design and model VAWT rotors up to 20 MW, with a focus on floating concepts; Develop models for simulation of VAWT wind farms, including optimization of wind farm performance.
10. Airborne Wind Energy: The pioneering kite power research group of TU Delft has developed a pumping kite power system which uses the traction force of an inflatable membrane wing to drive a generator on the ground. The current 20 kW prototype is using a wing of 25 m2 surface area. Objectives are Understanding and dealing with the mutual influence of involved disciplines; Improving the airworthiness, robustness and reliability of AWE technologies; Increasing the component life time of the critical airborne system components; and Developing guidelines, methods and instruments for effective optimization of AWE systems and AWE parks.

4.11.2.6 Environmental Research Impacting Urban Areas[70]

Wind is a key component of the urban climate due to its relevance for ventilation of air pollution and urban heat, wind nuisance, as well as for urban wind energy engineering. These winds are governed by the atmospheric boundary layer (ABL). the dynamics of the atmosphere closest to the surface, the atmospheric boundary layer Making use of a conceptual bulk model of the ABL, we find that for certain atmospheric conditions the boundary-layer mean wind speed in a city can surprisingly be higher than its rural counterpart, despite urban wind island (UWI) the higher roughness of cities. This urban wind island effect prevails in the afternoon and appears to be caused by a combination of differences in ABL growth, surface roughness and the ageostrophic wind, between city and countryside. Enhanced turbulence in the urban area deepens the ABL, and effectively mixes momentum into the ABL from aloft. Furthermore, the oscillation of the wind around the geostrophic equilibrium, caused by the rotation of the Earth, can create episodes where the urban boundary-layer mean wind speed is higher than the rural wind. By altering the surface properties within the bulk model, the sensitivity of the UWI to urban morphology is studied for the 10 urban local climate zones (LCZs). These LCZs classify neighborhoods in terms of building height, vegetation cover etc., and represent urban morphology regardless of culture or location. The ideal circumstances for the UWI to occur are a deeper initial urban boundary-layer than in the countryside, low-rise buildings (up to 12 m height) and a moderate geostrophic wind (~5ms). The UWI phenomenon challenges the commonly held perception that urban wind is usually reduced due to drag processes. Understanding the UWI can become vital to accurately model urban air pollution, quantify urban wind energy potential or create accurate background conditions for urban computational fluid dynamics models.[71]

The heterogeneous landscape of (large) cities causes a complex micro-climate, which can vary from street to street. The combination of ongoing urban expansion and climate change underlines the need to understand the dynamics behind this microclimate. The urban microclimate directly impacts citizen's health through additional heat, air pollution, and influences energy demand. While urban heat has been widely studied, knowledge about urban wind and its variability remains scarce. These winds are governed by

[69] https://iopscience.iop.org/article/10.1088/1748-9326/aad8ef/pdf
[70] Environmental Research "A Virtual Conference 2021, November, cited in https://iopscience.iop.org/article/10.1088/1748- 9326/aad8ef/pdf.
[71] https://iopscience.iop.org/article/10.1088/1748-9326/aad8ef/pdf

the dynamics of the atmospheric layer closest to the surface, i.e., the atmospheric boundary layer (ABL), see figure 4.28. This study aims to quantify the difference in wind dynamics between city and countryside, using the conceptual mixed-layer model for the daytime ABL. Urban wind has hitherto mainly been studied for individual buildings or street canyons, or over limited areas using computational fluid dynamics (CFD) models, though CFD models are increasingly able to model larger urban areas. However, knowledge regarding mean wind behavior at the scale of the ABL can be valuable for more general aspects, such as the mean wind load on buildings, wind potential for energy production or wind nuisance in urban planning. Differences between urban and rural wind dynamics can be caused by increased roughness, by enhanced surface heating due to the urban heat island (UHI) and changed ABL evolution. The aim is to identify and understand these differences in wind behavior through simulating contrasting wind regimes, by altering the surface parameters (roughness, displacement height, built fraction) and the initial conditions forcing the model (wind profile, geostrophic wind speed, boundary layer depth).[72]

The integration of wind energy systems in the urban environment is meaningful because wind energy would then be produced close to where it is needed. The disadvantages are that small wind turbines are less efficient, less economically viable and that mean wind speed in the urban environment is generally lower and turbulence higher than outside the urban environment, either offshore or onshore. For the potential of urban wind energy in the Netherlands, the first choice is vertical axis wind turbines (VAWTs) of the Darrieus type, the second choice being the type of integration. There are three categories of possibilities: (1) siting stand-alone wind turbines in urban locations; (2) retrofitting wind turbines onto existing buildings; and (3) full integration of wind turbines together with architectural form. Category 2 provides the most straightforward solution, i.e., to mount Darrieus VAWTs on the roof on masts that are high enough so that the rotors are situated above the areas of separated flow above the roofs, see figure 4.29.[73]

FIG. 4.29: A VAWT ON A ROOF TOP (SOURCE:[74])

For the assessment of the wind energy potential, we considered 85 Dutch cities and only buildings with a height equal to or above 35 m. The annual statistical mean wind speed distribution was obtained by Weibull distributions. Only the wind speed at roof height was used. It is a first-order approximation and for actual implementation on a specific building, a more detailed wind potential assessment by computational fluid dynamics (CFD) or by a wind tunnel study should be provided. Evidently, installations in the high wind speed area in the Netherlands will yield much higher energy output than in the medium or low wind speed area. The selected turbine was a high-performance VAWT with high power coefficients and high-efficiency (> 97%) direct drive brushless permanent magnet generator (PMG). It was a 2-bladed turbine with diameter 1 m, height 5 m, NACA0018 blade and operating in variable speed. The total number of VAWTs was 18,154, yielding a total Annual Energy Production (AEP) of about 170 GWh. The associated estimated costs, including rotor, tower, electrical components, fixation, grid connection, planning etc. were 310.24 million €, yielding a Levelized Cost of Energy (LCoE) of 90.96 €/MWh (~0.091 €/kWh). This cost is reasonable, but the number of turbines is very high, requiring an excessive installation and maintenance campaign. These 18,154 small VAWTs on the roofs of high-rise buildings would yield a similar AEP as 28 large 2.5 – 3 MW onshore horizontal axis wind turbines (HAWTs) and could cover the yearly electricity demand of about 42,500 average households. Technological developments that are needed to exploit this urban wind energy potential in the Netherlands first and foremost include the realization of highly-efficient and reliable Darrieus VAWTs in large numbers and at low costs. For slender and light-weight VAWTs, the horizontal loads on the building are very small compared to the overall wind loads on high-rise buildings. Noise and structural vibrations should not constitute major challenges if standard vibration dampers are installed. Individual and project-based integration of wind energy systems in iconic building projects can be considered, given the low LCoE, however not without overcoming the challenges and problems. Since 2012, the energy policy in the Netherlands has been focused on specific technologies. For Urban Wind Energy the potential for wind energy harvesting on top of high-rise buildings in urban areas exists. Which solutions are currently commercially available, and which are in development to harvest wind energy on high-rise buildings? Two different levels of solutions are considered: (1) solutions in terms of types of wind turbines and (2) solutions in terms of positioning around, in or on top of high-rise buildings. Among the small wind turbines, a distinction is made between horizontal-axis wind turbines (HAWTs) and vertical-axis wind turbines (VAWTs). The growing interest for VAWTs can be attributed to several advantages compared to HAWTs, such as Omni-directional capability and no yaw system is needed. This characteristic of VAWTs is very important for the urban environment where the wind direction exhibits larger and more frequent variations. Low noise, due to operating at lower tip speed ratios and smaller diameters, the blade tip speed is lower, which leads to lower levels of aerodynamic noise. Low manufacturing cost, due to simple blade profiles with no taper and twist as well as simplicity in the control system, i.e., no pitch and yaw control

[72] https://iopscience.iop.org/article/10.1088/1748-9326/aad8ef/pdf
[73] https://www.topsectorenergie.nl/sites/default/files/uploads/Urban%20energy/publicities'/Report%20Urban%20Wind%20Energy%20TKI%20RVO.pdf
[74] Ibid.

system. Low installation and maintenance costs, due to having the generator on the ground. Scalability, the turbine height-to-diameter ratio can scale up with minimal effect on performance.[75]

4.11.2.7 European Wind Atlas and the Netherlands Wind Energy Research

Complex terrain experiments in the New European Wind Atlas project create a freely accessible wind atlas covering Europe including Netherlands, Holland and Denmark to develop the model to perform experiments in many different kinds of complex terrain to validate the models. These experiments are described to focus on the flow properties that are relevant for wind turbines, so the main focus is the mean flow and the turbulence at heights between 40 and 300 m. Also, extreme winds, wind shear and veer, diurnal and seasonal variations of the wind are studied. Common to all the studies is the use of Doppler lidar systems to supplement and in some cases replace meteorological towers. Many of the lidars will be equipped with scan heads that will allow for arbitrary scan patterns by several synchronized systems. Synchronized scanning lidar, both in terms of the accuracy of the measurements and the atmospheric physical processes are studied and validated for use in various complex terrains, such as in Spain, Portugal and Netherlands.[76]

The New European Wind Atlas (NEWA) is a project concerned with (i) creation of a freely accessible wind atlas covering the European Union, Turkey and European coastal water within 100 km off the shore. The main content of the atlas is not only sitespecific wind resources, but also extreme winds, turbulence, shear and other parameters relevant for wind turbine siting; (ii) development of models or a model chain to produce the high-resolution wind atlas; (iii) atmospheric flow experiments in various kinds of complex terrain to validate the models. The main focus of the modelling activity is to downscale mesoscale meteorological models, so they can be used to predict wind in local terrain, which is done by computational fluid dynamics (CFD), typically Reynolds-averaged Navier–Stokes (RANS) models. These models are going to be applied to an entire continent, considerations about computational efficiency are important. Also, the much more computationally intensive large-eddy simulation is used to understand the flow over complex terrain. The largest part of NEWA, and the lasting contribution to our knowledge about atmospheric flow for wind energy purposes, is a series of complex-terrain experiments which is the focus of this contribution. Doppler scanning of flow over steep terrain has been done before, but rarely fully three-dimensional, which requires the coordination and synchronization of at least three lidars.[77]

Active radio wave back-scatter from satellites is used to map the wind resources offshore, but as the returned signal basically gives information on the drag on the surface it is important to know how to extrapolate that information up to turbine relevant heights. Careful analysis and extrapolation indicate that wind resources derived from satellites are comparable in uncertainty to mesoscale models. Additional uncertainties in both satellite wind and mesoscale models appear close to the shore. An experiment to map the wind resources the first 5 km from the coast with several scanning lidars was performed at the west coast of Denmark (see 'RUNE' in figure 4.30) close to the large turbine test station Høvsøre. An experiment to study a greater level of terrain complexity started in April 2016 near Hornamossen in Sweden (see figure 4.30). For the first time ever 12 long-range scanning Doppler lidars measured winds over a forested hill near Kassel in August 2016, see figure 4.30.[78]

Terrain inclination, roughness change, stratification, gravity waves, valley flows, recirculation, separated flow impinging on a hill are a plethora of atmospheric flow features requiring further study that can be encountered in a well-planned field experiment. The importance of these phenomena has raised the interest of US universities and research institutes in atmospheric sciences to participate in the 2017 campaign, Perdigão-2017. A large array of sensors was located on meteorological towers, 10 to 100 m high, distributed along the valley, the SW and NE ridges, and two transect lines perpendicular to the ridges (figure 4.31).[79]

An experiment was conducted to make use of the instrumentation and data management infrastructure of National Renewable Energy Centre (CENER's), Sarriguren, Spain test site for large wind turbines in Alaiz. The Alaiz mountain range is located in Navarra (Spain), around 15 km SSE from Pamplona, see figure 4.30. To the north, a large valley is found at around 700 m lower altitude limited by a 150 m high ridge (Sierra de Tajonar) that runs parallel to the Alaiz mountain range 5 km to the north. To the south, complex terrain is found with the presence of some wind farms operated by Acciona, the closest one situated 2 km behind the row of six wind turbines of the test site, see figure 4.32.

The conclusion of the series of atmospheric field tests conducted with the purpose of characterizing wind flow that is relevant for wind energy is that although wind resources is the primary concern, turbulence, extreme winds, shear, and diurnal and seasonal variations are all important as shown by the use of Doppler lidars in the experiments. Other conclusions concern the flow in physics and modelling: analysis of the RUNE experiment which shows a good agreement between lidar data and mesoscale modelling, but more varied results are found for satellite-derived winds in this near shore area. Indications from Hornamossen are that the boundary-layer height is an important parameter for the understanding of the turbulence profiles at wind turbine relevant heights, especially under stable atmospheric conditions. Two-dimensional flow fields measured at Østerild (see figure 4.31) at 50 m a.g.l. are related to the variation in the height of the underlying canopy. The ultimate goal of the experiments is to provide high quality measurement data for the validation of the models used to produce the New European Wind Atlas. Emphasis is not only on the distribution of mean wind speeds but also on turbulence, shear, extreme winds and other parameters important for wind turbine siting.[80]

---

[75] Ibid.

[76] https://royalsocietypublishing.org/doi/full/10.1098/rsta.2016.0101

[77] J. Mann et al.2017: Complex terrain experiments in the New European Wind Atlas. Phil. Trans. R. Soc. A 375: 20160101. http://dx.doi.org/10.1098/rsta.2016.0101, 23-November-2016, also cited in https://royalsocietypublishing.org/doi/full/10.1098/rsta.2016.0101

[78] https://royalsocietypublishing.org/doi/full/10.1098/rsta.2016.0101

[79] https://royalsocietypublishing.org/doi/full/10.1098/rsta.2016.0101

[80] https://royalsocietypublishing.org/doi/full/10.1098/rsta.2016.0101

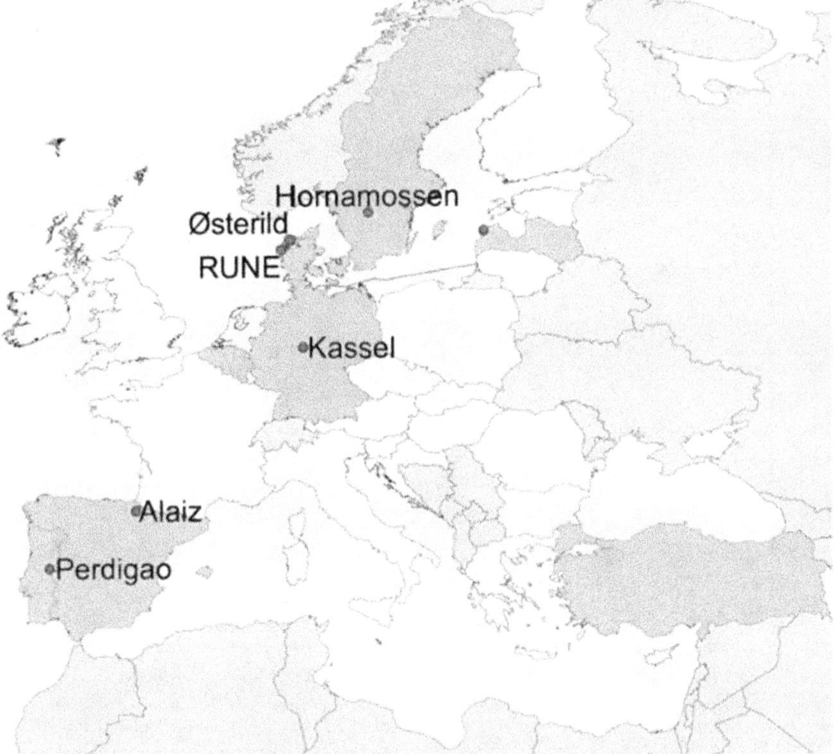

FIG. 4.30: THE MINIMUM ONSHORE COVERAGE OF THE ATLAS AND GREEN NUANCES, THE NEWA PARTNER COUNTRIES IN GREEN, OFFSHORE COVERAGE IN BLUE, MAIN EXPERIMENT SITES IN RED, AND FERRY-MOUNTED PROFILING DOPPLER LIDAR MEASUREMENTS AS A RED CURVE. (SOURCE: [81])

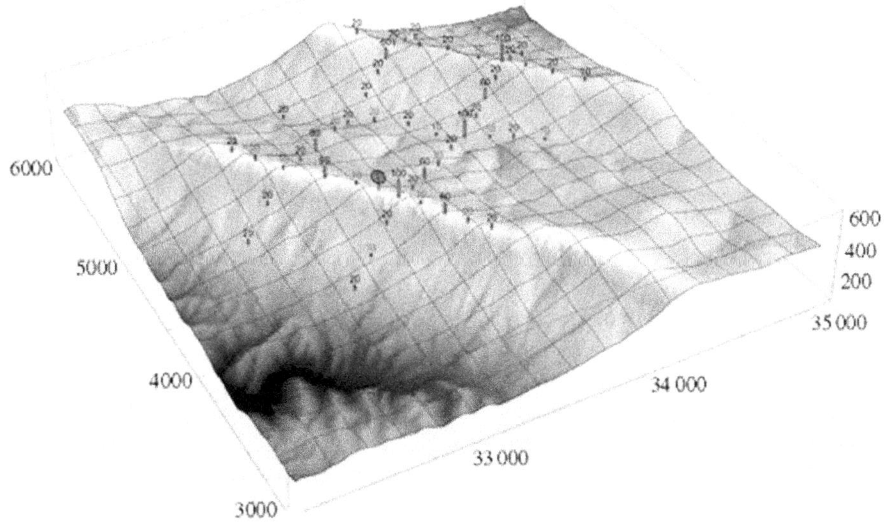

FIG. 4.31: TERRAIN AND TOWER OF THE PERDIGO, 2017. (SOURCE: [82])

[81] J. Mann et al.2017: Complex terrain experiments in the New European Wind Atlas. Phil. Trans. R. Soc. A 375: 20160101. "http://dx.doi.org/10.1098/rsta.2016.0101"http://dx.doi.org/10.1098/rsta.2016.0101, 23-November-2016, also cited in https://royalsocietypublishing.org/doi/full/10.1098/rsta.2016.0101

[82] https://royalsocietypublishing.org/doi/full/10.1098/rsta.2016.0101

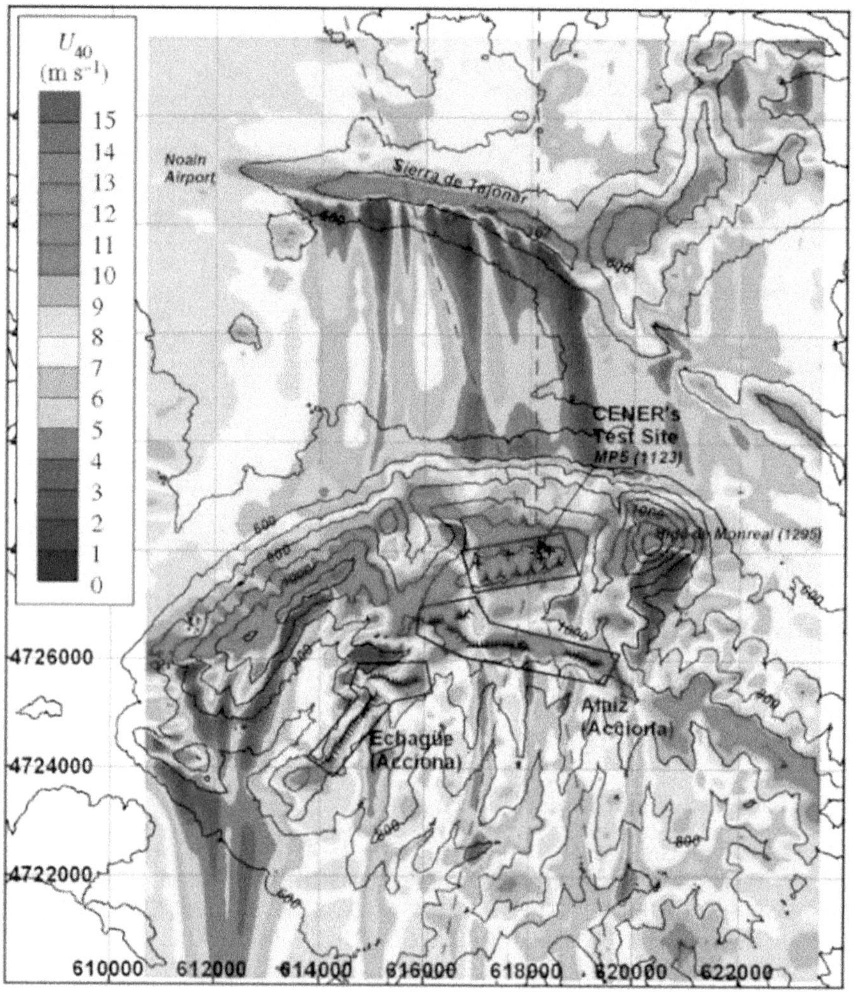

FIG. 4.32: CONTOUR MAP OF HORIZONTAL WIND SPEED AT 40 M ABOVE GROUND LEVEL (A.G.L.) SIMULATED WITH CFD WIND. WIND DIRECTION FROM THE NORTH. NORTHERN AND SOUTHERN SECTORS OF INTEREST FOR THE ALAIZ EXPERIMENT DELIMITED BY DASHED LINES AND CENTERED AT THE MP5 POSITION OF CENER'S TEST SITE. (SOURCE:[83])

4.11.2.8 Netherlands Wind Energy Association (NWEA) Research Policy Plan[84]

The goal of NWEA 2020, is towards at least 70% renewable electricity by 2030 in Netherlands. NWEA offers our members a powerful lobby, on a variety of topics regarding wind energy with participation in a network with national and regional authorities, with politicians, policymakers, scientists and academic institutions, businesses, environmentalists, journalists and other (social) and nongovernmental organizations (NGO's) in the Netherlands. NWEA maintains participation on committees, subcommittees and/or working groups on all the main topics, both nationally and internationally to maintain up-to-date sector intelligence, experience, knowledge, training and innovations. The challenges facing the Netherlands are ample with still a relatively high share of fossil fuels in the energy mix (92.6%) and a low degree of electrification (20%). The Dutch potential for wind energy is 15 GW onshore and 75 GW offshore. NWEA members their own voice in the national energy transition and climate discussions. NWEA represents the sector in international associations, including in particular Wind Europe, where knowledge is also exchanged with other national wind energy associations. The mission of NWEA was to contribute to a lowcarbon society within and outside of the Netherlands and a strong export position for the Dutch wind energy sector. NWEA strategy for Europe is to closely link it to WindEurope. For the upcoming years NWEA has outlined 5 focal points A Healthy Business Model, Community, No Supply Without Demand, Strengthening the Supply Chain, and Visibility.[85]

[83] https://royalsocietypublishing.org/doi/full/10.1098/rsta.2016.0101
[84] https://www.nwea.nl/nwea-beleidsplan-2020-2022/?lang=en
[85] https://www.nwea.nl/nwea-beleidsplan-2020-2022/?lang=en

FIG. 4.33: RWE BUILDS WIND TURBINES ON A SEA DIKE DAM IN THE NETHERLANDS (SOURCE: [86])

4.11.3 Programs AND Progress Of Wind Energy In Netherlands Since 2011[87]

As of October 2020, wind power in the Netherlands has an installed capacity of 4,990 MW, 19% of which is based offshore. The Dutch government has set a target of 6,000 MW of onshore wind power by 2020 and 4,450 MW of offshore wind power by 2023. This will contribute towards the Dutch target of 14% renewable energy use out of total energy use by 2020 and 16% by 2023. For offshore wind a new system of tendering is under development. The two Dutch wind farms that have contracted to sell power to NS (Noordoostpolder and Luchterduinen) are both connected directly to the Dutch grid, along with all the other power plants in the country, and NS draws its power from the grid.[88]

4.11.3.1 RWE, NL Wind Turbines on Dutch Dams

The Oostpolderdijk Wind Farm in the Dutch province of Groningen consists of three turbines on dykes. This is the first time that such a project has been undertaken. Due to limited space at the site of the dike a climbing crane is used to install the top turbine components, see figure 4.33. The foundations and the lowest turbine components of the wind turbines have already been built. One wind turbine has been fully completed, and the climbing crane has begun the further construction of the second wind turbine. RWE and Water Board Noorderzijlvest (the owner of the sea dike dam) conducted meticulous research into the construction process, dike safety, and the impact of the project on the local environment, which resulted in a highly specialized technical design for implementation. The most important aspect is to guarantee the safety of the dike and its dam function. Wind Farm Oostpolderdijk, with a total installed capacity of 7.5 MW, is expected to be operational by the end of 2021. Sustainable production portfolio in the Netherlands shows that RWE is a globally leading player in the energy transition sector. RWE is currently operating seven onshore wind farms with a capacity of 268 MW, including Oostpolderdijk Wind Farm and is now constructing four new onshore wind farms in the Netherlands with a total capacity of 116 MW. In addition, RWE is also working on the expansion of CO_2-free production capacity. Oostpolderdijk Wind Farm is the first of its kind worldwide with wind turbines on a sea dike dam and is committed to expand the opportunities in the field of sustainable electricity production. By combining knowledge, innovation and determination, RWE has added a new function to a sea dike dam in the Netherlands.[89]

4.11.3.2 Infrastructure for Wind Energy in The Netherland

A €250 million (£215m) loan has been agreed by the European Investment Bank (EIB) to finance the upgrade of an energy infrastructure project in the north of the Netherlands. Grid operator TenneT Holding will use the loan to build a 40-kilometre long electricity transmission corridor from Eemshaven to Vierverlaten in the province of Groningen, enabling the transmission of offshore wind energy to users in the Netherlands and beyond. The upgraded energy infrastructure will support the connection of wind farms in the North Sea to the electricity grid and ensure a high security of supply. Once the new transmission lines are operational in 2024, the existing smaller capacity lines along the same route will be decommissioned and the project will result in an overall doubling of capacity of the existing 220kV circuit. The Netherlands' ambitious targets for renewable energy

[86] https://www.bing.com/search?q=dutch+wind+power&qs=HS&pq=dut&sc=8-3&cvid=6363464B9200487E9C10D118E2AE0927&FORM=QBLH&sp=1
[87] https://en.wikipedia.org/wiki/Wind_power_in_the_Netherlands
[88] https://www.evwind.es/2021/10/20/rwe-builds-wind-turbines-on-a-sea-dike-dam-in-the-netherlands/82883
[89] https://www.evwind.es/2021/10/20/rwe-builds-wind-turbines-on-a-sea-dike-dam-in-the-netherlands/82883

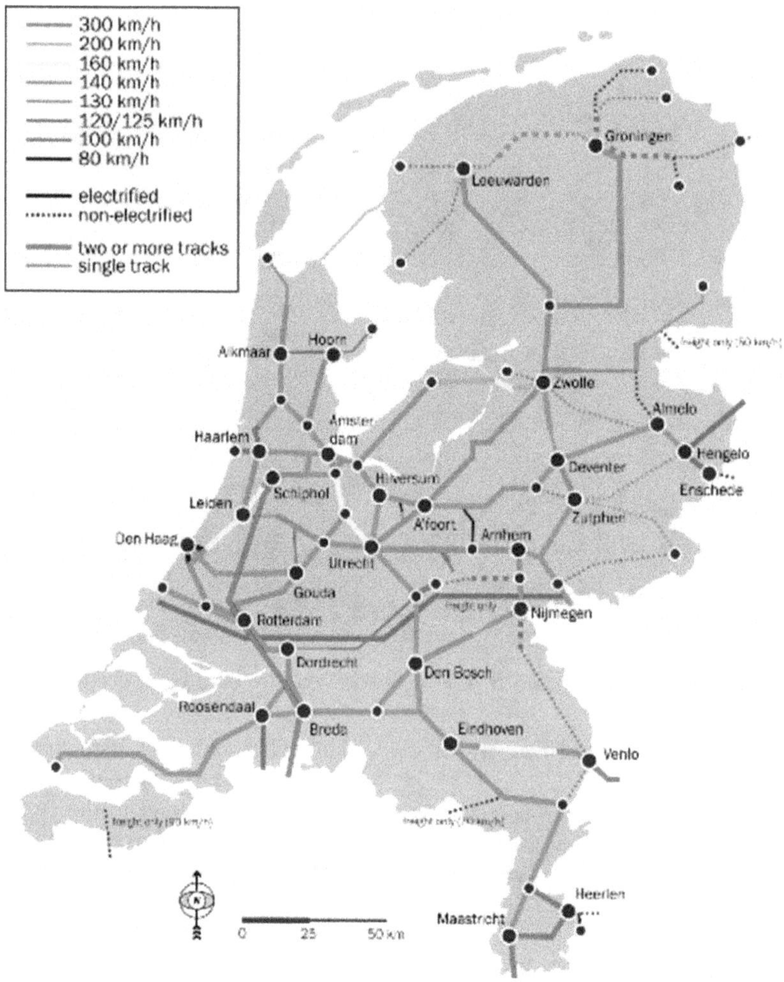

FIG. 4.34: THE NETHERLANDS' RAILWAY NETWORK (SOURCE:[90])

generation would require more investment in the transmission network and EIB provides financing to contribute for achieving this goal. This project secures sustainable electricity to users across the Netherlands.[91]

4.11.3.3 Trains by Wind Energy in The Netherland

Nearly 10-15% of the electricity consumed annually by NS's electric trains will come from wind. NS clarified that only the electrified portion of its system runs completely on wind power. This is about two-thirds of the system.[92] The electrified portion of the Dutch railway network (*Nederlandse Spoorwegen*, or NS) runs on grid electricity that comes dominantly from fossil fuel generation (natural gas and coal). NS claims 100% wind power because it has a contract with various wind farms to produce enough energy to power its rail system, but this is just an accounting transaction. Only a small fraction of the power delivered to its trains actually comes from wind. In The Netherlands' electricity sector, installed capacity is dominantly fossil fuel, with natural gas making up 61% of total installed capacity and coal 15%, wind contributes 4,117MW, representing 13% of the capacity mix. According to Cleantechnica wind power in the Netherlands generates 7.4 billion kWh (7.4TWh) of electricity annually, and according to BP the Netherlands' total electricity generation in 2015 was 109.6TWh. However, wind power consumption in the Netherlands in 2015 was 12.5TWh, indicating that about 5TWh of wind power was imported during the year. While wind contributes about 7% to the Netherlands' electricity generation it contributes about 11% to the country's electricity consumption. The amount of electricity NS uses to power its electric trains is 1.2 or 1.4TWh/year. The Netherlands imports wind power basically because it's falling behind its EU renewable energy targets. Eneco, which contracts to supply NS with wind power, gets a "Guarantee of Origin" from the exporter under which the exporter confirms that the power came from wind and assigns the rights to it to NS. For NS to use only wind power from wind farms to power its rail system the wind farms must be connected directly to NS's railways. In Figure 4.34 the dotted lines show the non-electrified track. According to LJ Electrical only 2,231km of NS's total 3,223km of track is electrified:[93]

---

[90] https://www.energylivenews.com/2021/09/21/dutch-energy-infrastructure-upgrade-granted-e250m-eu-loan/

[91] http://energyfairness.org/dutch-trains-really-running-completely-renewable-power/

[92] http://euanmearns.com/do-the-netherlands-trains-really-run-on-100-wind-power/

[93] http://euanmearns.com/do-the-netherlands-trains-really-run-on-100-wind-power/

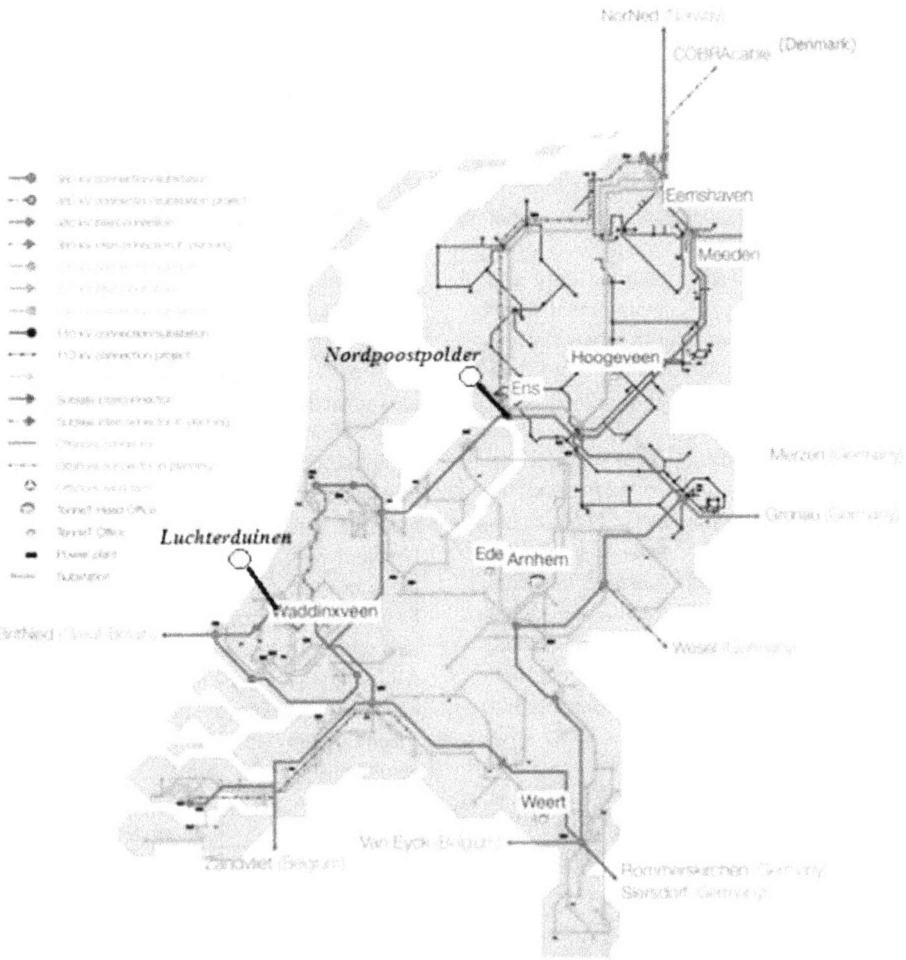

FIG. 4.35: THE NETHERLANDS' ELECTRICITY GRID. GRID CONNECTIONS FOR THE LUCHTERDUINEN AND NORDPOOSTPOLDER WIND FARMS ARE SHOWN IN BLACK.(SOURCE:[94])

The two Dutch wind farms that have contracted to sell power to NS (Noordoostpolder and Luchterduinen) are both connected directly to the Dutch grid, along with all the other power plants in the country and NS draws its power from the grid. When wind power is fed into a grid it becomes inextricably mixed with all the vibrating electrons from other generation sources to the point where there is no way of knowing where any power taken from the grid came from. Grid power in fact reflects the overall generation mix, which in the case of the Netherlands it is dominantly gas and coal with only a small contribution from wind. Over the course of this year the average will be around 11%, equal to wind power's share of the Netherlands' annual grid electricity consumption. And only half of the wind power NS has contracted comes from the Netherlands. The other half comes from "newly built wind farms in …. Belgium and Finland".

Only 10-15% of the electricity consumed annually by NS's electric trains will come from wind, with the rest a mixture that includes mostly Dutch gas and coal plus a small amount of Belgian and German coal, nuclear and lignite and maybe even a little German solar. The supply of wind power to the Dutch grid will also not be constant. With an installed capacity of around 1850 MW in February 1, 2017 the overall wind capacity factor was 11% and there were a number of occasions on which wind generation fell effectively to zero for hours on end. During these periods wind generation in the Netherlands would also have fallen to low levels. Were these conditions to repeat themselves now, and if NS's trains were powered exclusively by wind, they would almost certainly come to a halt, although "wind farm portfolio guarantees sufficient capacity to cover such eventualities" . NS actually claims that only the electrified portion runs on 100% wind. According to Railway Technology NS has a: "green energy contract – thought to be among the largest yet signed in Europe – between power supplier Eneco and VIVENS, an energy procurement joint venture comprising Netherlands Railways (NS), Veolia, Arriva, Connexxion and rail freight firms", and because "NS and Eneco have

[94] http://euanmearns.com/do-the-netherlands-trains-really-run-on-100-wind-power/

carefully selected a list of wind farms that fulfil their criteria of being traceable, sustainable or renewable energy", and because "this partnership ensures that new investments can be made in even newer wind farms, which will increase the share of renewable energy. In this way, the Dutch railways aim to reduce the greatest negative environmental impact caused by CO_2 in such a way that its demand actually contributes to the sustainable power generation in the Netherlands and Europe." by purchasing wind power that would otherwise have gone elsewhere NS is leaving the door open for more wind projects and more CO_2 reductions, the only one that offers any tangible benefits. But there is no guarantee that the unfilled demand will be met by renewables, and in any event the 1.2-1.4 TWh/year consumed by NS represents barely more than 1% of the Netherlands' annual electricity consumption.[95]

4.11.3.4 The German-Dutch Wind Tunnels DNW

As of January 1, 2016, DNW is one of Europe's most advanced and specialized organizations for wind tunnel testing. DNW NWB transferred to new business unit Braunschweig / Göttnigen / Köln (BGK), formerly GUK. DNW's six wind tunnels include subsonic, transonic and supersonic facilities, and provide experimental aerodynamic simulation capabilities. DNW provides techniques for aerodynamic, aero-acoustic or aeroelastic simulations and tests of scale models in a controlled environment. Its sophisticated variety of wind tunnels and varying wind speeds required for experimental simulation, DNW also provides the expertise and support for designing the appropriate measurement techniques. The techniques use the most sophisticated tools for obtaining quantitative information needed to describe the phenomena dealt with in wind tunnel testing rotating windmill blades.[96]

The Large low-speed facility, (LLF) in Marknesse, NL is an industrial wind tunnel for the low-speed domain. Type of wind tunnel is a closed circuit, atmospheric, continuous low-speed wind tunnel with one closed wall and one configurable (slotted) wall test section and an open jet.[97] Next is the supersonic and transonic wind tunnel TST-27 is a blowdown wind tunnel with a test section 280 mm wide of height varying from 250 mm to 270 mm.[98] The transonic wind tunnel (HST) in Amsterdam is an industrial wind tunnel for the subsonic, transonic and low supersonic domain. Type of wind tunnel is a closed circuit, variable density, continuous wind tunnel with slotted top and bottom test section walls (12% open).[99] The transonic wind tunnel (TWG) in Göttingen, Germany is a facility for research and development tests in the transonic and supersonic domain. The Type of wind tunnel is a closed circuit, continuous, variable density sub-transand supersonic wind tunnel with three exchangeable test sections.[100] The low-speed wind tunnel (LST) in Marknesse, NL is an industrial wind tunnel for the low-speed domain. Type of wind tunnel is a continuous, atmospheric, low-speed wind tunnel with exchangeable test sections.[101] The NWB in Braunschweig, Germany is a closed-circuit low speed wind tunnel of the atmospheric type, located on the premises of the DLR, Germany. From 2009 to 2010, the NWB underwent substantial refurbishment to become an aero-acoustic wind tunnel of excellent acoustic properties, while at the same time its well-known high aerodynamic qualities were preserved. Type of wind tunnel is an industrial wind tunnel for the low-speed domain with cutting edge technology and with capabilities for aero-acoustic and dynamic testing.[102]

As noted above, DNW innovations in wind tunnel testing and new moving belt system for Large Low speed Facility are in the German-Dutch Wind Tunnels DNW which is one of

FIG. 4.36: TRAINS ELECTRIFIED BY WIND ENERGY IN THE NEDERLAND (SOURCE:[103])

[95] http://euanmearns.com/do-the-netherlands-trains-really-run-on-100-wind-power/

[96] https://i1.wp.com/www.nationofchange.org/wp-content/uploads/2017/01/nsmolen01- e1484168837811.jpg?fit=800%2C532&ssl=1

[97] https://www.dnw.aero/services/general/

[98] https://www.dnw.aero/wind-tunnels/

[99] https://www.tudelft.nl/en/ae/organisation/departments/aerodynamics-wind-energy-flight-performance-and-propulsion/facilities/high-speed-wind-tunnels/tst-27-transonicsupersonic-windtunnel

[100] https://www.dnw.aero/wind-tunnels/hst/

[101] https://www.dnw.aero/wind-tunnels/twg/

[102] https://www.dnw.aero/wind-tunnels/lst/

[103] https://www.dnw.aero/wind-tunnels/nwb//

Europe's most advanced and specialized organizations for wind tunnel testing. DNW's wind tunnels include subsonic, transonic and supersonic facilities, and provide experimental aerodynamic simulation capabilities to a global user community at large. DNW provides techniques for aerodynamic, aero-acoustic or aeroelastic simulations and tests of scaled models in a controlled environment. Its experimental simulation techniques capture the essence of the issues to be investigated. The Large Low Speed Facility (LLF) in Marknesse (the Netherlands) is an industrial wind tunnel for the low-speed domain. It is a closed circuit, atmospheric, continuous low-speed wind tunnel with one closed wall and one configurable (slotted) wall test section and an open jet. Low speed means testing of aircraft in take-off and landing flight configurations and therefore DNW focusses its investments for the LLF on safety (ground proximity, powered and unpowered) and environmental issues (acoustics) related testing capabilities. Recent DNW-LLF upgrade programs focused on ground proximity simulation (procurement of a new moving belt system) and reducing of wind tunnel circuit background noise level to improve its capabilities and market attractiveness. The main drive for the latter initiatives is a clear trend in aircraft characteristics, i.e., continuous reduction of aircraft noise levels. Funding support was provided by the Ministry of Economic Affairs (the Netherlands), the German Aerospace Center DLR and the European Commission through EU 7th Framework European Strategic Wind Tunnels Improved Research Potential ESWIRP. This reference details the various development steps taken for the new moving belt system and elaborate on the calibration activities conducted.[104]

4.11.3.5 Recent Developments of Wind Farm Design[105]

The Dutch are trying to meet the EU-set target of producing 14% of total energy use from renewable sources by 2020, and 16% by 2023. In this attempt many of these smaller companies were acquired by the larger wind turbine manufactures such as Vestas, Siemens and Lagerwey. Since 2015 there has been a trend towards the deployment and planning for large wind farms, both onshore and offshore, with a view to the approximate tripling wind power capacity from 2015 levels by 2023. These newer and larger wind farms are making use of some of the largest wind turbines available, in particular in the case of Noordoostpolder the Enercon E-126 7500 kW wind turbine, the largest onshore turbine available at that time.[106]

The Dutch government has set future targets of 6,000 MW of onshore wind power by 2020 by 2023. This will contribute towards the Dutch target of 14% renewable energy use out of total energy use by 2020 and 16% by 2023. Some of the newer larger wind farms currently under construction are replacing smaller turbines previously installed at given locations. One such example is the largest wind farm in Holland, Noordoostpolder, which is installing industrial scale wind turbines such as the new Enercon onshore wind turbines, at the time the largest onshore turbines in the world rated at 7500 kW. The turbines have a hub height of 135m, and each new turbine can generate as much electricity as all 50 turbines of the old wind farm combined. The wind farm demonstrates the growth in scale in wind power as between 1987 and 1991 was the largest of its kind in Europe with 50 turbines and a total capacity of 15 MW. The new Zuidwester site at the wind farm will have just 12 turbines and a total capacity of 90 MW. Timeline of developments are:[107]

2011: In November 2011, the Dutch government decided to no longer fund €6 billion per year to maintain subsidized wind kWh at €0.18. It sharply cut subsidies down to €1.5 billion, leaving private sector to carry over wind turbine investments should these be beneficial.

2013: By December 2013, 1,975 wind turbines were operational on land in the Netherlands, with an aggregate capacity of 2,479 MW. An additional 228 MW of capacity was installed at sea.

2015: Whilst there are a growing number of large onshore wind farms in Holland, most of the onshore farms in 2015 consisted of a large number of small wind farms, including many single installation turbines as well as farms of between 1–10 turbines. In total there were 3,004 MW of onshore wind power at the end of 2015. Most wind power is produced in the province of Flevoland.
2018: In March 2018 the Dutch government announced that it will build between 2023 and 2030 several new windfarms with a total installed capacity of 7 gigawatt.[14] Where the parks will come

TABLE 4.3: CUMULATIVE LAND-BASED WIND CAPACITY (MW) IN THE NETHERLANDS[108]

Year	2010	2011	2012	2013	2014	2015	2016	2017	2018	2019	2020
Land-based Capacity (MW)	2,009	2,088	2,205	2,485	2,637	3,034	3,300	3,245	3,436	3,572	4,278
Installed Capacity (MW)	2,237	2,316	2,433	2,713	2,865	3,391	4,257	4,202	4,393	4,484	6,719
Electricity Generated (GWh)	4,503	4,725	4,939	5,368	5,810	6,917	8,364	9,642	10,030	10,774	13,896
% Electricity Consumption	3.69	3.84	4.10	4.47	4.92	5.81	6.97	7.98	8.22	8.84	11.62

[104] Christophe Hermans and Sinus Hegen in CEAS Aeronautical Journal (2018) 9:283–290 https://doi.org/10.1007/s13272- 018-0285-4; https://link.springer.com/content/pdf/10.1007%2Fs13272-018-0285-4.pdf
[105] https://en.wikipedia.org/wiki/Wind_power_in_the_Netherlands
[106] https://en.wikipedia.org/wiki/Wind_power_in_the_Netherlands
[107] https://en.wikipedia.org/wiki/Wind_power_in_the_Netherlands
[108] https://en.wikipedia.org/wiki/Wind_power_in_the_Netherlands

and how they will be built is still being discussed. However, the government wants 40% of the total electricity used in 2030 in the Netherlands to be supplied by wind.

The largest wind farm in the Netherlands is Noordoostpolder with a capacity of 429 MW. The farm consists of three smaller wind farms owned by separate organizations with local community links. The wind farm's 86 wind turbines will produce 1.4 billion kWh of electricity, enough power for over 400,000 households every day, a figure roughly equivalent to the number of households in the Northern Netherlands. The next largest onshore wind farm is the Westereems wind farm located in Eemshaven in Groningen, in the north of the Netherlands. Up until the Westereems Wind Park was commissioned the largest onshore wind farm was the Princess Alexia Windpark in Zeewolde (122MW), consisting of 36 REpower turbines and a 3.2MW/33MWh battery storage power station of BMW i3 batteries. Other large wind farms are located in Delfzijl-zuid (43 MW), Lelystad (46 MW), Terneuzen, (Koegorspolder, 44 MW), and Biddinghuizen (WP Kubbeweg, 34 MW). The Rachel Carson wind farm (also known as the Eemmeerdijk Wind Park) consists of wind turbines with just 2 blades, manufactured by Nedwind, the 18 turbines have a combined nominal power of 18 MW.[109]

4.11.3.6 Recent Developments of Offshore Wind Farm Design[110]

The Netherlands is well prepared for a significant rise in the production of intermittent power from wind energy by good linkages to its neighbors via high voltage cables enabling power to be imported or exported as required. These include the 580 km NorNed submarine cable (700 MW) link to Norway, the 1,000 MW BritNed cable link to the United Kingdom and the COBRA cable link to Denmark (700 MW). The 600 MW Gemini offshore wind farm was commissioned in 2017. For offshore wind a new system of tendering is under development. New wind farm deployment is based on the SER agreement that describes a plan for five years of tendering 700 MW per year. Under this system the government chooses locations and organizes tenders for projects of 350 MW, and project developers can offer bids for each farm. These sites are also surveyed centrally by the government allowing developers to avoid costly and multiple surveys.[111]

Some of the newer larger wind farms currently under construction are replacing smaller turbines previously installed at given locations. Timeline for Offshore developments:[112]

2015: The Dutch government has expressed the aspiration to build 4.450 MW of offshore wind power by 2023 which will have a considerable impact on the Dutch electricity grid, operated by TenneT. As a first step, the government has determined 65 sites for offshore wind farms in the North Sea and IJsselmeer.

2016: In the first week of December 2016, all 150 turbines of the new 600 MW Gemini offshore windfarm began producing power, and the project is expected to be fully commissioned by mid-2017.] In July 2016, the first two stages of offshore wind farm development for a combined 700 Megawatt capacity in a water area near Borssele was awarded to DONG Energy at a price of 7.27 Euro cent per kilowatt hour for 15 years. Transmission costs of 1.4 eurocent/kWh is to be added as TenneT is required to take power from sea to shore.

In 2050 all energy used in the Netherlands will come from renewable sources. Offshore wind is key to the transition to a zero-carbon energy supply. Offshore wind energy plays a key role in the energy transition. The North Sea is a good place to install wind turbines due to its shallow waters; its favorable wind patterns; good access to ports and energy-consuming industries. The costs of offshore wind energy have fallen significantly in recent years, making it the cheapest large-scale source of renewable energy. The total installed capacity of offshore wind power in the Netherlands is around 2.5 gigawatts (GW) in 2021 and should increase to at least 4.5 GW by 2023. This commitment is enshrined in the Energy Agreement for Sustainable Growth. Offshore wind turbines will then supply 3.3% of the Netherlands' total energy needs. The Coalition Agreement and the Climate Agreement (2019) include a commitment to continue the successful offshore wind energy policy. This will result in a total offshore wind power capacity of 11 GW by 2030, which is enough to supply 8.5% of all the energy in the Netherlands and 40% of our current electricity consumption. The construction of offshore wind farms is an economic opportunity. Thanks to the large domestic market, Dutch offshore and wind power companies can further develop their expertise, which also strengthens their position on the international market. Already, Dutch companies have a share of around 25% of the total European market for offshore wind energy. Measures for Offshore Wind Energy are based on extensive consultation with a large number of stakeholders, and the government has presented the plans in the 'Offshore Wind Energy Roadmap'. This sets out, among other things, where and when the new wind farms will be built. By doing so, the roadmap provides clarity to all stakeholders and ensures certainty for wind farm developers. The area map below provides an overview of the locations of the wind farm zones.[113]

FIG. 4.37: ONSHORE WIND TURBINES AT THE WESTEREEMS WIND PARK (SOURCE: [114])

[110] https://en.wikipedia.org/wiki/Wind_power_in_the_Netherlands
[111] https://en.wikipedia.org/wiki/Wind_power_in_the_Netherlands
[112] https://www.government.nl/topics/renewable-energy/offshore-wind-energy
[113] https://www.government.nl/topics/renewable-energy/offshore-wind-energy
[114] https://en.wikipedia.org/wiki/Wind_power_in_the_Netherlands

TABLE 4.4: SELECTION OF LARGER ONSHORE WIND FARMS IN THE NETHERLANDS, FEB 2021[115]

Wind farm	Province	Capacity (MW)	Turbines	Commissioning
Windplan Groen	Flevoland	400-500	90	Planned 2023
Noordoostpolder	Flevoland	429	38 × Enercon E-126 7.5MW 48 × Siemens 3.0DD-108 3MW	2017
Wieringermeer	North Holland	360	82 × Nordex N117 3.6MW × Vestas V126 3.6MW × Other	2020
Zeewolde	Flevoland	320	91	Planned 2022
Windplan Blauw	Flevoland	250	37 × EnVentus V162-5.6MW 24 × Other	Planned 2023
De Drentse Monden en Oostermoer	Drenthe	175.5	45 × Nordex N131 3.9MW	Planned 2021
Westereems	Groningen	171	50 × Enercon E82 3MW 2 × Senvion M6 6.15MW 2 × Lagerwey L136 4.5MW	2009/2012/2020
N33	Groningen	147	35 × Siemens Gamesa DD-230 R19 4.2MW	2021
Princess Alexia Windpark	Flevoland	122	36 × Senvion 3.4M104	2013
Windpark Krammer	Zeeland	102	34 × Enercon E-115 3MW	2018/2019
Windpark Oostpolder	Groningen	94.5	21 × Enercon	Planned 2021
Windpark Kroningswind	South Holland	80	19 × Vestas	Planned 2022
Windpark Oosterhorn	Groningen	77.4	18 × Vestas V136	Planned 2021
Windpark Delfzijl Zuid	Groningen	75	34 × Enercon E70	2006
GroWind	Groningen	63	21 × Vestas	2008
Windpark Delfzijl Noord	Groningen	63	19 × Nordex	2015
Windpark Geefsweer	Groningen	60.2	14 × Vestas 136	Planned 2021
Slufterdam	South Holland	50.4	14 × Vestas V112 3.6MW	2018
Deil	Gelderland	46	11 × Vestas V136 4.2MW	2020
Delfzijl Zuid	Groningen	43	10 × Enercon E70 2MW 10 × Enercon E70 2.3MW	2008
Koegorspolder	Zeeland	42	21 × Vestas V80/2000	2007
Delfzijl Zuid	Groningen	32	14 × Enercon	2008
Rachel Carson (Eem-meerdijk)	Flevoland	18	18 × NedWind 1MW	1998

[115] https://en.wikipedia.org/wiki/Wind_power_in_the_Netherlands

TABLE 4.5: CUMULATIVE SEA-BASED WIND CAPACITY IN THE NETHERLANDS (MW)[116]

Year	2010	2011	2012	2013	2014	2015	2016	2017	2018	2019	2020
Sea-based capacity (MW)	228	228	228	228	228	357	957	957	957	957	2,441
Installed capacity (MW)	2,237	2,316	2,433	2,713	2,865	3,391	4,257	4,202	4,393	4,484	6,719
Electricity generated (GWh)	4,503	4,725	4,939	5,368	5,810	6,917	8,364	9,642	10,030	10,774	13,896
% Electricity consumption	3.69	3.84	4.10	4.47	4.92	5.81	6.97	7.98	8.22	8.84	11.62

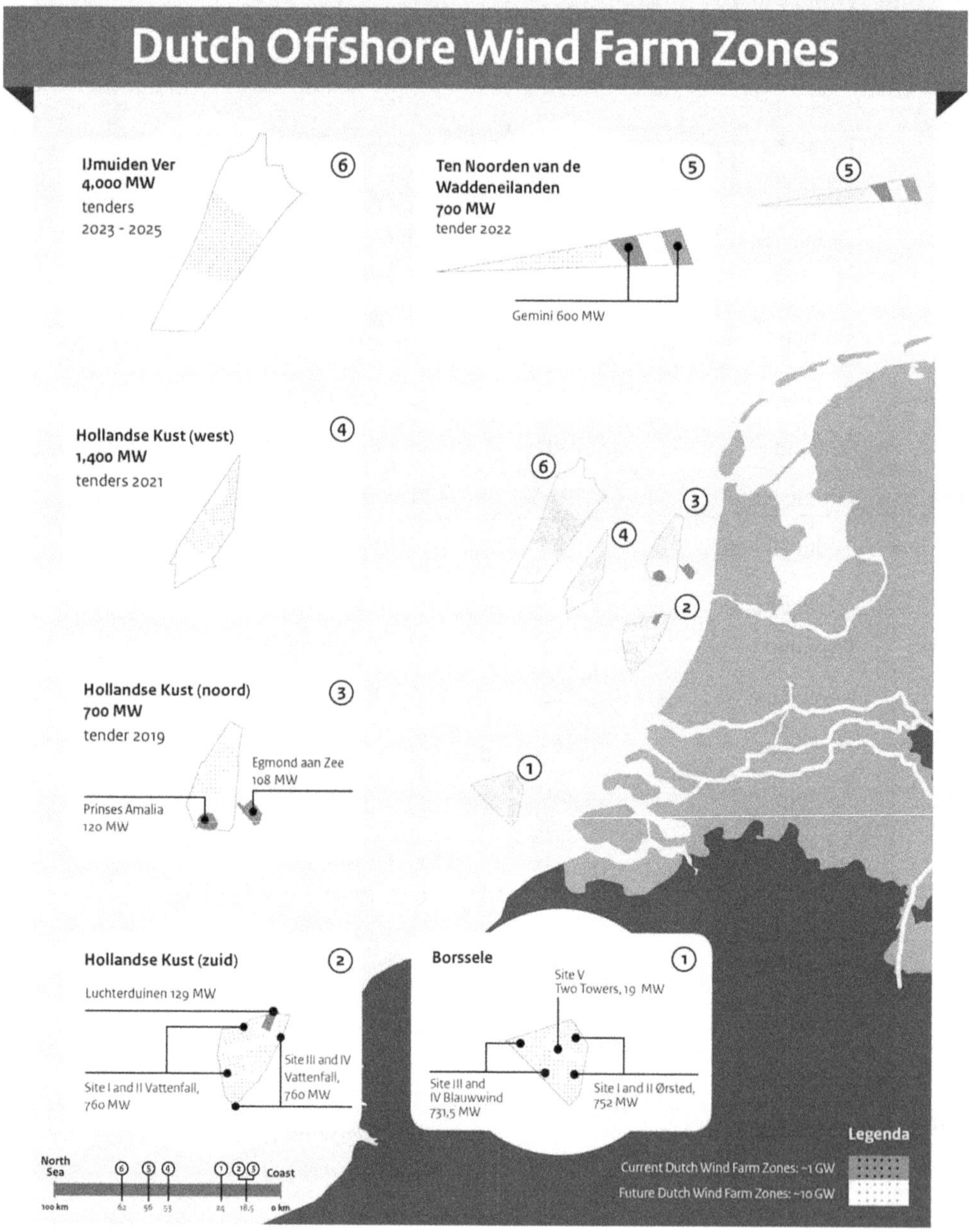

FIG. 4.38: DUTCH OFFSHORE WIND FARMS (SOURCE:[117])

[116] https://en.wikipedia.org/wiki/Wind_power_in_the_Netherlands

[117] https://www.government.nl/topics/renewable-energy/offshore-wind-energy

TABLE 4.6: WIND FARMS IN THE NORTH SEA: (SOURCE [118]) IN CONTRAST TO ONSHORE WIND FARMS, ALMOST ALL TURBINES BY JANUARY 2017 WERE SITUATED IN JUST A FEW LARGE WIND FARMS.

Wind farm	Location	Capacity (MW)	Turbines	Commis-sioned	Build-ing Cost	Cap. fac.	Depth range (m)	km to shore	Owner
Borssele 1 & 2	51°41′44″N 3° 4′56″E	752	94 × Siemens Gamesa 8 MW	2020			14–40	22	Ørsted
Borssele 3 & 4	51°41′50″N 2° 55′22″E	731.5	77 × MHI Vestas V 164 9.5MW	2021			14–38	22	Shell, Van Oord, Eneco, Partner s Group, DGE
Egmond aan Zee (OWEZ)	52°36′22″N 4° 25′8″E	108	36 × Vestas V90-3MW	2008	€200m		15–18	13	Nuon, Shell
Eneco Luchter-duinen	52°24′18″N 4° 09′43″E	129	43 × Vestas V112/3000	2015	€450 million		18–24	24	Eneco, Mit-subishi
Gemini	54°2′13″N 5°5 7′54″E	600	150 × Siemens SWT4.0–130	2017	€2.8 b		28–36	55	Northland Power, Siemens, Va n Oord, HVC Groep
Hollandse Kust (Noord) V	52°43′10″N 4° 14′14″E	759	69 × Siemens Gamesa 11.0-200 DD	Planned 2023			15–28	18.5	Shell, Eneco
Hollandse Kust (Zuid) I–IV	52°19′15″N 4° 2′36″E	1540	140 × Siemens Gamesa 11.0-200 DD	Planned 2023			18–28	18–36	Vattenfall/ BASF
Princess Amalia	52°35′16″N 4° 13′23″E	120	60 × Vestas V802MW	2008	€350m		19–24	26	Eneco En-ergie
Friesland	53°2′26″N 5°1 6′59″E	320	89 × Siemens SWT-3.6–107	2021			3–4	2	Windpark Fryslân B.V.
IJsselmeer Bui-tendijks (Wind-plan Blauw)			24 ×	Planned					
Irene Vorrink	52°35′53″N 5° 35′20″E	17	28 × Nordtank NTK600/43	1996	GBP 19m		2–3	0.1	Nuon
Lely (decommis-sioned 2016)	52°47′49″N 5° 7′8″E	2	4 × Nedwind 500 kW/41	1994	GBP 4.4m		3–4	0.8	Nuon
Westermeerwind (Noordoostpo-lder)	52°43′50″N 5° 34′50″E	144	48 × Siemens SWT-3.0-DD	2016			4–7	0.7–1.2	Westermeer-wind BV

TABLE 4.7: FUTURE OFFSHORE WIND FARMS (SOURCE:[119])

Wind farm[52]	Capacity (MW)	Planned Commisioning
Hollandse Kust (West) VI	700	2025/2026
Hollandse Kust (West) VII	700	2025/2026
Ten noorden van de Waddeneilanden I	700	2027
IJmuiden Ver I–IV	4000	2028/2029

[118] https://en.wikipedia.org/wiki/Wind_power_in_the_Netherlands
[119] https://en.wikipedia.org/wiki/Wind_power_in_the_Netherlands

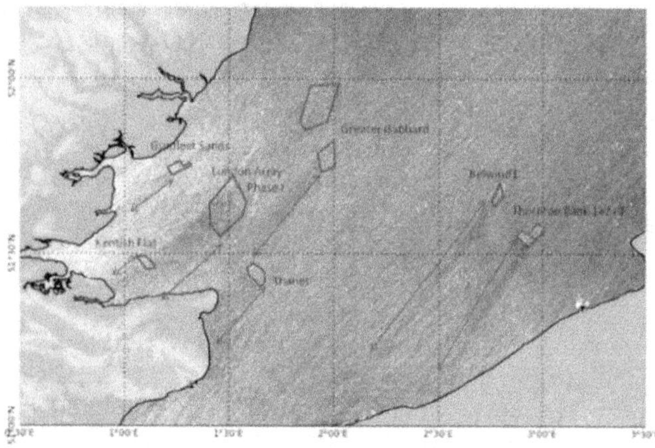

FIG. 4.39: RADARSAT-2 INTENSITY MAP OF THE SOUTHERN NORTH SEA OBSERVED 30 APRIL 2013 AT 17:41 UTC. THE BLUE LINES UTLINE WIND FARMS AND THE RED ARROWS THE WIND FARM WAKE. THE SAR-WIND PROCESSING CHAIN WAS SET UP BY COLLECTE LOCALISATION SATELLITES (CLS). (SOURCE [120])

FIG. 4.41: PRINCESS AMALIA WIND FARM IN THE NORTH SEA (SOURCE:[122])

An Overview of Offshore Wind Farm Design indicates that the offshore wind energy to be viable, the design of wind turbines is not the only an important factor but in fact it is crucial. Wakes are not only an issue within a wind farm, but they also extend for many km downstream of the wind farm. The wind speed near the sea surface can be measured using Synthetic Aperture Radar (SAR), e.g., mounted on satellites. In certain weather situations, the area of reduced production can extend several tens of kilometers. At Belwind wind farm the wake is around 55 km long, at Thornton Bank 45 km, at London Array 15 km, at Thanet 14 km and at Kentish Flat 10 km, and probably continues inland. It is the intensity image where the darker area is due to lower wind speed (Figure 4.39). The European Energy Research Alliance (EERA) integrated many of the institutes' software's into a common Design Tool for Offshore Clusters (eera-dtoc.eu) in a project sponsored by the EU in 2012–2015 with a project budget of 4 million euro, with the main emphasis for offshore wind farm design model in 2016, DTU Vindenergi. This integrated the workflow and the models from different planning aspects, i.e., wind climate, wakes and electrical models from grid to turbine plus a LCOE model, which currently embodies the cost function of an offshore developer. The design of the grid inside the farm and the connection to the shore were integrated in 2015 for the Design Tool for Offshore Clusters (DTOC). The EERA-DTOC tool was designed based on input from end users. Its aim is to support the optimization of LCOE by comparing different variants for the farm layout. Scenarios can be cloned or duplicated and inherit the settings from the higher level scenario. The DTOC software supports the generation and comparison of the calculation results of many design scenarios. The workflow to optimize through comparing LCOE is shown in Figure 4.40. In order to calculate the LCOE, the sub models for e.g., cabling, AEP and grid compliance have to be called for each scenario. GIS

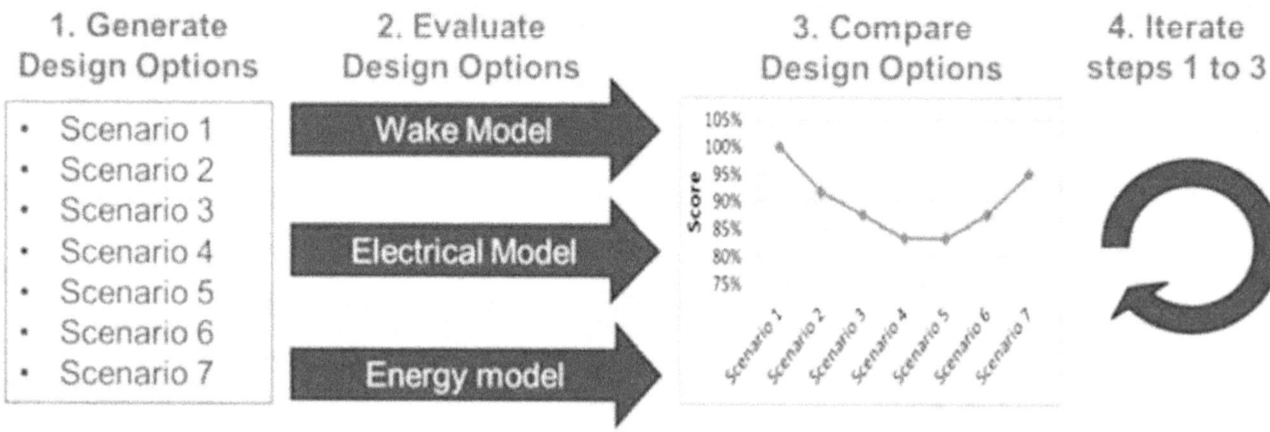

FIG. 4.40: WORKFLOW FOR THE DTOC-TOOL BASED OPTIMIZATION PROCESS, (SOURCE: [121])

[120] Hasager et al. (2015), cited in https://www.researchgate.net/publication/307522478_An_Overview_of_Offshore_Wind_Farm_Design
[121] https://www.researchgate.net/publication/307522478_An_Overview_of_Offshore_Wind_Farm_Design
[122] https://www.rvo.nl/sites/default/files/2015/03/Offshore%20wind%20energy%20in%20the%20Netherlands.pdf

FIG. 4.42: DESIGNATED WIND FARM ZONES IN THE NETHERLANDS CONTINENTAL SHELF. (SOURCE:[123])

FIG. 4.43: WIND TURBINES AT THE WESTEREEMS WIND PARK (SOURCE: [124])

data is also integrated in the tool. A timeline can be established when the production is going to drop due to wake effects from new wind farms. Since this wind climate is notionally the same for all wind farm configurations, it is run only once. This wind climate is then put into different farm layout options, which are compared according to Figure 4.40. Therefore, instead of having to convert data from one program to the next, the user can handle many more scenarios before the bidding process begins.

The ENECO Luchterduinen wind farm was officially opened in September 2015 and is a 50/50 joint venture between Dutch energy company ENECO and Japan's Mitsubishi corporation. The 129 MW ENECO Luchterduinen wind farm is located 23 km off the coast between Zandvoort and Noordwijk cities in the Netherlands.

TABLE 4.8: SCHEDULE OF TENDERS FOR OFFSHORE WIND FARMS 2015 – 2019 (SOURCE:[125])

Year	Capacity (MW)	Wind Farm Zone	Becoming Operational
2016	700	Borssele Wind Farm Zone Sites I and II	2020
2016	700	Borssele Wind Farm Zone Sites III, IV and V	2020
2017	700	South Holland Coast Wind Farm Zone Sites I and II	2022
2019	700	South Holland Coast Wind Farm Zone Sites III and IV	2023
2020	700	North Holland Coast Wind Farm Zone Site V	2024
2021	700	West Holland Coast Wind Farm Zone Site VI	2025-2026
2021	700	West Holland Coast Wind Farm Zone Site VII	2025-2026
2022	700	Ten noorden van de Waddeneilanden, Site I	2027
2023	1000	IJmuiden Ver, Site I	2028
2023	1000	IJmuiden Ver, Site II	2028
2025	1000	IJmuiden Ver, Site III	2029
2025	1000	IJmuiden Ver, Site IV	2029
Total	**9,600**		

[123] https://www.rvo.nl/sites/default/files/2015/03/Offshore%20wind%20energy%20in%20the%20Netherlands.pdf
[124] https://en.wikipedia.org/wiki/Wind_power_in_the_Netherlands#/media/File:Wind_farm_Westereems.JPG
[125] https://www.government.nl/topics/renewable-energy/offshore-wind-energy

The farm consists of 43 Vestas V112 turbines of 3 MW capacity each. A further 228 MW of installed capacity will be added when completed in 2015. In addition, The Gemini wind farm will be constructed with 150 Siemens 4 MW turbines on the twin locations Buitengaats and Zee-Energie, 85 km off the coast. Gemini is owned by Northland Power, Siemens, Van Oord and HVC. This wind farm is expected to be fully operational in 2017. The road map towards 4,500 MW offshore wind power foresees an annual tendering of 700 MW in the period 2015 – 2019, it is a precondition that the cost of offshore wind power will decrease by 40% in the coming years. The Government has decided that three offshore wind farm zones will be used for the deployment of the 3,500 MW new offshore wind capacity: Borssele (1,400 MW), South Holland coast wind farm zone (1,400 MW) and North Holland coast wind farm zone (700 MW).The tenders to select parties for realizing the offshore wind farms will be done according to the following schedule: [126]

4.11.3.7 Climate Control in Netherlands with Impact on Wind Energy

The Climate Agreement is part of the Dutch climate policy. It is an agreement between several organizations and companies in the Netherlands to combat climate change. The government's central goal with the National Climate Agreement is to reduce greenhouse gas emissions in the Netherlands by 49% by 2030 compared to 1990 levels.[127] A letter to the President of the House of Representatives of the State's General Binnenhof dated Date 28 June 2019 regarding the subject "Draft Climate Agreement" states that the "The climate is changing". In the Netherlands, responsibility for doing everything to combat climate change, to ensure the country remains habitable for future generations. For environmental benefits, an ambitious climate policy provides opportunities for the economy, for prosperity and for earning capacity and positions Netherlands to decrease the greenhouse gas emissions. With this in mind, the government set an ambitious climate target in the "Coalition Agreement" to implement measures for a 49% reduction of greenhouse gas emissions by 2030 compared to 1990. On a European level, the government is advocating a 55% reduction of greenhouse gas emissions by 2030. Should such a target prove unfeasible within the EU, the Netherlands will strive to achieve more ambitious agreements with like-minded North-Western European countries. Climate change goes beyond national borders. A significant first step for the long-term was taken by the European Council on 20 June 2019, with a large majority of Member States endorsing the goal of climate neutrality for 2050. In the Coalition Agreement, the Dutch government announced measures to achieve the national reduction target. A commitment was also made to realize a Climate Agreement. In the past year, more than one hundred parties have jointly worked on a cohesive set of proposals with the aim of achieving the carbon reduction; (Parliamentary Paper 32813, No 163), while simultaneously defining the political frameworks with which the Climate Agreement must comply with "Priorities" for the government The government launched talks on a Climate Agreement on 23 February 2018. Affordable items were to fulfil the task in the most cost-efficient way possible to ensure the transition remains affordable. For the costs of the transition the government has one central goal, which is to reduce carbon emissions by 49% by 2030 compared to 1990. This will allow the society to choose from the best possible and most cost-effective combination of measures and technologies thereby minimizing costs. A "Fair" step would be for the transition to be managed in a "just manner" with "Measures to be adopted for each sector" which are:[128]

- Built environment Attractive financing and subsidies; Affordable energy bills; The need for cost reduction; Built environment target to be within the reach. Transition vision for heat and for Non-residential buildings.
- Mobility Citizens should be able to travel from safely, quickly and easily. Companies want their goods to reach their destination quickly and reliably. This will not change, but the modes of transportation will. Transport accounts for a quarter of all emissions. The ambition is for all new cars to be emission free by 2030. Hydrogen and electric cars produce zero greenhouse gas emissions and keep the air clean and cause less noise pollution. In the Netherlands, new cars have an average life of 18 years. The government's ambition is for all new cars to be emission free by 2030 at the latest.
- Industry Additional investment in green hydrogen; Reasonable and objective carbon levy; Limitation of Carbon Capture and Storage (CCS) subsidies; Surcharge for Sustainable Energy (ODE) and SDE with fair distribution of costs. The government believes that everyone should make a fair contribution towards the transition. To this end, the government will increase the share that businesses contribute towards the Surcharge for Sustainable Energy (ODE) from 1/2 to 2/3 with effect from 2020. The cost of this increase will be borne by major consumers, including those in industry. Industry will therefore pay €550 million in 2030.
- Electricity Where the electricity system is concerned, the long-term goal is to have a zero CO_2 with reliable, affordable and safe system by 2050.
- Agriculture and Land Use In the draft Climate Agreement, the sector platform for Agriculture and Land Use has stated that it expects to be able to reduce emissions by 6 Mt instead of the 3.5 Mt stipulated in the Coalition Agreement.
- Governance of the Climate Agreement Concurrence for implementation is most important.
- Follow-up process by Parliament. The implementation of the Climate Agreement will be an adaptive process in which allowance will be provided for responding to relevant new developments. Further decisions will therefore need to be made during this process. However, with this Climate Agreement, the government looks forward to this process with confidence.

4.11.4 Summary

The primary focus of Dutch energy policy is carbon reduction through a so-called Energy Transition. While the Netherlands is

[126] https://www.rvo.nl/sites/default/files/2015/03/Offshore%20wind%20energy%20in%20the%20Netherlands.pdf

[127] file:///C:/Users/Rao/Downloads/Letter+to+the+House+of+Representatives+about+the+proposal+for+a+National+Climate+Agreement%20(7).pdf

[128] https://www.government.nl/topics/renewable-energy/documents/reports/2019/06/28/climate-agreement

TABLE 4.9: RENEWABLE ELECTRICITY PRODUCTION IN THE NETHERLANDS IN MILLIONS OF KWH (SOURCE: [129])

	2016	2017	2018
Total Renewable Production	15,082	16,670	18,142
Onshore wind	6,041	6,267	6,578
Offshore wind	2,323	3,375	3,452
Hydropower	98	94	94
Solar	1,602	2,204	3,201
Biomass	5,018	4,729	4,817
Renewables as a percentage of total electricity consumption	12.7	13.9	15.0

not on track to meet the 2020 renewable energy goals set by the EU, it is one of the first EU countries to announce plans to eliminate natural gas from its energy mix. The current government coalition is committed to a 49 percent reduction in carbon emissions by 2030, which would surpass the current EU target. This goal will be reached primarily through the construction of new offshore wind farms in the North Sea. Other important aspects of Dutch government policy include closing the Netherlands' sole nuclear power plant by 2024 and encouraging increased use of carbon capture and storage (CCS) technologies by factories, power plants, and waste incinerators. The share of renewable energy in total energy consumption in 2018 was 7.4 percent, up from 6.6 percent in 2017. Biomass processing accounts for 61 percent of renewable energy production and 23 percent is from wind energy. The contribution of solar energy grew to 8 percent in 2018. Hydropower, geothermal energy and heat from the outside air contributed 8 percent. In terms of production statistics, in 2018, more than eighteen billion kilowatt-hours of electricity were produced from wind energy, hydropower, solar energy and biomass, equating to 15 percent of total electricity usage. By 2025 this proportion will have increased to around half, and it will be close to two-thirds by 2030. Conventional production from gas, and later also from coal, will come under pressure. Under these circumstances, the Netherlands will increasingly become a net power exporter. An additional roll-out of offshore wind over the period 2023-2030 and a continued increase in the contribution of solar power will lead to growth in the proportion of renewable electricity in national electricity production. Obstacles to growth of renewable energy include the absence of large-scale hydro generation, solar usage, and onshore wind due to population density. [130]

4.11.4.1 Wind Farms in Netherlands During 2011-through-2030

9.6 percent of average annual electricity demand in the Netherlands is provided by wind energy, and due to ongoing wind farm creation particularly in the North Sea, wind power has experienced the fastest growth of any renewable sector. In 2019, 2,000 onshore wind turbines generated 8,000 MW of electricity. The Netherlands successfully issued its first and subsidy-free tender for offshore wind energy in 2017 and 2019, signs of a booming renewable energy sector. Currently, there are six offshore wind farms off the coast of the Netherlands: Gemini (600 MW), Eneco Lucterduinen (129 MW), Princess Amalia (120 MW), Egmond aan Zee (108 MW), Irene Vorrink (17 MW), Lely (2 MW). The 320 MW Windpark Friesland is scheduled to come online by 2021, and the 760 MW Hollandse Kust Zuid 3 and 4 farms will come online by 2023. Opportunities abound for offshore wind power. In 2019, offshore wind turbines generated one gigawatt of power in total, and this will increase to 4.5 GW in 2023. This commitment is enshrined in the Energy Agreement for Sustainable Growth. Offshore wind turbines will then supply 3.3% of all the energy in the Netherlands. Dutch firm TenneT will develop five or six 2GW offshore platform grid connection systems for integrating offshore wind parks into the Dutch and German power grid over the next ten years. Each offshore grid connection will convert 2GW of electricity generated by offshore wind from AC to DC, and will transport this to onshore, where the DC is converted back to AC. 2GW offshore high voltage direct current (HVDC) systems have not yet been developed. TenneT's objective is to standardize the 2 GW platform design. They have launched the design phase with five HVDC suppliers based on an innovative partnership. Participants are GE Renewable Energy, ABB Power Grids, Consortium Global Energy Interconnection Research Institute, Siemens and Xian Electric. These projects will be realized in 2028. Long-term plans are for a total of 11.5 GW of offshore wind by 2030, which will supply 8.5 percent of all the energy in the Netherlands, which is 40 percent of current electricity consumption. The Dutch government is accelerating collaboration on smart grids with its Intelligent Grids Innovation Program. To provide a boost to large-scale application, this program has supported ninety-four pilot projects in residential districts, city centers, office parks, industrial estates, and agricultural areas. There is a growing number of local energy cooperatives that focus on energy production (particularly collective solar and wind projects) and energy conservation. In 2015, these cooperatives represented only one percent of the total peak power of all solar panels in the Netherlands and three percent of onshore wind energy capacity. These percentages are expected to rise sharply in the coming years, with installed capacity of collective wind projects expected to grow from 82 MW in 2015 to 232 MW by the current period, 2020-10-10, publication of this paper.[131]

[129] Statistics Netherlands (cbs.nl) cited in https://www.trade.gov/knowledge-product/netherlands-energy

[130] https://www.trade.gov/knowledge-product/netherlands-energy

[131] www.trade.gov/knowledge-product/netherlands-energy

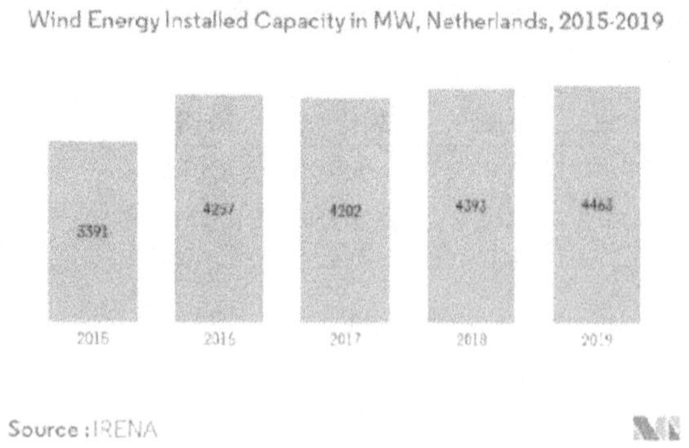

FIG. 4.44: CAPACITY OF WIND ENERGY INSTALLATIONS IN MW BETWEEN 2015-2019 (SOURCE:[132])

4.11.4.2 Futuristic Wind Farms in Netherlands

Netherlands wind energy market growth, trends, Covid-19 impact, and forecasts (2021 2026) has been studied with 2020 as the base year for this study. The Netherlands is working on a transition to a sustainable, reliable, and affordable energy supply. The Netherlands wind energy market is expected to grow at a CAGR of more than 9.5% during the forecast period 2020-2025. Wind energy in the Netherlands powers about 5.7 million homes and also cuts the CO_2 footprint of electricity by 12%. Factors like increasing demand for renewable energy, rising investments in wind farms, and reducing CO_2 emissions are driving the wind energy market in the country. Offshore wind is expected to witness significant growth in the wind energy market in the Netherlands during the forecast period. The Netherlands has an overly ambitious renewable-energy plan in the works. The country is expected to build the world's largest offshore wind farm by 2027, along with a 2.3-square-mile artificial island to support it. The wind farm is expected to be capable of producing 30 GW of power, which is likely to provide an opportunity to a growth in the deployment of wind energy in coming future. The Netherlands is making strides toward the country's renewable energy targets i.e., 16% renewable energy sources (RES) by 2023 of total energy demand. The conditions for offshore wind energy in the Netherlands are excellent. The factors like shallow waters, good wind resources, good harbor facilities, experienced industry, and a robust support system are driving the offshore wind energy market. The active offshore wind farms in the North Sea are Gemini (600 MW), Luchterduinen (129 MW), Prinses Amalia (120 MW), and Egmond aan Zee (OWEZ) (108 MW). Offshore wind farms which are under construction are built at a number of locations in the Dutch part of the North Sea and they are Borssele Wind Farm Zone Sites I and II, Borssele Wind Farm Zone Sites III and IV, Hollandse Kust (zuid) Wind Farm Zone Sites I and II, Borssele Wind Farm Zone Innovation Site V, and Hollandse Kust (zuid) Wind Farm Zone Sites III and IV. The upcoming offshore wind farm zones for the deployment of the 3,500 MW new offshore wind capacity in Netherlands are Borssele (1,400 MW), South Holland coast wind farm zone (1,400 MW) and North Holland coast wind farm zone (700 MW). Netherlands holds 4463 MW of wind energy installed in 2019 and the wind-generated electricity in Netherlands accounted for 10.5 TWh in 2018. As demand for energy is rising, Netherland is turning toward the adoption of renewable energy for sustainable and clean form of energy. The adoption of offshore wind energy along with higher wind potential and energy demand attracts the companies for high investment. The Netherlands wind energy market is moderately consolidated. Key players in this market include ENERCON GmbH, Siemens Gamesa Renewable Energy SA, Mitsubishi Corp, General Electric Company, and Lagerwey Wind BV.[134]

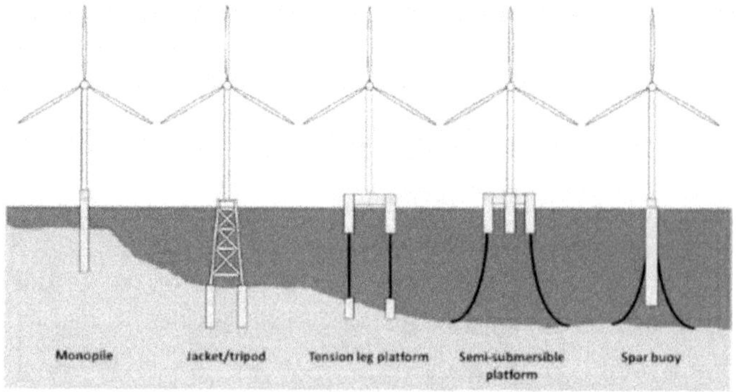

FIG. 4.45: OFFSHORE PLATFORMS (SOURCE: [133])

The futuristic and emerging technologies in the wind power sector from European Perspective is worth looking at to understand how Netherlands is going to assimilate EU experience. The notable features to assimilate and capturing the essence are the areas of concentration for the foreseeable future in EU countries for wind energy:

- Airborne Wind Energy
- Offshore Floating Concepts
- Smart Rotors
- Wind-Induced Energy-Harvesting Devices
- Blade Tip-Mounted Rotors
- Unconventional Power Transmission Systems
- Multi-Rotor Turbines
- Alternative Support Structures
- Modular High Voltage Direct Current Generators
- Innovative Blade Manufacturing Techniques
- Diffuser-Augmented Turbines and Small Turbine Technologies.

The future role of advanced multiscale modelling and data availability is a significant feature. The EU fears are that there is a need to identify synergies between fundamental and industrial research by correctly targeting public and private funding in these emerging technology areas as industrial development may outpace more

[132] https://www.mordorintelligence.com/industry-reports/netherlands-wind-energy-market

[133] https://www.sciencedirect.com/science/article/pii/S1364032119304782

[134] www.trade.gov/knowledge-product/netherlands-energy

fundamental research faster than anticipated. The development of floating wind structures has grown out of existing fixed structure technology (see figure 4.45). There is therefore potential for optimization of floating offshore systems. This optimization would be possible through the integrated design of the platform and the wind turbine. The more specific, tailored design of wind turbines for offshore floating platforms, including downwind rotors, high tip speed ratio operation and two bladed rotors, could have an impact on the cost of the platform and the whole floating system.[135]

4.11.5 References

1) Energy and Power Generation Handbook – Established and Emerging Technologies, Editor K. R. Rao published by ASME, 2011.
2) Meeting the Challenge of Practical Implementations, Volumes 1 and 2, by K. R. Rao, Springer Publications, Switzerland, 2019

[135] https://www.sciencedirect.com/science/article/pii/S1364032119304782

CHAPTER

5

ROLE OF WIND ENERGY TECHNOLOGY IN INDIA AND NEIGHBORING COUNTRIES

5.1 INTRODUCTION

Wind power has soared in terms of installed capacities and market share over the last decades. It has also become a source of sizeable economic activity. Its project activities in terms of manufacture both for domestic and international markets, installation & commissioning have increased. Operational wind power has provided a sizeable number of high technology jobs in remote areas. Several positive impacts observed in terms of major electrical grid expansions have resulted in many centres of economic importance. The only threat for wind power, if we may call it, is from the solar electricity. Notwithstanding the fact that wind power technologies originally evolved in Europe and to an extent in United States, major commercial exploitation is seen in Asia, predominantly China. As a matter of fact, China leads the pack with India occupying fifth position. China changed its stance regarding Wind Power in 2007-08. Chinese approach has been to concentrate on "building big" and there was no hesitation to employ technology resources including expertise from Europe and the United States to set up comprehensive technology base. India, in contrast, concentrated more on deployment of technologies from overseas, even in terms of component development. There have been considerable efforts by the Government and private sector players in terms of large-scale wind projects with meteorological towers up to 160 m height. Well documented and informed assessment reports are produced for future wind farms and wind solar hybrid projects.

Regarding expertise in terms of design capability, it was again Europe-centric or US centric with somewhat restricted expertise transferred to global players having their R&D branch offices established in India. Most of the global players do have their out reaches established in India and are looking at possible end to end manufacturing facilities as an alternative.

However, existing fully indigenous design capabilities cannot compete in international markets because there were never any carefully thought-out policies, efforts and plans in that direction. However, with an emphasis on local manufacture and serious interest by global players looking for an alternate location, India is well positioned in getting into manufacturing in a big way. Solar power, which is in abundance over the Indian subcontinent, has caught the imagination of policy makers and is perhaps the fastest growing Renewable Energy power deployment in India. Sri Lanka has a more tropical wind climate and its ability to accommodate more wind power in addition to the major and mini hydro generation, which can combinedly turn it into a decent market for wind power. However, the required bulk energy levels being limited, it's not going to be a bottomless market. Much is happening in Pakistan in terms of planning which has barely 1.25 GW which is due to a late start, and there is a good possibility of looking at a growing presence of wind power on the Pakistan grid.

5.2 INDIA

Under the Indian Constitution, power is a concurrent subject and the responsibility for distribution and supply of power to rural and urban consumers rests with the States. The centre provides support on all matters under various schemes and programs that encompasses the entire value chain. Due to its pre-eminent position Local Governments have significant say in the way these programs and schemes get implemented. Due to this there can be considerable differences in the way projects take shape in different parts of the country. As of 30.11.2021, the total installed power in the country was 392 GW, of which 157.3 GW is of renewable power (including hydro) and the goal is for the installed RE capacity in the country reach about 500 GW by 2030. The peak demand met with this size of installed power was in the range of 203 GW in year 2021-22[1]. This demonstrates that the power supply in the country is in a reasonably healthy position. However, by 2030 with a large chunk of conventional (311 GW) and renewable power (including hydro) of (506 GW) there will be very stiff competition to be the preferred generator. Obviously conventional power has a better appeal because of the controllability it affords and the familiarity of decades of operation of grids with very large percentage of thermal power.

Growth of wind power in India was originally driven mainly by tax incentives and to a smaller extent based on Independent Power Producers. Figure 5.1 gives the annual capacity additions [2] of wind power over the years from 1986. The 60 GW by 2022 target has to be extended due to various factors including Covid-19.

The year 2017 saw a record number of installations due to the rush to get into the Feed-in Tariff based power purchase agreements. By then auction based power determination had taken shape with Solar Energy Corporation of India. Subsequently the installations have shown a sharp fall.

The industry saw a major shift in terms of holdings of few tens of MW for self consumption to supplying energy to distribution companies (DISCOM) at pre-fixed preferential tariffs to moderately bigger Independent power producers (IPPs) with the sole aim of supplying power to the DISCOMS at a price determined by

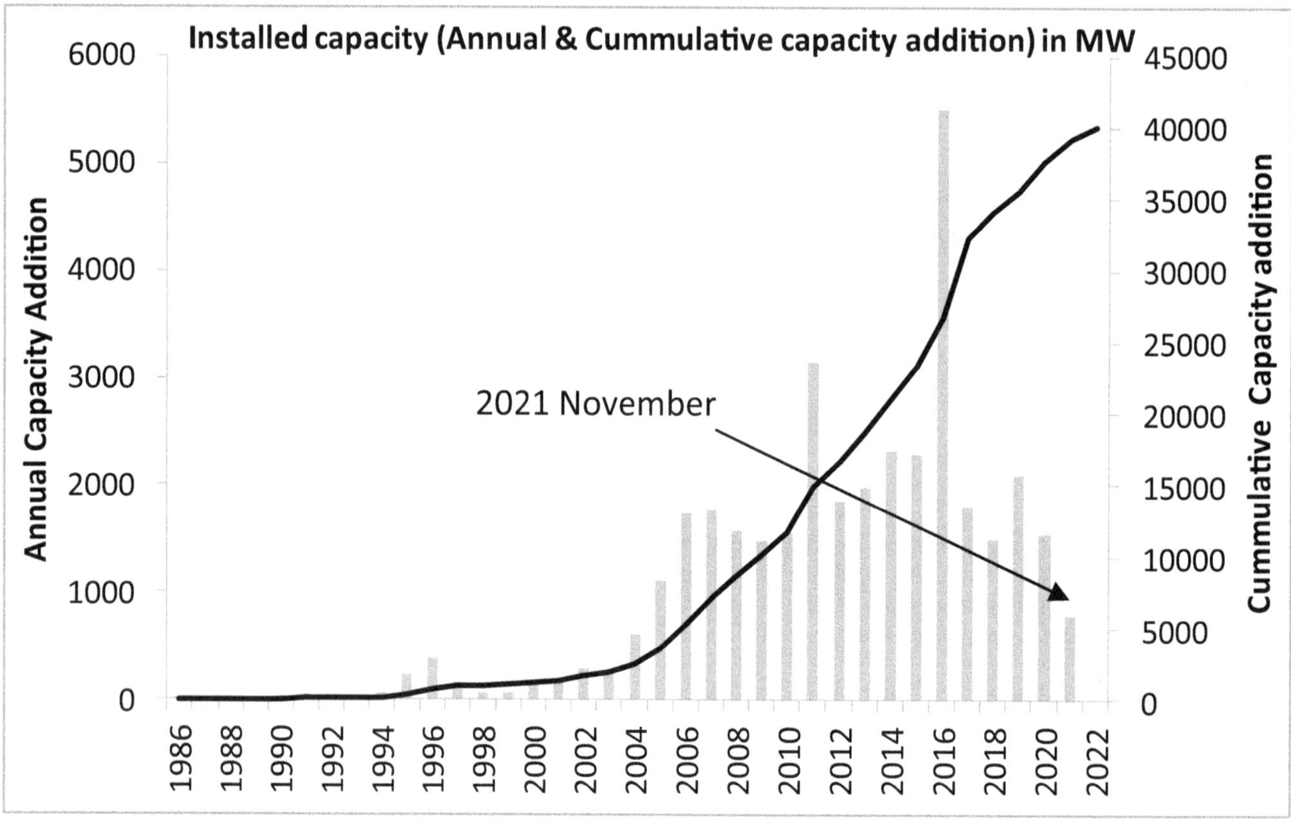

FIG. 5.1: GROWTH OF WIND POWER INSTALLATIONS SINCE 1986 (MNRE ANNUAL REPORTS)

reverse bidding. The business of investing part of the earnings into Wind turbines for tax benefits was completely done away with. Solar Energy Corporation of India (SECI) starting the process of reverse bidding with Solar Photo Voltaic power stations extended the process for wind power procurement in 2016.

The reverse bidding gave rise to a very new and interesting business model. Very imaginative and innovative approaches were taken. Those with fair knowledge of wind business and good ideas of procuring low-cost funds from overseas formulated viable projects of the sizes needed. They were in a good position to negotiate with OEMs on one hand and on the other, the necessary financial power to follow through to effect financial closure. In the transition period of Feed-In-Tariff (FIT) to auctioned pricing of electricity, OEMs were on a weaker wicket in terms of stock piled inventories and continued manufacturing. The market was no longer driven by the OEMs who were in the business of providing end to end service with somewhat older turbine models. It was taken over by the new Independent Power Producers (IPPs) without the limitation of having to go with one or another manufacturer for everything starting from resource assessment to the turbine selection to installation and commissioning. What may have been missing was the connection to the ground realities of setting up wind power projects. It was perhaps essential to change the scales of operation to bring on a different level of deployment.

5.2.1 Technology Trends

There has been a fairly big change in the way wind power has been looked at in India. For nearly two decades the potential for deployment was kept at 50 GW at 50 m agl in the country. This was based on the available evacuation capability with the then and future grid sizes. There were various other estimates ranging from 100 to 3000 GW based on several desk top studies and loosely validated assumptions. National Institute of Wind Energy, a Government of India laboratory made fresh estimates at 100 m agl and pegged the number at 350 GW as the theoretical upper limit. Based on this, the ceiling for setting up wind power projects saw a seven-fold increase. This became an important driver for the new approach.

The wind turbines deployed until recently were sub-MW range except for Suzlon's 1.25 MW and a few others with less significant market shares. These turbines were with rotor diameters in the range of 60 to 97 m with hub heights in the range of 70 to 80 m. By and large, the turbines were of IEC Class II/III (IRC 61400-1 ed. 2) catering to moderate wind regimes. The plant load factors achieved by this class of turbines started showing a decline essentially because of lower wind potential at the practiced hub heights. With land becoming scarce over a period in moderately windy sites the need to look at higher hub heights and larger rotor diameters became imperative. The turbine sizes have grown in terms of higher hub heights and rotor diameters with relatively smaller generators and are able to deliver much higher plant load factors by 10 to 12% over older turbines. Unlike in the past, the new potential areas are largely in the plains. There are a few reasons for this. The complex terrain sites that could provide high plant load factors are no longer available. In plain terrain, with increased hub heights and good power law indices it is now possible to look at 35% plant load factors with hub heights in the range of 120 m and above. On a parallel, it is now possible to transport very large components with much improved infracture. With the business model changed entirely, it is now possible to look at much lower capital costs although it may not be sustainable. There may have been some incremental improvements in the designs evolved to meet the market requirements. Other design driver has been better understood loads and ease of tailoring an optimised product suitable for the moderate wind sites.

5.2.2 SECI and other wind power project bids

Government of India made a major departure from the well-established negotiated FIT which was roughly based on the cost-plus model to an open market mode system. The underlying idea was that keeping all things equal, let us get the lowest procurement price independent of the involved costs from the point of view of the customer. One major advantage here is that the project developer is free to choose the best suited turbine for a given wind regime which was not feasible under the manufacturer cum developer model. Some of the salient features of these bid invitations were:

The bidders were not required to use the grid networks owned and managed by the State authorities. Power Grid Corporation of India (PGCIL) under Ministry of Power would provide evacuation facilities for the wind farms established under the bids. Independent entites called Central Transmission Utilities (CTUs) were made responsible for providing evacuation and transmission.

The projects shall be completed in 18 months from date of award.

The determined tariffs were fixed for 25 years.

Power Purchase Agreements could be signed with any state or private entities for supply of electricity.

This possibility helps in state utilities to meet their Renewable Power Obligations.

It will be the responsibility of the bidder to obtain a financial closure of the project keeping in view the stringent project parameters including timelines. The first SECI bid of 1000 MW obtained a final average price of INR 3.63 (February 2016). Figure 5.2 gives an idea as to how the competitive bidding resulted in a continuously falling bid prices till the third auction in February 2018.

Thereafter there was a market stabilisation and the bid prices started rising. To keep control over the upper limits, a ceiling price above which bids shall not be permitted. For example, for the sixth bid, a ceiling price of INR 2.88 was fixed, and downward bidding brought the numbers to something in the range of INR 2.85.

On a parallel, few State Governments also attempted their own state level exercises for projects ranging from 250 to 1000 MW and determined prices were in comparable ranges. There was one aberration to the trends in case of a 500 MW auction from the State of Maharashtra. Original expectations were that the wind farms were to be connected to Inter State Transmission systems governed by PGCIL in case of the SECI bids. This was to minimise problems with state level evacuation and the need to overcome associated constraints. It also helped in facilitating the possibility of renewable energy transmission occurring across state boundaries. As the PGCIL transmission networks can evacuate power on 220/400/730 kV multi-circuit lines there would be hardly any capacity constraints on one hand and higher grid availability at these voltage levels.

A number of bids for Hybrid have been floated and won. Two rather important bids which encouraged bundling with Solar/Storage/Thermal power was auctioned and bids were won at modest prices. Attempts have been made to get bids that ensure Round The Clock and to supply Peak Power supplies based largely on renewable energy. Several projects are in various stages of implementation. Of about 10 GW projects under the auction route about 3.2 GW have been commissioned. Delays are anticipated due to various factors such as quick policy shifts, mismatch between Centre and State understanding, interpretation, and implementation processes.

5.2.3 Central Grid Extension Initiatives

One of the most daunting tasks with wind power (as also the Solar power) is the evacuation of generated power to load centres. The country though is crisscrossed with Central and State transmission lines and sub stations, evacuation limitations are very serious and a limiting factor. Firstly, the State grids are limited by the

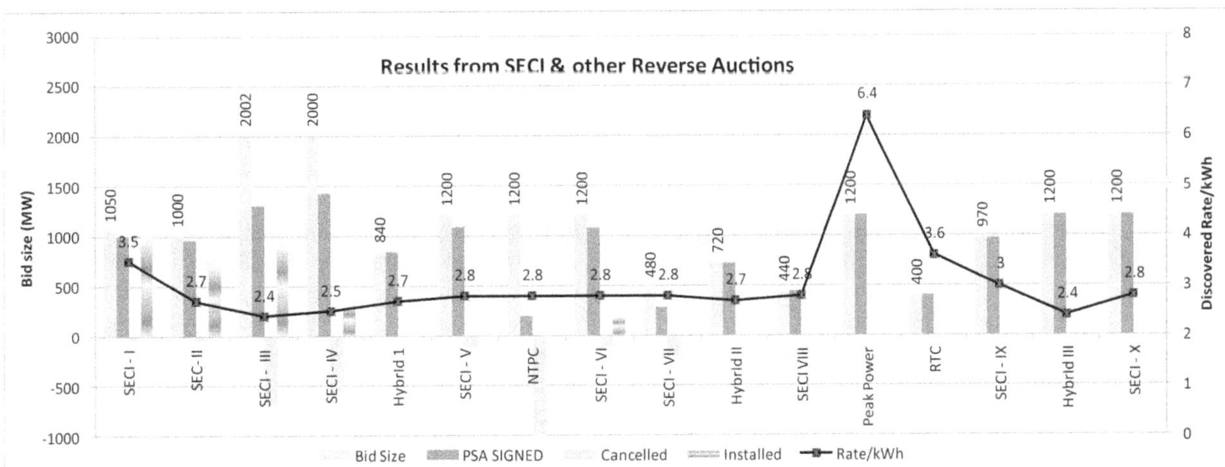

FIG. 5.2: SOLAR ENERGY CORPORATION OF INDIA AND OTHER WIND AUCTION RESULTS (SECI, MEC+ GWEC)

voltage levels employed. Subsequently the interstate exchanges were managed between utilities and there were no possible rules under which a private power producer could sell power in other than the home state for the wind farms. The severity of congestion was so much that very often the grid managers must ask wind farms to run at sub optimal levels and many times even to back down. Considering these aspects PGCIL initiated in 2012 a holistic action plan that aimed at creating an evacuation & transmission infrastructure in two phases of 33 & 22 GW. The basis for this estimate was the available information at that point of time and looking back in 2019 the estimates were reasonably good except that there was not much on Solar power. The central push for Solar energy was far more effective. Falling Solar Photo Voltaic (SPV) costs further helped in pushing solar power deployment to 30 GW while wind power installation showed a sharp decline since 2017.

In 2012 a new concept called Green Corridor for use of renewable energy was envisaged [3]. However, the overall transmission planning is considering just 10 % of available RE power to contribute during the evening power market (6 pm to 8 pm). The Green corridor project also envisaged several enabling policies, institutional arrangements such as Renewable Energy Management Centers (REMC), real time forecasting and scheduling and creating frameworks for interactions among various stake holders.

5.2.4 Central initiatives to improve grid availability for RE power:

A few initiatives apparently for improving the absorption of renewable energy were proposed. The underlying expectation was that with a history of over twenty years, the wind technology was mature enough to be treated on par with conventional power. The policy makers, largely from conventional background went with a rather simplistic requirement. They needed scheduling wind power generation with 15-minute time blocks and it was envisaged as a penal system. If the scheduled power falls outside the ± 10 or 12 % there will be a penalty to be paid which will result in reduced financial returns to the generators thereby compelling them to find ways of getting the right forecasting system in place.

In the matter of grid disturbance anywhere new demands on the low voltage and high voltage ride through the ability to ride through such events at the point of common connection level. Present regulations have kept things vague and have revised the LVRT and HVRT regulations to more stringent levels.

The regulations are announced after due consultations with various stake holders leaving the accountability aspects open. Though there are possibilities for redress, it is always a long and painful path which has the potential of failure on account of financial reasons.

Formation of vague management systems such as aggregators and REMCs without clearly spelling out the institutional arrangements required.

Looking at these limitations the Power ministry formulated a much-needed relief measure. It was decided to create REMCs in RE rich states which would coordinate with RE generators, LDCs and help in better integration of RE power.

5.2.5 State level interventions

At the state level, things get considerably more interesting. Every state has some form of renewable energy department or an autonomous institution with the mandate to facilitate and promote use of renewable energy. In many cases it also acts as the single window clearing house for project permissions. However, the need to very actively engage with all other departments to get the permits and clearances will not be obviated. Intervention of the renewable energy department will mostly be limited to providing guidance about the requirements and keeping track of renewable energy projects for the state.

The central decisions though appear to be mandatory; States have sufficient autonomy to control most of the factors in association with implementation and management. An illustration of this is seen where in the price discovery by central auctions for wind farming has made state run DISCOMS wanting to cancel existing power purchase agreements entered into under negotiated feed in tariffs and seek to purchase power at the discovered prices by auctions with entirely different parameters. DISCOMS perpetually run under losses due to various reasons some of which are the agricultural loads to be provided under heavy subsidy, inefficient distribution grids, and power theft. The State Governments chose to write off than pay when the burden reaches a certain level. Under these circumstances there will be a natural tendency to transfer some of the debts to other entities such as IPPs.

5.2.6 New capacity additions

Many factors have been considered from the present and past situations and are expected to yield positive growth patterns over next years. For example, once the PGCIL's plans are executed, evacuation stops being an issue. Therefore, it is possible to expect large number of projects commissioned. In terms of integration of RE power, REMCs will take control of maximising RE power on grid. On a parallel, State Governments are also into auctioning power procurements along the same lines as the SECI auctions. The virus break out in March 2020 will certainly impact an already seemingly stiff target. The SECI bids for wind power have not yielded the anticipated growth. To encourage more and more installations, hybrid projects consisting of Wind – Solar and other resources including coupling the projects with thermal power also have been put under bidding route.

5.2.7 Offshore Wind

After considerable deliberations and near shore measurements along the coastline of India starting from Kutch in Gujarat to Puri in Odisha there was an uneasy understanding that Indian offshore potential was rather limited to a few MW at Dhanushkoti near Rameshwaram in Tamil Nadu mainly from evacuation point of view. However, a recent investigation in Gujarat using Lidar and other advanced equipment there seems to be some sizeable potential. Accordingly, Government of India has sought expression of Interest globally for a demonstration project for 1000 MW to be established in Gujarat. The LIDAR based wind speed at 150 m asl is at 7.98 m/s. [4] It is interesting to note that the demonstration project not funded by Government is to be established in Gulf of Khambhat off Gujarat coast about 25 km into the sea. Presently

5.2.8 Future

With changed scenario from the cost-plus models to competitive bidding financial models there are new challenges. Further, with grid size becoming bigger, management of variable power availability by grid managers accustomed to operate grid with controllable generation will be a challenge. It will require considerable changes in the mind sets. There are engineering challenges that could be overcome. India has a clear vision that by 2030 it shall have about 440 GW of RE power majorly coming from Solar and wind. This will be part of 836 GW of total power capacity [5]. The original target of achieving 175 GW by 2022 may be dislocated to an extent due to the fact that the policy deployment in reality has not been as smooth as was expected. Added, there has been the impact of the Covid – 19 pandemic that has had a very negative impact on economy and business as usual scenarios. However, this may merely postpone the plans and not shelve them. Though repowering is often suggested as an option to get high wind sites to accommodate higher capacity turbines, in reality over the next five years just 3.5 GW would complete the 20-year service life. With moderate winds in such sites, the actual service lives of most of the turbines can be as much as thirty or more years. When the turbine sizes change, the entire infrastructure will have to be completely redone.

5.3 CHINA

Everything about China has to be on a mega scale. It is not any different in wind power. For a country that was in a few GW till 2007 China now has 216.75 GW in terms of installed power. There are no parallels to this exponential growth anywhere else.

Figure 5.3 provides an overview of the country wide installations in China.

5.3.1 Technology development

China boasts a mega scale manufacturing capability. With five wind turbine manufacturers on the list of top ten companies in the world are engaged in the manufacture of wind turbines with market share of 37.8% in the onshore wind business [7].

5.3.2 Growth Drivers

In China, as can be seen from figure 5.4, wind power was nowhere on the radar until 2006 when few GW's were installed. Subsequently the installations doubled every year and until 2015 the industry added sizeable capacities both from manufacturing and deployment angle. There were some issues about evacuation from far flung areas to load centres and this was a dampener for the growth. Over the next few years, the electrical infrastructure developed and the forced backing down of large wind farms were considerably reduced. On a parallel, centralised forecasting and scheduling were undertaken by the grid managers to give priority access to the Wind and Solar power.

Some of the most important factors that were helpful were:

Holistic policy framework with good clarity supported by clear Renewable Energy Law enacted.

Purpose was the first consideration and commerce came later.

Long term policies that remained in force with little or no tweaking.

Each province had a double-digit renewable energy purchase obligations that were enforced in earnest.

In the matter of giving priority to RE power there were no conflicts between different provinces.

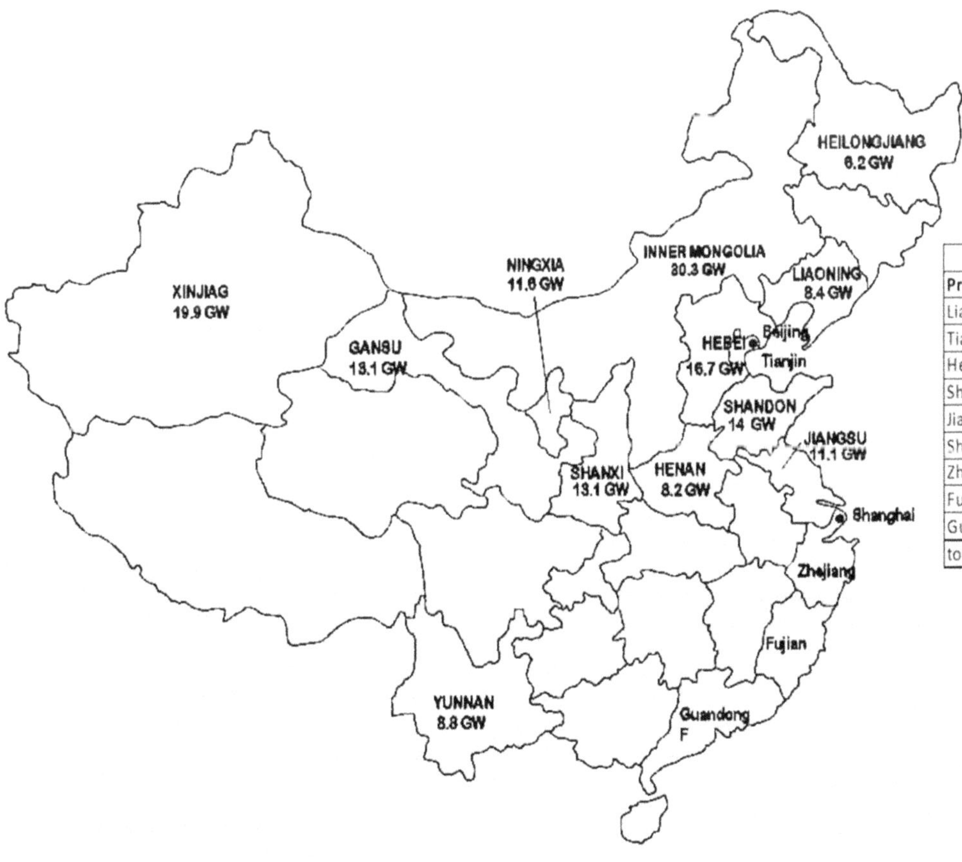

China offshore Wind	
Province	MW
Liaoning	245
Tianjin	117
Hebei	292
Shandong	15
Jiangsu	4725
Shanghai	417
Zhejiang	265
Fujian	490
Guangdong	458
total	7024

FIG. 5.3: TOP TWELVE PROVINCES LEADING IN WIND POWER DEVELOPMENT [6]

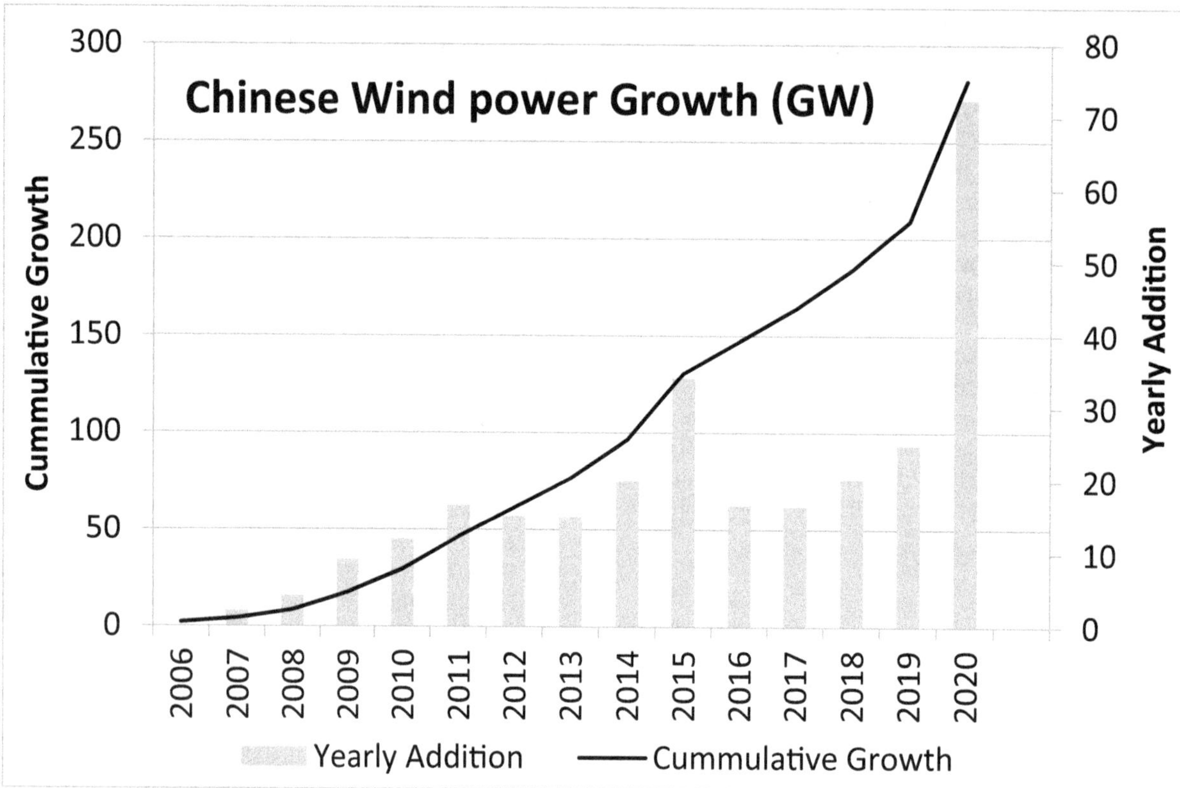

FIG. 5.4: CHINESE WIND POWER GROWTH OVER LAST 15 YEARS (SOURCE: CHINAENERGYPORTAL.ORG)

Grid extension and infrastructure development were taken up automatically in a planned manner.

Export of wind turbine hardware was fully supported by the Government. Though most of the production was consumed internally, several Chinese wind turbine manufacturers have set up shops abroad including India.

China's Standards addresses the type certification aspects and till date produced fourteen publications including applicable standards, recommendatory standards, and guidelines. While keeping internationally accepted standards, the localisation helps in encouraging indigenous production.

With such enabling policy environment, the wind farms were established with a feed in tariff assured for 20 years, the installations were quite aggressively followed. The idea was to charge a RE-cess on energy sold and pooled centrally. The generation was paid for at the coal equivalent by the distribution company and the difference in the FIT and the sale price is paid by the RE-fund. This went off quite well in the initial years. As the number of GW deployed grew at a very rapid rate, the RE fund started depleting too quickly and it was becoming increasingly difficult to meet the commitments. The idea of green certificate in lieu of the payment was introduced and has had a limited success. Even as the green certificates were introduced, without a strict Renewable Energy Portfolio enforcement, it becomes difficult to find willing takers for the certificates. Though polluting industrial consumers are being targeted, the realisation of revenue is posing a major challenge. China, finding the complexities of providing fixed feed in tariffs for twenty years recently moved over to competitive bidding route. With much capacity already permitted, there FIT regime will end by 2021. Therefore, there is a rush to install and commission as many projects as was feasible to be in this regime. The Covid Pandemic has disrupted the supply chain quite considerably. Notwithstanding this it is possible to see record capacity addition in 2020 and 2021.

Another major bottleneck was about the grid access to the wind farms. The national average of curtailment for the year 2016 was as much as 17% and some districts like Gansu had to do with 57% grid availability while Xinjiang managed just 62% grid availability. In most cases, grid extension did not keep pace with the wind farm establishment. The situation has remedied to some extent over last few years.

5.3.3 Off-shore Wind

China, with a vast eastern coastline is moderately windy with the southeast coastline. It varies between 30% near Shandong in the north, increasing to around 35 to 40% near Shanghai. An estimated 45%-56% is available off Fujian in the Southwest. It drops to about 30% along the southern coastline [9].

China started its offshore experiments with a modest 1.5 MW in 2007 spent over a decade to reach 6.8 GW [10]. Up to 44 GW of offshore wind farms have been permitted under FIT, providing that the OEMs and the EPC contractors can meet the challenge.

Just as in Europe the Chinese have been developing offshore wind turbines with a rating of over 6 MW in the recent years. Consistent with the policy of possible export together with domestic consumption, China has developed capabilities of designing, deploying, and operating offshore wind farms.

5.4 SRI LANKA

Sri Lanka, being an island nation has modest power requirements. The 2018 statistics from Ceylon Electricity Board [Statistical Digest 2018 CEB accessed on20.05.2020] shows the total installed capacity of grid power is at 4046 MW shared between thermal (coal and oil based) and Hydro electric stations (figure 5.5). Key stake holder is the Ceylon Electricity Board owning major hydro and thermal generation under its control. To a small extent there is private participation in Renewable energy generation which include mini hydel, solar and wind power installations. A 10 % of Private Public Partnership (PPP) in terms of oil based thermal plants is also operating in Sri Lanka.

The first demonstration project of 3 MW (1999) at Hambantota owned by CEB was decommissioned in 2018 due to very high operating costs. In the meanwhile, under PPP 128 MW wind and 88 MW of solar capacity has been added. Solar capacities are largely roof top systems.

The Sri Lanka energy policy [11] envisages a sizeable role for renewable energy and expects to have on ground 1600 MW of RE power operating by 2023. The CEB perspective plan till 2039 shows a steady growth of RE Power in the energy mix (figure 5.8). The envisaged Wind installation is at about a GW most of which is to come through competitive bidding.

Sri Lanka proposes to add more and more renewable energy sources in its energy mix. The perspective plans anticipate the inclusion of hydro and other renewable to be in the ratio of 51 % by 2039 [12]. The plan also envisages creating pumped hydro power stations to serve as peaking suppliers. This will also serve as spinning reserves.

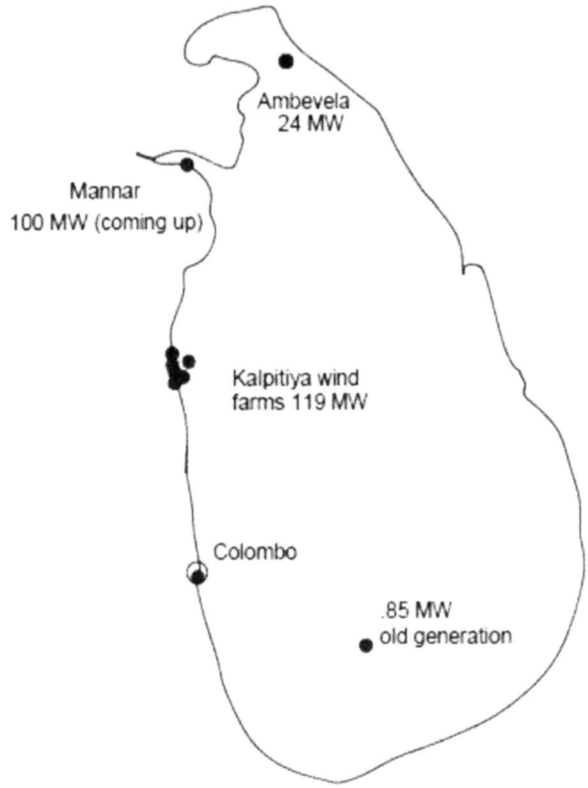

FIG. 5.6: CURRENTLY RUNNING WIND FARMS IN SRI LANKA

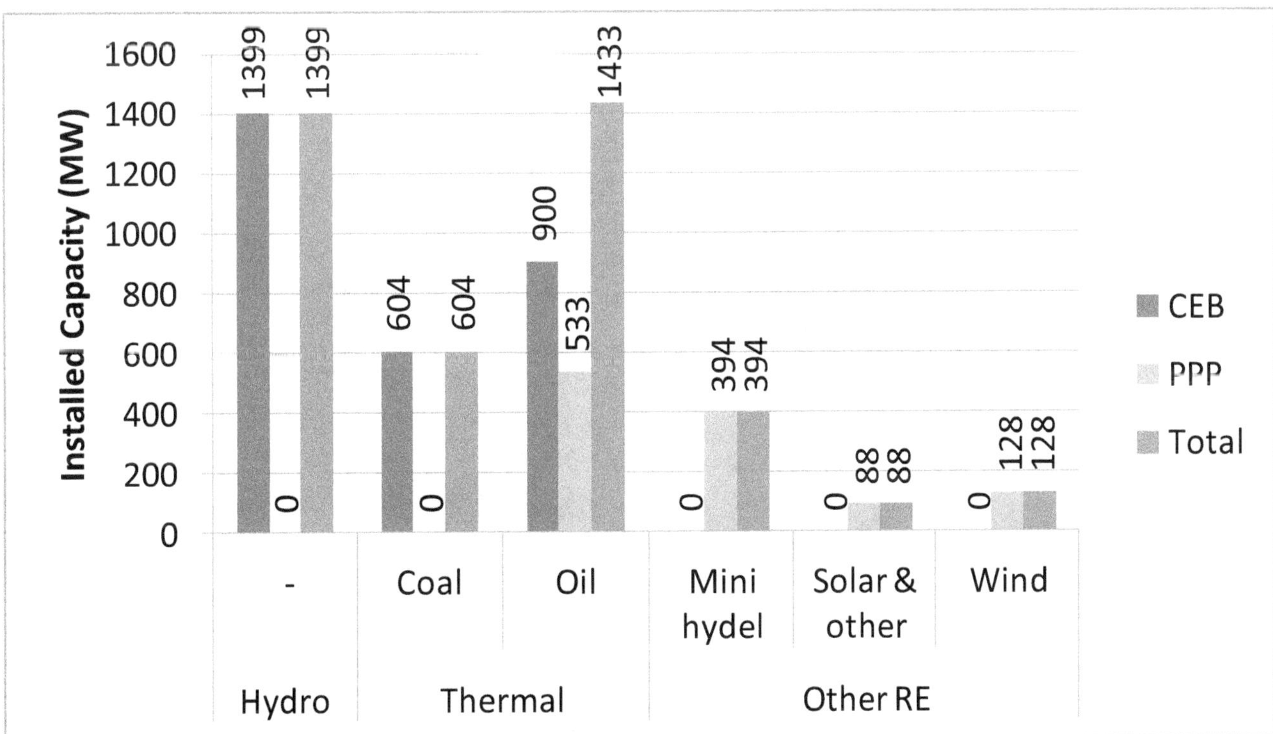

FIG. 5.5: GENERATION MIX FOR GRID POWER IN SRI LANKA

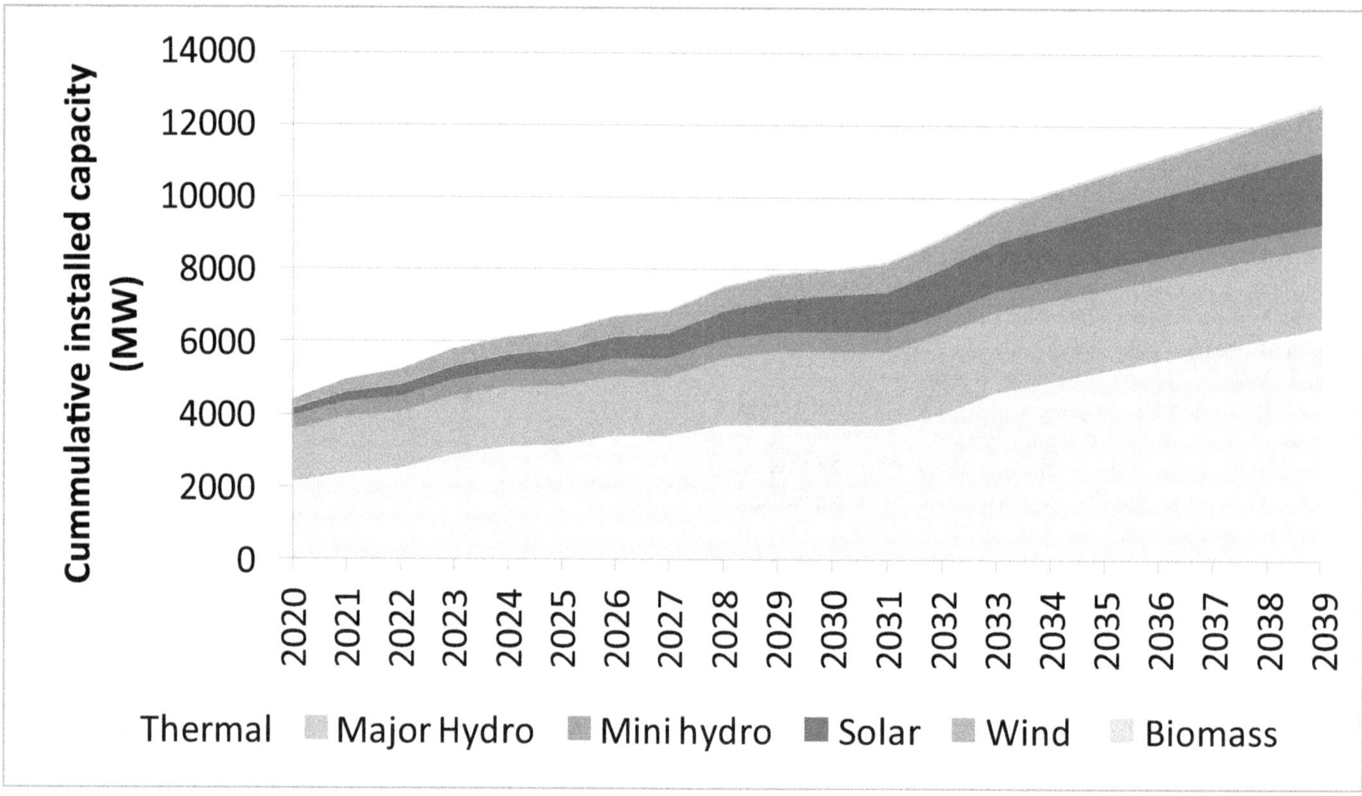

FIG. 5.7: PLANNED DEVELOPMENT OF GRID POWER

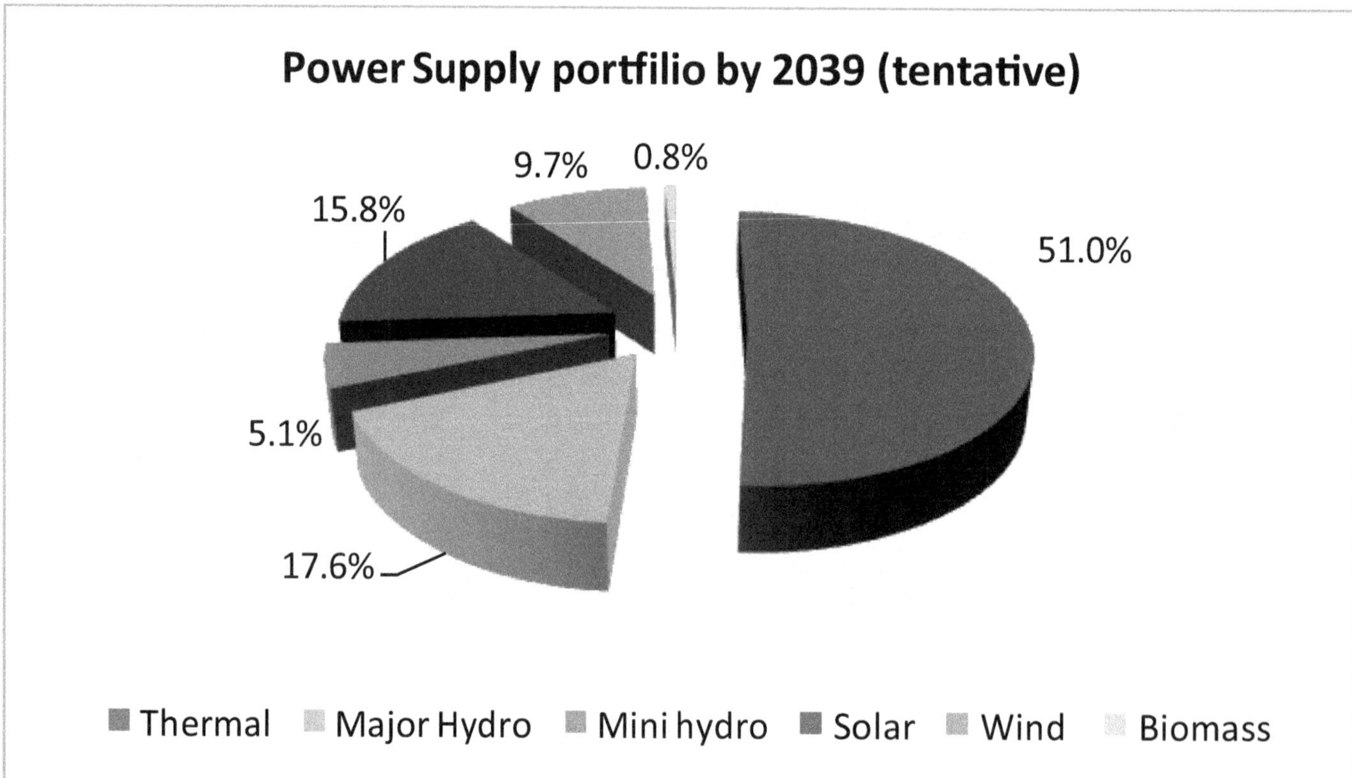

FIG. 5.8: ANTICIPATED POWER SCENARIOS IN SRI LANKA IN 2039

TABLE 5.1 POWER SUPPLY MIX IN PAKISTAN AS OF 6-30-2020

Source	Capacity (MW) as of 30 June 2020	% of installed capacity
Thermal	24817	64.1%
Hydro	9861	25.5%
Wind	1248	3.2%
Solar	530	1.4%
Baggasse	369	1.0%
Nuclear	1467	3.8%
Small& Captive	427	1.1%
Total	38719	

The planned capacity addition year on year is in the range of 4 to 5% and the thermal component shows a gradual reduction still maintaining a major part of the energy mix. The inclusion of RE technologies show a slightly higher rate as can be seen from figure 5.7. As per projections the thermal component is a little over fifty percent. Major hydro is about 17.6 %.

Solar power would reach 16 % while wind power installations may reach around 10 % in terms of installed capacity. As of the moment there is no proposal of venturing into off-shore wind power. Looking at the small exposure to wind, there may be no incentive to get into manufacturing activity either through licence agreements or through indigenous design and development.

5.5 PAKISTAN

A study by NREL back in 2007 identifies some of the possible wind farming sites based on a number of assumptions estimates about 50 GW wind potential for Pakistan[13]. However these estimates depend on a number of assumptions and they have to be clearly understood. One of the most important aspects is that it ignores evacuation related issues. The installed power generation capacity of Pakistan as on 30th June 2020 stands at 38,719 MW. Table 5.1 gives the shares of each of the resources[14]

Pakistan has been slow on uptake of renewable energy projects and was looking for high level of private sector involvement. The need to use less and less fossil fuels has resulted in establishing conductive investment atmosphere for wind power projects. The renewable energy has a pre fixed feed in tariff that makes it an attractive option and with very high cost of fossil fuel based generation; it is an advantageous position for the distribution companies to purchase wind or solar power.

5.5.1 Wind resources in Pakistan

Wind mapping in Pakistan has been originally based on a mapping program under US-AID and the Meteorological Department has identified some high wind zones in the Gharro-Keti Bunder wind corridor. Presently on-site measurements are also underway in 12 locations across the country to validate the potential.

Figure 5.9 shows the windy regions of Pakistan. Though some hills to Northwest of Kech and Gilgit in the North may present very high winds, the area is rather complex and would present challenges in deployment. Essentially therefore the Sind province is where most of the projects have been getting established. The wind farms established in been operating in Jamphir, Thatta district of Sind province. Future projects also are planned in this area.

There are serious issues of grid extension for very large projects and this has inhibited rapid growth.

5.5.2 Energy Mix and a case for Renewable Energy

Solar and wind power in Pakistan offer one of the most economic resources with vast areas that are entirely suit-

FIG. 5.9: AREAS WITH USEFUL WIND POTENTIAL IN PAKISTAN

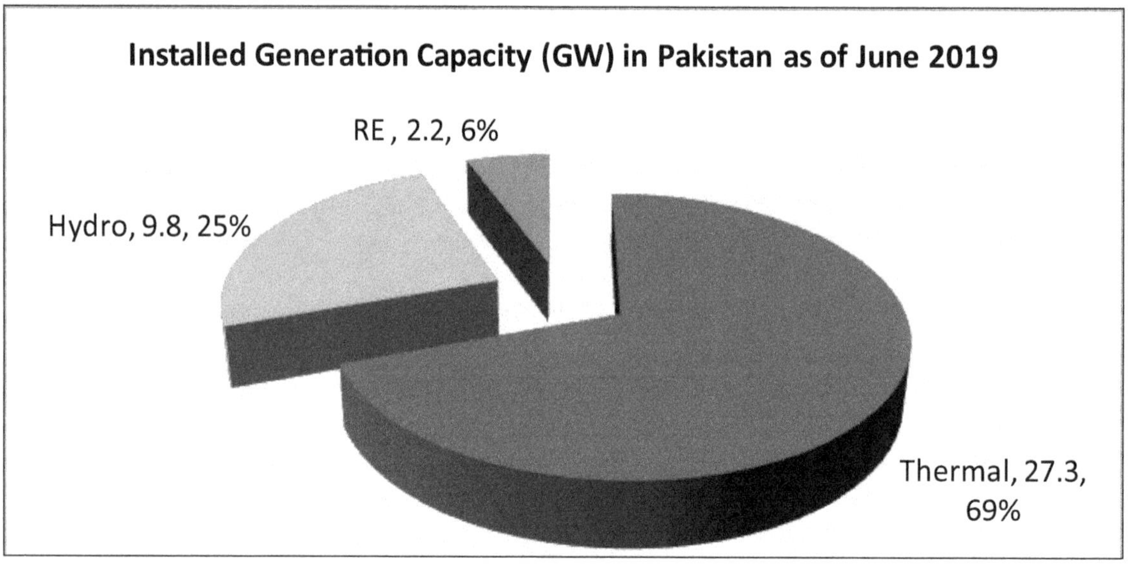

FIG. 5.10: GRID POWER CAPACITY IN PAKISTAN, 2019

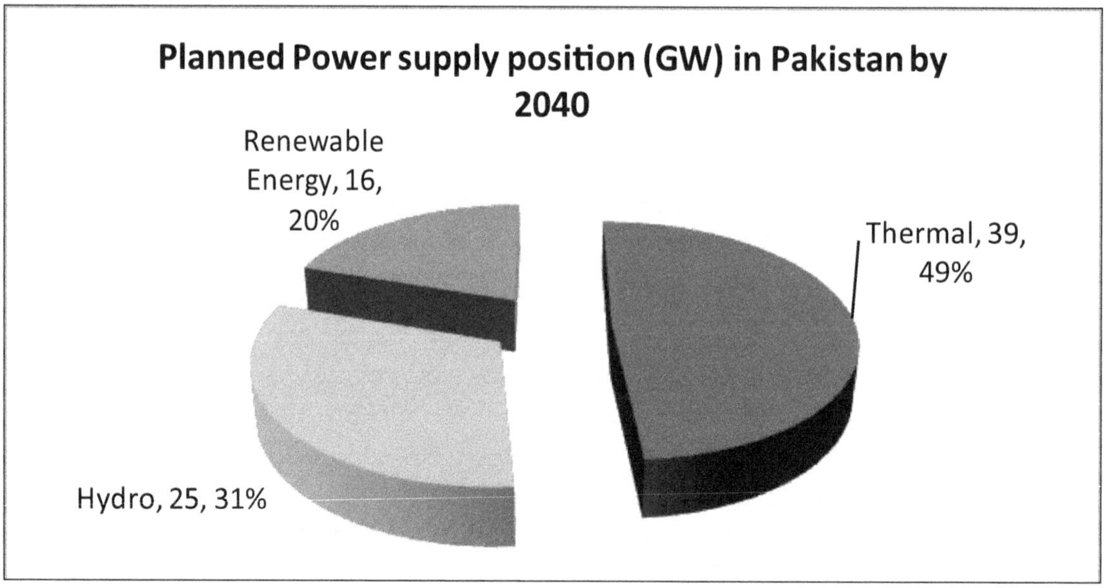

FIG. 5.11: PLANNED POWER MIX IN PAKISTAN, 2040

able for large scale deployment [14]. There are challenges in terms of grid extension which are posing serious dislocation due to local area congestion and frequent grid outages. However, these are known issues and future plans have taken into account the need to have a strong electrical network. The role of renewable energy as part of the energy mix is taken into consideration.

It can be seen that the hydro and RE power account for 31% in terms of installed capacity. Though there is an increase in the use of fossil fuels in absolute terms, inclusion of RE power is definitely part of expansion. The policy requires a 30% penetration of RE technologies by 2025 which seems very optimistic. There are perceived conflicts among various stake holders including current and future investors. Over a period of time it will be inevitable to move towards renewable energy technologies. Though the planning envisages over 10 GW of wind power alone and 6 GW from Solar Power (figure 5.11), realisation may face several hurdles. The Local governments normally have the last word and central planning to implementation always remains a complex issue. Pakistan is also looking at off grid solutions mainly powered by SPV. It makes sense in several cases where grid extension would prove to be uneconomic both in terms of investments and maintainability. Some common traits can be seen in this with India.

5.6 SUMMARY AND CONCLUSIONS

Large scale exploitation of wind power in the last decade has entered hundreds of GW league and is now considered as an essential element for energy security. There are several important aspects that should be highlighted.

With increased penetration levels, grid managers have huge issues with balancing.

One is the engineering limitations that can be addressed with some imagination.

Other is the commercial angle. Presently, in the merit order grid access, renewable energy has, in most cases a must run status. With multiple options from all other IPPs and state run corporations, there is severe competition for grid access. Since it is a question of returns on investment, there will be various pressures; well intentioned regulations may not be deployed in an even-handed manner.

It will, in many cases result in increased number of Non-Performing Assets (NPAs) Over last few years, the Solar Photovoltaic has found favour with planners due to the perceived predictability and falling prices. The SPV is crowding out wind power to a considerable extent. However, due to the complementary nature of these two resources and storage technologies, the combinations are being looked at as viable alternative to firm power. There are issues of commercial viability of such combinations. The integration of such systems and successful commercial deployment is yet to happen. One primary issue here is that the renewable energy production from each individual resource is almost entirely in private commercial space. Due to the fact that the procurement is by reverse bidding route, the prices have been really very low. Most often the bidders assume idealistic future:

- Projects are envisaged, smooth passage to establishment of the projects is anticipated where the federal and local governments play their constructive role,
- Full grid access is provided without any preconditions for the project lifetime.
- The DISCOMs would honour Power Purchase Agreements and pay on time for the project life time.

With such a backdrop, the temptation to out bid competitors is very strong. As a matter of fact most of the numbers are arrived at based on sweeping assumptions. In most cases the bidders would just have over optimistic energy estimates. They would also not have a clear picture of where the projects would be established.

In order to prevent gaming, bank guarantees are taken for non performance at various levels. Further a Ceiling price is also mandated for the bidding which cannot be breached. These factors still do not deter certain amount of reckless price discovery. Therefore, workable project configurations have smaller chances of staying afloat.

In reality each such expectation needs to be recalibrated based on experience.

Now, to be in the firm power league and still remain cost competitive is a huge challenge. To set up a fully RE powered firm power plant, one has to take into account the variability of different resources, plan for the no power availability situations. This will result in having to have larger capacity wind power plants, solar plants and have access to storage. Based on a very theoretical approach, bidding gets done and it will be found quite quickly that it will be almost impossible to get financial closure in many cases.

Notwithstanding these challenges, the outlook remains positive. It is a question of time and learning to get to the right price bands that make wind projects get back on its feet and thrive. What is essential is that all stake holders need to understand that they are working for the common good.

5.7 ACRONYMS

agl	above ground level
CEB	Ceylon Electricity Board
CTU	Central Transmission Utility
DISCOM	Distribution Company
EPC	Engineering Procurement Construction
FIT	Feed in Tariff
GW	Giga Watts
INR	Indian Rupee
IPP	Independent Power Producer
LIDAR	Light Detection and Ranging
NPA	Non Performing Asset
NREL	National Renewable Energy Laboratory
OEM	Original Equipment Manufacturer
PGCIL	Power Grid Corporation of India Limited
PPP	Private Public Partnership
RE	Renewable Energy
REMC	Renewable Energy Management Centre
SECI	Solar Energy Corporation of India
SPV	Solar Photo Voltaic

5.8 REFERENCES

1) https://powermin.gov.in/en/content/power-sector-glance-all-india Accessed on 20.12.2021
2) https://mnre.gov.in/wind/current-status/ accessed on 30.05.2020.
3) https://powermin.nic.in/en/content/green-energy-corridor accessed on 01.05.2020
4) https://niwe.res.in/assets/Docu/Dr.Rajesh_Katyal-Offshore_wind_energy_02Jun2017.pdf
5) http://cea.nic.in/reports/others/planning/irp/Optimal_generation_mix_report.pdf
6) https://chinaenergyportal.org/en/2020-q2-wind-power-installations-and-production-by-province/ accessed on 28.12.2020
7) https://www.windpowerengineering.com/business-news-projects/vestas-gains-impressive-lead-as-top-wind-turbine-manufacturer/ downloaded on 19.05.2019
8) https://www.eia.gov/opendata/qb.php?category=2134745&sdid=INTL.37-7-CHN-MK.A accessed on 28.12.2020

9) https://www.evwind.es/2019/12/29/china-offshore-wind-energy/72824 accessed 30.12.2020
10) GWEC annual report 2019
11) The Gazette of the Democratic Socialist Republic of Sri Lanka PART I : SECTION (I) MINISTRY OF POWER, ENERGY AND BUSINESS DEVELOPMENT National Energy Policy and Strategies of Sri Lanka,08th August, 2019. NATIONAL ENERGY POLICY AND STRATEGIES OF SRI LANKA MINISTRY OF POWER, ENERGY AND BUSINESS DEVELOPMENT - GOVERNMENT OF SRI LANKA
12) Long term Generation Expansion plan 2020 -2039, CEB draft
13) https://www.trade.gov/knowledge-product/pakistan-renewable-energy accessed on 27.05.2020
14) State of Industry Report 2020.pdf (nepra.org.pk) accessed on 23.12.2020

CHAPTER

6

Artificial Intelligence in Wind Energy

Weifei Hu

ABSTRACT

Despite the increasing development of wind energy (WE) which has notably become the leading renewable energy now, there are still persistent barriers to wider implementation related to technology, cost, and environmental and social impacts, e.g., super large-scale (≥10 megawatt) wind turbine (WT) technology, WT deployment at low wind speed and/or residential-closed areas, and offshore wind farm under severe wind-wave conditions. As one of the emerging technologies, artificial intelligence (AI) could take some of these challenges and assist in achieving the goal of further lowering the levelized cost of energy from wind. Recently AI has been tremendously investigated in an extremely wide range from financial data analytics and industrial production design to novel biometric and forensic applications and various renewable energy development. This chapter provides the fundamental introduction of four categories of state-of-the-art AI methods including (i) neural learning methods, (ii) statistical learning methods, (iii) evolutionary learning methods, and (iv) hybrid learning methods in a lucid way, then carry out a comprehensive survey of various AI methods used in WE including (i) wind speed prediction, (ii) wind power estimation, (iii) WT design and optimization, and (iv) WT condition monitoring.

Keywords: Wind Energy; Artificial Intelligence; Wind Speed Prediction; Wind Power Estimation; Artificial Neural Network; Statistical Methods, Evolutionary Methods

6.1 INTRODUCTION

Wind energy (WE) has become the fastest growing renewable energy source to meet the ever-increasing world energy demand. Global installed wind power capacity has been tremendously increased over the last 16 years from 23,900 MW in 2001 to 539,581 MW in 2017 [1]. There are more than 314,000 wind turbines (WTs) now operating around the world and supply more than 4.3% of 2015 global electricity demand [2, 3]. However, this significant development of WE also brings challenges, e.g., exploring new available wind resources on land and offshore, designing reliable and cost effective titanic WTs (e.g., 10 MW WTs), and developing new monitoring approaches that can reduce the operational and maintenance (O&M) cost. Artificial intelligence (AI) techniques, emerging from other fields (e.g., computer science), are becoming promising tools to overcome these challenges and boosting the further development of WE.

AI is a kind of intelligence demonstrated by machines in contrast to the natural intelligence displayed by humans [4]. It has been performing more complex tasks than straightforward automation in different domains and applications, such as robotics [5], cyber-physical systems [6], and renewable energy [7]. It consists of many branches including artificial neural network (ANN), genetic algorithm (GA), expert system (ES), fuzzy logic (FL), and various hybrid systems, which are combinations of two or more of the branches, to produce efficient and effective computing solutions [8]. Even though some current AI techniques still require interaction with human intelligence, the appropriate use of AI techniques leads to smart systems with improved performance that cannot be achieved by traditional methods.

A large number of AI techniques have been proposed and applied in WE. This chapter distinguishes various AI techniques into four categories, neural learning methods, statistical learning methods, evolutionary learning methods, and hybrid learning methods, which are further separated into different specific AI methods as explained in Section 6.2. Section 6.3 provides surveys of various AI methods used in WE focusing on wind speed prediction, wind power prediction, WT design and optimization, and WT condition monitoring (WTCM). Concluding remarks are provided in Section 6.4. This chapter aims to offer non-exhaustive surveys due to the extremely wide topic (the combination of AI and WE) and present readers an overall but clear picture of AI in WE.

6.2 FUNDAMENTALS OF STATE-OF-THE-ART AI

This section provides the fundamental introduction of four categories of state-of-the-art AI methods including neural learning methods, statistical learning methods, evolutionary learning methods, and hybrid learning methods with their demonstrating usages in WE.

6.2.1 Neural Learning Methods

Neural learning methods apply various ANN algorithms, inspired by the biological neural networks (NNs) that constitute human brains, for solving a wide range of WE problems, for example, pattern classification, clustering, function approximation, forecasting, optimization, association, and control [9]. WE problems could involve multidisciplinary theories. For example, the physical model used for WT blade fatigue optimization include random wind field simulation, aerodynamic analysis, WT dynamic and blade structure analyses, and composite fatigue analysis. Consequently, the whole physical-based analyses of WE problems could be extremely complex. Fortunately, ANNs have been demonstrated that they can efficiently solve WE problems (e.g., wind speed and wind power forecasting) that have complex physical processes [10].

A basic ANN consists of several simple processing units, named neurons, with weighted connections between neurons (Fig. 6.1). The output of each neuron is the sum of some non-linear functions of its inputs. The weights between the neuron connections are used to increase or decrease the strength of the signal. Each connection can transmit a signal from one artificial neuron to another if the aggregated signal exceeds a threshold. Neurons are aggregated into layers which are often classified as input layer, hidden layer(s), and output layer in the ANN architecture. An ANN with one hidden layer can be expressed as

$$\hat{P}(t) = f\left[\sum_{i=1}^{m} a_i f\left(\sum_{j=1}^{n} A_{ij} x_j(t)\right)\right] \tag{1}$$

where a represents a weight vector, A is a weight matrix, f is a transfer function, and $x_j(t)$ and $\hat{P}(t)$ are the input values and the prediction at time t, respectively. Multiple hidden layers are commonly used to solve complex nonlinear and implicit WE problems. For instance, Fig. 6.1 provides a typical two-hidden-layers ANN that uses historical meteorological data (e.g., wind speed, air density, and humidity) and WT power to train the ANN model for WT power prediction (e.g., [11, 12]).

The input dataset is fed into an ANN which starts a training process by adjusting the ANN parameters, e.g., weights. The training is called supervised training if the output is known, otherwise it is named as unsupervised training. As a closed-loop feedback system, supervised learning directly compare the actual network output and the predicted output. In contrast, unsupervised learning draws inferences from datasets consisting of input data without labeled responses. The difference between the network output and the training sample output is used to training the learning process, e.g, adjusting the network parameters. The accuracy is often measured by the mean square error (MSE), root mean square error (RMSE) and mean absolute percentage error (MAPE).

As the significant development of neural learning methods during the past decades, more than 50 types of ANN have been investigated including multilayer perceptron (MLP), backpropagation network (BPNN), convolutional neural network (CNN), wavelet neural network (WNN), recurrent neural network (RNN), adaptive resonance theory neural network (ARTNN), to name a few [9]. More neural learning theories and practical NN design methods in depth are seen in [13, 14].

6.2.2 Statistical Learning Methods

Statistical learning methods, also named as the direct random time-series methods, create mathematical models for WE problems using pattern identification, parameter estimation, and model checking based on historical data. The mathematical models are often classified as moving average (MA) model, autoregressive (AR) model, autoregressive moving average (ARMA) model, and autoregressive integrated moving average (ARIMA) model [15].

A MA model is often used to smooth noises and remain the long-term trends in measured data. Given a series of data, the moving mean is calculated over a sliding window across neighboring samples with a window size depending on the application. For example, a series of 10-Hz wind speed data measured by a sonic anemometer applies 3-second window size to obtain a moving average (i.e., 3-second moving mean wind speed), which is often used to identify wind gust events as demonstrated in Fig. 6.2.

An ARMA model is a powerful tool to describe the dynamics and predict the forthcoming value of an individual time series in WE, e.g., wind speed, wind direction, and WT power. The prediction is calculated as a linear combination of previously observed values and a series of noise. An ARMA model ARMA(p, q) with pth-order AR process and qth-order MA process is expressed as [17]

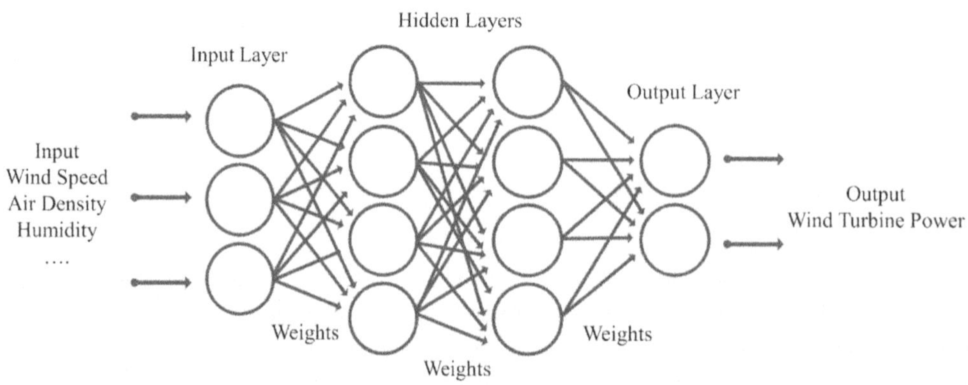

FIG. 6.1: AN ILLUSTRATION OF USING A TYPICAL ANN TO PREDICT WT POWER BASED ON METEOROLOGICAL DATA, E.G., WIND SPEED, AIR DENSITY, AND HUMIDITY.

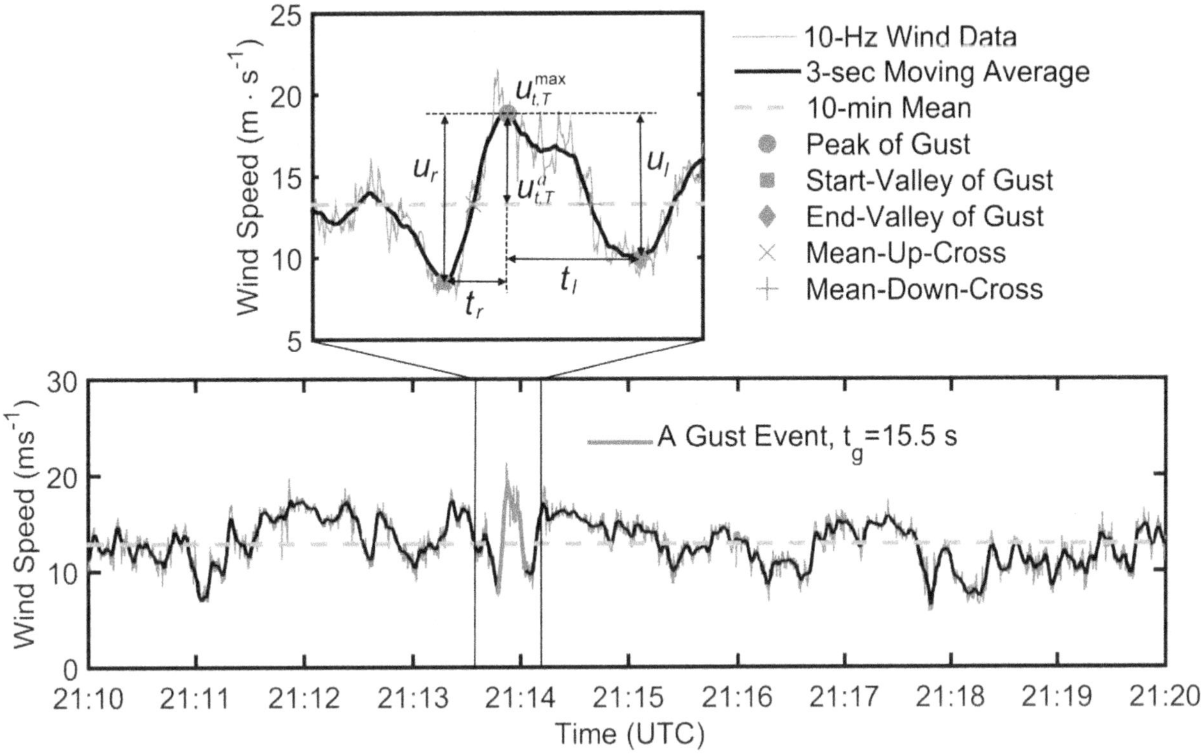

FIG. 6.2: AN ILLUSTRATION OF USING MA TO IDENTIFY WIND GUSTS. A 3-SECOND MA IS USED TO IDENTIFY A WIND GUST EVENT FROM A 10-HZ TIME SERIES OF MEASURED WIND DATA FOLLOWING THE GUST CRITERIA IN [16].

$$X_t = \sum_{i=1}^{p}\phi_i X_{t-i} + \sum_{j-1}^{q}\theta_j \varepsilon_{t-j} + \varepsilon_t \tag{2}$$

where X_t represents the time series to be described, X_{t-i} is a given testing time series, and coefficients ϕ_i and θ_j are both n by n matrices for the AR and MA processes, respectively. The noise terms ε_t is usually assumed to be white noise (i.e., a random variable with mean zero and a covariance, and for every $t_1 \neq t_2$ ε_{t1} and ε_{t2} are uncorrelated) or independent identically distributed normal variables with zero mean. If $q = 0$, the ARMA model reduces to an AR model; if $p = 0$, the ARMA model reduces to a MA model. The training process of an ARMA model is to determine the coefficients ϕ_i and θ_j which can be calculated by computing either nonlinear least square estimator or maximum likelihood estimator [18].

An ARIMA model can be derived from the ARMA model ARMA(p, q) as explained in the following. Equation (2) can be reorganized as

$$X_t - \phi_1 X_{t-1} - \cdots - \phi_p X_{t-p} = \varepsilon_t + \theta_1 \varepsilon_{t-1} + \cdots + \theta_q \varepsilon_{t-q} \tag{3}$$

or equivalently by

$$\left(1 - \sum_{i=1}^{p}\phi_i L_i\right) X_t = \left(1 + \sum_{i=1}^{q}\theta_i L_i\right)\varepsilon_t \tag{4}$$

where L_i is the lag operator. Assuming the polynomial has a unit root of multiplicity d, it can be rewritten as $\left(1 - \sum_{i=1}^{p}\phi_i L_i\right)$

$$\left(1 - \sum_{i=1}^{p}\phi_i L_i\right) = \left(1 - \sum_{i=1}^{p-d}\phi_i L_i\right)(1-L)^d \tag{5}$$

Then Eq.(4) can be expressed as

$$\left(1 - \sum_{i=1}^{p-d}\phi_i L_i\right)(1-L)^d X_t = \left(1 + \sum_{i=1}^{q}\theta_i L_i\right)\varepsilon_t \tag{6}$$

Now using a new "p" to replace the "$p-d$" in Eq. (6), an ARIMA(p, d, q) is expressed as

$$\left(1 - \sum_{i=1}^{p}\phi_i L_i\right)(1-L)^d X_t = \left(1 + \sum_{i=1}^{q}\theta_i L_i\right)\varepsilon_t \tag{7}$$

Thus, an ARIMA model is actually a special case of ARMA (p+d, q) model having the autogressive polynomial with d unit roots. In the case of $d = 0$, the ARIMA(p+d, q) model is reduced to the ARMA(p, q) model. Modified ARIMA methods have been proposed to address the stochastic characteristic of WE parameters. For example, a novel LARIMA model is developed by introducing a limiter into a standard ARIMA model to represent the upper and lower bounds of wind power generations [19].

6.2.3 Evolutionary Learning Methods

Evolutionary learning methods are generic population-based metaheuristic methods using mechanisms inspired by biological evolution, such as reproduction, mutation, recombination, and

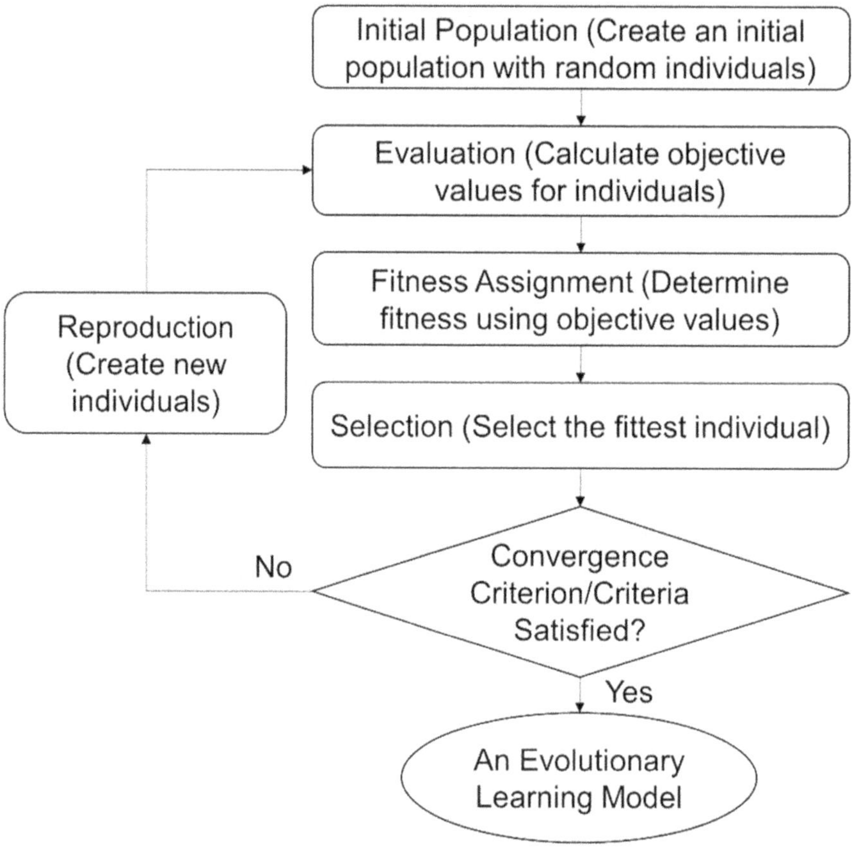

FIG. 6.3: A GENERAL FLOWCHART OF CREATING AN EVOLUTIONARY LEARNING MODEL

selection. They have been becoming important learning techniques implemented in a wide range of WE problems. This kind of methods belongs a set of general stochastic searching methods and is a subset of evolutionary computation methods. Typical evolutionary learning methods include GA, genetic programming (GP), evolutionary programming (EP), and evolution strategy (ES). There are also some extensions of traditional evolutionary learning methods, for example memetic algorithm and distributed evolutionary algorithm [20]. A general flowchart of creating an evolutionary learning model is shown in Fig. 6.3.

Because of the flexibility and robustness inherited from evolutionary computation, evolutionary learning methods become efficient means to identify global optimums in WE problems. For example, they are used for finding the optimal WTs siting [21], the optimal WT blade design [22], the optimal wind farm pattern for maximum power production with limited number of turbines installed [23], and the automated specification of the configuration parameters used for hybrid learning methods [24].

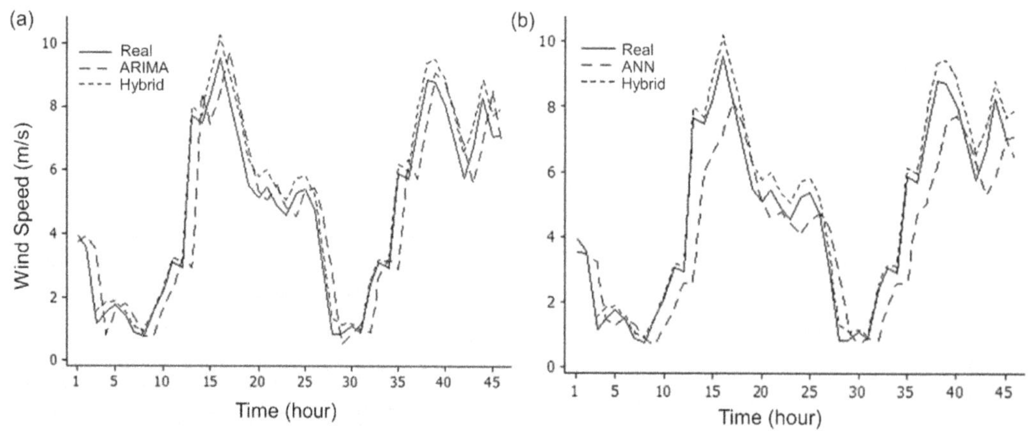

FIG. 6.4: AN ILLUSTRATION OF FORECASTING HOURLY WIND SPEED USING AN ARIMA-ANN HYBRID LEARNING MODEL, AN ARIMA MODEL, AND AN ANN MODEL [25].

6.2.4. Hybrid Learning Methods

The above three types of learning methods do not strictly repel each other. Actually use of a marriage of different types of learning methods, i.e., hybrid learning methods, could potentially generate more accurate prediction of WE parameters than use of a single method [25]. For example, hybrid models consisting of ARIMA models and ANN models have been developed in which the ANN models are built taking into account the nonlinear tendencies of errors that the ARIMA models cannot identify [25]. A comparison of wind speed forecasting using an ARIMA model, an ANN model, and an ARIMA-ANN hybrid model is demonstrated in Fig. 6.4.

6.3 APPLICATIONS OF AI IN WE

AI techniques have been proved to be versatile means for a wide range of problems which can be summarized from the following seven categories: classification, clustering, approximation, forecasting, optimization, association, and control [15, 26, 27]. This section provides an non-exhaustive survey of applications of AI in WE focusing on four major aspects: (i) wind speed prediction, (ii) wind power estimation, (iii) WT design and optimization, and (iv) WT condition monitoring.

6.3.1 Wind Speed Prediction

Wind speed is of critical importance considering the following usages: WT control and load tracking, WT preload sharing, and wind power system management and energy trading, as well as maintenance and repair of WTs [28]. Commonly used wind speed prediction methods include physical methods (e.g., numerical weather prediction (NWP)), neural learning methods (e.g., ANNs), statistical learning methods (e.g., ARIMA), and hybrid learning methods. Physical methods are often used for predicting long-term wind speed, while the neural learning methods and the statistical learning methods are efficient for wind speed prediction in short term [9]. Based on the predicted time intervals, wind speed prediction methods can be classified into four types, very-short term, short-term, medium-term, and long-term, as shown in Table 6.1. The main applications of different types of wind speed prediction are also summarized in Table 6.1.

The four types of learning methods explained in Section 5.2 are widely used for wind speed prediction as illustrated in the following. The neural learning methods have been developed for all four types of wind speed prediction methods according to the predicted time intervals. For example, in order to reduce the predicted errors and uncertainties Kani and Ardehali conducted a very-short-term wind speed prediction by developing a ANN-Markov chain hybrid learning model [30]. Li et al. compared the predictions of 1-hour ahead (short-term) wind speed using three different ANNs (linear element network, BPNN, and RBFNN), and concluded that one unique ANN cannot provide the best prediction in all cases. Since the very-short term and short-term wind speed predictions are often used for WT control, the computational time is a crucial factor to the prediction to be used in online application [9]. Regarding the medium-term wind speed prediction, Lawan et al. developed a feed forward neural network (FFNN) by trial and error techniques and found that the obtained MAPE was 12% less than those obtained by traditional methods [29]. However, the medium-term wind prediction usually causes less accurate forecasting due to the increase in the predicted time interval comparing with the very-short and short-time predictions. For a long-term wind speed prediction, Moustris et al. proposed an ANN model for a 24-hour ahead wind speed prediction for the optimum operation of hybrid power stations [31]. Azad et al. developed a feed-forward BPNN model for predicting the monthly mean wind speeds that are 30 days ahead and 1 year ahead [32].

Conventional statistical learning methods have also been extensively applied in the wind speed predictions [15]. For example, Ding et al. proposed an information criterion for an ARMA model that was employed to calculate the 1-month average wind speed and the 1-hour averaged wind speed [33]. Kavasseri and Seetharaman found that use of a fractional-ARIMA model to predict one-day-ahead and two-day-ahead wind speeds could generate significant improvements compared with the persistence model using wind speed records in North Dokota [34]. Other statistical learning methods used for wind speed prediction are seen in [35-37].

TABLE 6.1: WIND SPEED PREDICTION METHODS CLASSIFIED ACCORDING TO THE PREDICTED TIME INTERVALS [9, 29]

Wind speed prediction methods	Predicted time interval	Main applications
Very-short term	Few seconds – 30 minutes	Electricity market clearing, regulatory actions, WT control and load tracking
Short-term	30 minutes – 6 hours	Wind power dispatch to meet customer need
Medium-term	6 hours – 1 day	Wind generator on/off decision, operational security, electric market trading, power system management
Long-term	1 day – 1 week	Maintenance and repair of WTs

Evolutionary learning methods are not used extensively for direct wind speed prediction as the neural learning methods and the statistical learning methods. However, evolutionary learning methods are often used as optimization algorithms for finding the optimal parameters that could be tuned for other wind speed prediction models, e.g., ANN [38, 39] and support vector machines [40].

Many researchers have demonstrated that hybrid learning methods could obtain better wind speed prediction than their constituted single learning methods. Wang et al. developed a hybrid learning method (an integration of Support Vector Regression (SVR), seasonal index adjustment (SIA), and Elman recurrent neural network (ERNN)) to achieve more accurate forecasting of medium-term wind speed prediction than other models utilizing daily wind speed records at three different sites in the Xinjiang region of China [41]. A concise summary of different AI methods for wind speed prediction, as well as the wind power prediction, is provided in Table 6.3.

6.3.2 Wind Power Prediction

Wind power has a significant intermittent structure and variability comparing with other types of power systems, due to the inherent large uncertainty of wind. Accurate wind power prediction could benefit for dispatching, scheduling, and unit commitment of other power sources, e.g., thermal generators, hydro plants, energy storage plants, as well as energy market trading [42]. Since the wind power is theoretically calculated as

$$P = \frac{1}{2}\rho A V^3 \tag{8}$$

where ρ, A, and V are the air density, the cross sectional area of an air flow, and the wind speed, respectively, the accuracy of wind power prediction is highly related to the accurate forecasting of these input three variables as well as considering other factors, for instance local terrain and obstacle.

The wind power prediction methods can be classified into different types based on varied criteria. According to the predicted time horizons, wind power prediction can also be classified into very short-term (few seconds – 30 min), short-term (30 min – 6 h), medium-term (6 h – 1 day), and long-term time scales (1 day – 1 week) [28]. Foley et al. divided the wind power prediction into physical methods, traditional statistical or 'black box' methods, AI or 'gray box' methods, and hybrid methods [42]. In this chapter, the AI-based wind power prediction has a more general definition including the neural learning, statistical learning, evolutionary learning and hybrid learning methods. Among these four types, statistical learning, neural learning and hybrid learning methods are widely used for wind power prediction.

Several wind power prediction methods are surveyed as follows. Ma et al. compared a generalized dynamic fuzzy with BPNN and found that the prediction accuracy obtained by the former is 5.33% more accurate than that obtained by the latter [43]. A com-

TABLE 6.2: SOME SOFTWARE MODELS FOR WIND POWER PREDICTION [42]

Model Name	Developer(s)	AI Learning Methods
WPPT	Eltra/Elsam collaboration with Informatics and Mathematical Modeling at Danmarks Tekniske Universitet (DTU), Denmark	Statistical
Zephyr	Risø & IMM ay DTU, Denmark	Hybrid
Previento	Oldenburg University	Hybrid
e Wind™	True Wind Inc., USA	Hybrid
Sipreólico	University Carlos III, Madrid, Spain & Red Eléctrica de Espana	Statistical
WPMS	Institut für Solare Energieversorgungstechnik (ISET), Germany	Statistical
WEPROG	J. Jørgensen & C. Möhrlen at University College Cork	Hybrid
GH Forecaster	Garrad Hassan	Statistica
AWPPS	École des Mines, Paris	Statistical
LocalPred & RegioPred	M. Perez at Centro Nacional de Energias Renovables (CENER) and Centro de Investigaciones Energéticas, Medioambientales y Tecnalógicas, Spain (CIEMET)	Hybrid
Alea Wind	Aleasoft at the Universitat Polytécnica de Catalunya, Spain (UPC)	Statistical
EPREV	Instituto de Engenharia de Sistemas e Computadores do Porto (INESC), Instituto de Engenharia Mecânica e Gestão Industrial (INEGI) and Centro de Estudos de Energia Eólica e Escoamentos Atmosféricos (CEsA) in Portugal	Statistical
Scirocco	Aeolis Forecasting Services, Netherlands	Hybrid

bination of FFBPNN, RBFNN, and Broyden Fletcher Goldfarb Shanno NN has been developed by Zameer et al. to predict short-term wind power at several wind farms [44]. Torre et al. found their ARMA model could produce 20% error reduction compared to a persistence method when forecasting hourly mean wind speed at 10 h ahead [36]. ARIMA-family forecasting models are adopted to have an direct and indirect wind power prediction of an offshore 2-MW WT by Shi et al.[45]. Giorgi et al. utilized ANNs, Adaptive Neruo-Fuzzy Inference System, and ARMA models to predict short-term wind power in which the normalized mean error and the statistical distribution are used as accuracy metrics [46]. It is impossible to determine a best learning method that can universally solve all the wind power prediction problems since the forecasting is usually constrained by the nature of the data, the time horizon, and the computational time [9]. Foley et al. presented a non-exhaustive list of software models for wind power prediction, as reproduced in Table 6.2. [42].

Due to the intimate relationship between wind speed and wind power (Eq. (8)), the prediction of wind power is highly related to the prediction of wind speed. These two predictions are often sharing the sample principle [15]. Thus, literature works about wind speed prediction and wind power prediction are non-exhaustively selected and organized into the four categories of AI methods (as explained in Section 6.2), and are concisely summarized in Table 6.3.

6.3.3 WT Design and Optimization

AI methods have been used for design and optimization of different WT systems, e.g., blades [22, 78, 91, 92]. This section distinguishes different AI algorithms applied for WT design and optimization into two aspects, surrogate algorithms and optimization algorithms.

Design and optimization of WTs are complex processes that may involve random wind field simulation, aerodynamic wind load calculation/computational fluid dynamics (CFD), multibody dynamic analysis, structural stress/strain analysis, fatigue damage evaluation, and reliability analysis. Thus, the computational time is extremely expensive if all the aforementioned analyses are conducted for WT design and optimization. In order to reduce the computational time, meanwhile guarantee the accuracy, AI methods (e.g., neural learning and statistical learning methods)

TABLE 6.3: A LITERATURE SUMMARY OF AI METHODS FOR WIND SPEED PREDICTION AND WIND POWER PREDICTION.

Category of AI Methods	Specific AI Methods	Wind Speed Prediction	Wind Power Prediction
Neural Learning Methods	MLP	[31, 47-49]	[50, 51]
	BPNN	[32, 52-54]	[44, 55]
	RBFNN	[54, 56]	[44, 57-59]
	RNN	[41, 60, 61]	[60-63]
	WNN	[64]	[65]
	CNN	[62]	[62]
	ENN	[41, 66]	[67]
	FFNN	[32, 68]	[11, 44, 65]
	ART	[54]	[69]
Statistical Learning Methods	AR	[70-72]	[73]
	ARMA	[36, 74-77]	[77]
	ARIMA	[34, 78]	[42, 46]
Evolutionary Learning Methods	GA	[39, 79]	[80, 81]
	GP	[44]	[44]
	EP	[82]	[83]
Hybrid Learning Methods	NN-Fuzzy	[53]	[84, 85]
	NN-Bayesian	[56]	[86]
	NN-GA/GP	[38, 87]	[44]
	NN-Physical	[88]	[89]
	ANN-ARIMA	[25]	[90]

are widely used as surrogate modeling algorithms that replace the computationally expensive physical-based simulation processes. For example, Rubio proposed an analytic NN to model the WT dynamic behavior and obtained smaller RMSE than using the analytic model [93]. To reduce the time for performance analysis and minimize the prediction error, an BPNN method was used to approximate the hydrodynamic and structural performance of the substructure of floating offshore WT [94]. Statistical learning methods, such as ARIMA, Markov Chain, Kalman filter, are used to forecast a long-term wind speed pattern for calculating aerodynamic loads and fatigue damage of a horizontal-axis WT [78].

Due to the complex analysis procedure, the performance functions for WT design and optimization (e.g., blade deflection, hotspot stress of hub, drivetrain vibration, and WT power output) are highly nonlinear and implicit with respect to design variables. Thus, traditional sensitivity-based optimization algorithms may have difficult to identify the global optimum. Evolutionary learning methods, that usually do not require explicit gradient information of the performance functions, are the versatile optimization tools to drive the initial WT design to the optimum design. For example, an evolutionary-based GA is incorporated into CFD for the optimization of Savonius WT's blades and produce 33% improvement in power coefficient at the optimal blades design [95]. EA algorithms are used to solve the multi-objective structure optimization of composite WT blades based on ultimate limit state analysis [22] and fatigue analysis [92]. It is worth noting that the evolutionary optimization algorithms may require more optimization iterations and simulation samples than traditional sensitivity-based optimization algorithms (e.g., sequential-quadratic programming) in order to find the global optimum. Thus, AI methods (e.g., evolutionary learning methods), as WT optimization tools, are often accompanied by efficient and accurate surrogate models which help to overcome the computational burden.

6.3.4 WT Condition Monitoring

WTCM techniques are differentiates into WT subsystems CM techniques (e.g., vibration analysis, stress/strain measurement, acoustic emission) and WT global system CM techniques (e.g., power signal analysis, signature analysis, SCADA data analysis) [96]. AI-based CM techniques for WTs have been widely applied for both the WT subsystems and global system CM. Among the four types of AI learning methods, the neural learning methods are the most frequently used WTCM techniques due to their strong capability to represent complex nonlinear relationships. The neural learning methods often employs SCADA data to indicate the occurrence of a fault [97].

AI-based WTCM methods can be applied into different WT components including gearbox and bearings, generator, power electronics and electric controls, rotor blades and hydraulic controls, and false-alarm rate reduction. Taking the CM of WT gearbox as an example, gearbox failure caused by cracks is one of the most frequency failures that occur in modern WTs [98]. Li et al. proposed a new gearbox CM approach that is based on wavelet transform and an ARTNN method using the vibration signals collected from a gearbox to diagnose a gear crack fault (Fig. 6.5) [99]. The proposed ARTNN method is based on unsupervised learning and is able to recognize the changing trend of crack faults in gearbox. Kusiak and Verma use SCADA data from 24 WTs to train five MLP algorithms to detect bearing faults from over-temperature events [100]. According to the authors, the beset MLP configuration could obtain a prediction accuracy of more than 97% with a leading time of 1.5 h [100]. Marugán et al. carry out an exhaustive survey of neural learning methods for CM of WT gearbox and other components [9].

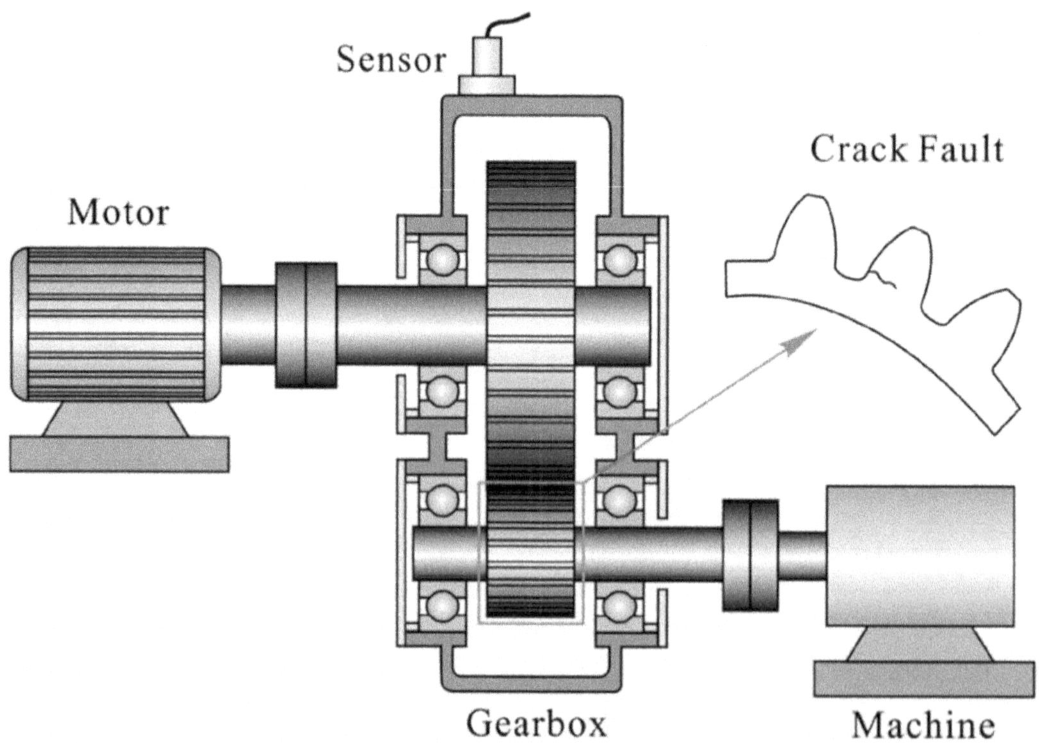

FIG. 6.5: AN ILLUSTRATION OF CRACK FAULT IN THE EXPERIMENT GEARBOX [99]

6.4 CONCLUSIONS

This chapter presents the fundamentals of state-of-the-art AI methods and a non-exhaustive survey of their wide applications in WE. These AI methods provide new avenues toward the next-generations of modern WT systems and wind farms that become increasingly complex, smart, and robust. Concluding remarks are provided as follows:

- The neural learning methods, e.g., MLP, RNN, and BPNN, are proven to be versatile tools for wind speed prediction, wind power prediction, WT design and optimization, and WTCM due to their strong flexibility and robustness.
- The statistical learning methods, e.g., ARMA and ARIMA, are widely used in time series prediction of wind speed and wind power.
- The evolutionary learning methods, e.g., GA, are often used as optimization algorithms for WT design and optimization, and are also applicable for other WE fields, e.g., wind speed prediction, power prediction, and WTCM.
- The hybrid learning methods that consist of multiple learning methods potentially generate better performance in WE applications than a single method does.
- Development and selection of specific AI methods for WE applications need to understand both the fundamentals of the AI methods and the characteristics of the WE applications.

6.5 ABBREVIATIONS

ANN	Artificial Neural Network
AR	Autoregressive
ARIMA	Autoregressive Integrated Moving Average
ARMA	Autoregressive Moving Average
ART	Adaptive Resonance Theory
ALENN	Adaptive Linear Element Neural Network
BPNN	Back Propagation Neural Network
CNN	Convolutional Neural Network
ENN	Elman Neural Network
EP	Evolutionary Programming
ES	Evolution Strategy
FFNN	Feed Forward Neural Network
FFBPNN	Feed Forward Back Propagation Neural Network
GA	Genetic Algorithm
GP	Genetic Programming
MA	Moving Average
MAE	Mean Absolute Error
MAPE	Mean Absolute Percentage Error;
MLP	Multilayer Perceptron
NN	Neural Network
RBFNN	Radial Basis Function Neural Network
RMSE	Root Mean Square Error
RNN	Recurrent Neural Network
WNN	Wavelet Neural Network

6.6 ACKNOWLEDGMENTS

This work is supported by the National Natural Science Foundation of China (Grant No. 51905475).

6.7 REFERENCES

1) Global Wind Energy Council. Global Wind Statistics 2017. 2018; Available from: https://gwec.net/wp-content/uploads/vip/GWEC_PRstats2017_EN-003_FINAL.pdf.
2) Hu, W., R.J. Barthelmie, F. Letson, and S.C. Pryor, A new seismic based monitoring approach for wind turbines. Wind Energy, Published online in October 2018.
3) Wiser, R., K. Jenni, J. Seel, and et al., Expert elicitation survey on future wind energy costs. Nature Energy, 2016. **1**(10): p. 16135.
4) Wikipedia. Artificial Intelligence. 2019; Available from: https://en.wikipedia.org/wiki/Artificial_intelligence.
5) Wachter, S., B. Mittelstadt, and L. Floridi., Transparent, explainable, and accountable AI for robotics. Science Robotics, 2017. **2**(6): p. eaan6080.
6) Leitao, P., S. Karnouskos, L. Ribeiro, J. Lee, T. Strasser, and A.W. Colombo, Smart agents in industrial cyber–physical systems. Proceedings of the IEEE, 2016. **104**(5): p. 1086-1101.
7) Jha, S.K., J. Bilalovic, A. Jha, N. Patel, and H. Zhang, Renewable energy: Present research and future scope of Artificial Intelligence. Renewable and Sustainable Energy Reviews, 2017. **77**: p. 297-317.
8) Medsker, L.R., Microcomputer applications of hybrid intelligent systems. Journal of Network and Computer Applications, 1996. **19**(2): p. 213-234.
9) Marugán, A.P., F.P.G. Márquez, J.M.P. Perez, and D. Ruiz-Hernández, A survey of artificial neural network in wind energy systems. Applied Energy, 2018. **228**: p. 1822-1836.
10) Olaofe, Z.O., A 5-day wind speed & power forecasts using a layer recurrent neural network (LRNN). Sustainable Energy Technologies and Assessments, 2014. **6**: p. 1-24.
11) Jain, V., A. Singh, V. Chauhan, and A. Pandey. Analytical study of wind power prediction system by using feed forward neural network. in Computation of Power, Energy Information and Commuincation (ICCPEIC). 2016. 2016 International Conference on, pp. 303-306. IEEE.
12) Hamed, C., N. Amjady, and H. Zareipour, Wind power forecast using wavelet neural network trained by improved clonal selection algorithm. Energy Conversion and Management 2015. **89**: p. 588-598.
13) Hagan, M.T., H.B. Demuth, M.H. Beale, and O.D. Jess, Neural network design, 2nd Ed. 2014: Martin Hagan
14) Du, K.-L. and M.N.S. Swamy, Neural networks and statistical learning. 2014, London: Springer.
15) Ma, L., Shiyan Luan, Chuanwen Jiang, Hongling Liu, and Y. Zhang, A review on the forecasting of wind speed and generated power. Renewable and Sustainable Energy Reviews, 2009. **13**: p. 915-920.
16) Hu, W., F. Letson, R.J. Barthelmie, and S.C. Pryor, Wind gust characterization at wind turbine relevant heights in moderately complex terrain. Journal of Applied Meteorology and Climatology, 2018. **57**: p. 1459-1476.
17) Chen, J. and C. Rabiti, Synthetic wind speed scenarios generation for probabilistic analysis of hybrid energy systems. Energy, 2017. **120**: p. 507-517.
18) Hogg, R.V., J.W. McKean, and A.T. Craig, Introduction to Mathematical Statistics, Six Edition. 2005, Upper Saddle River, NJ: Peason Education International.
19) Chen, P., T. Pedersen, B. Bak-Jensen, and Z. Chen, ARIMA-based time series model of stochastic wind power generation. IEEE Transactions on Power Systems, 2010. **25**(2): p. 667-676.
20) Vikhar, P.A. Evolutionary algorithms: a critical review and its future Prospects. in 2016 International Conference on Global Trends in Signal Processing, Information Computing and Communication. 2016. IEEE.
21) Huang, H.S. Distributed genetic algorithm for optimization of wind farm annual profits in The 14th International Conference on Intelligent System Applications to Power Systems. 2007. November 4-8, 2007, Kaohsiung, Taiwan.

22) Hu, W., I. Han, S.-C. Park, and D.-H. Choi, Multi-objective structural optimization of a HAWT composite blade based on ultimate limit state analysis. Journal of Mechanical Science and Technology, 2012. **26**(1): p. 129-135.
23) Charlie Vanaret, N.D., Jean-Marc Alliot. Windmill farm pattern optimization using evolutionary algorithms. in GECCO 2014, Genetic and Evolutionary Computation Conference. 2014. July 2014, Vancouver, Canada.
24) Jursa, R. and K. Rohrig, Short-term wind power forecasting using evolutionary algorithms for the automated specification of artificial intelligence models. International Journal of Forecasting, 2008. **24**(4): p. 694-709.
25) Cadenas, E. and W. Rivera, Wind speed forecasting in three different regions of Mexico, using a hybrid ARIMAeANN model. Renewable Energy, 2010. **35**: p. 2732-2738.
26) Basheer, I.A. and M. Hajmeer, Artificial neural networks: fundamentals, computing, design, and application. Journal of microbiological methods 2000. **43**(1): p. 3-31.
27) Chang, W.-Y., A literature review of wind forecasting methods. Journal of Power and Energy Engineering, 2014. **2**: p. 161-168.
28) Hu, W., Advanced wind turbine technology. 2018, Switzerland: Springer.
29) Lawan, S.M., W.A.W.Z. Abidin, W.Y. Chai, A. Baharun, and T. Masri., Different models of wind speed prediction: a comprehensive review. International Journal of Scientific and Engineering Research, 2014. **5**(1): p. 1760-1768.
30) Kani, S.P. and M.M. Ardehali, Very short-term wind speed prediction: a new artificial neural network-Markov chain model. Energy Conversion and Management, 2011. **52**(1): p. 738-745.
31) Moustris, K.P., D. Zafirakis, D.H. Alamo, R.N. Medina, and J.K. Kaldellis, 24-h ahead wind speed prediction for the optimum operation of hybrid power stations with the use of artificial neural networks. In Perspectives on Atmospheric Sciences, 2017: p. 409-414, Springer, Cham.
32) Azad, H.B., S. Mekhilef, and V.G. Ganapathy, Long-term wind speed forecasting and general pattern recognition using neural networks. IEEE Transactions on Sustainable Energy, 2014. **5**(2): p. 546-553.
33) Ding, M., L.-j. Zhang, and Y.-c. Wu, Wind speed forecast model for wind farms based on time series analysis. Electric Power Automation Equipment, 2005. **25**(8): p. 32-34. in Chinese.
34) Kavasseri, R.G. and K. Seetharaman, Day-ahead wind speed forecasting using f-ARIMA models. Renewable Energy, 2009. **34**(5): p. 1388-1393.
35) Erasmo, C. and R. Wilfrido, Wind speed forecasting in the south coast of Oaxaca, Me´xico. Renewable Energy, 2007. **32**(12): p. 2116-2128.
36) Torres, J.L., A. Garcı´a, M.D. Blas, and A.D. Francisco, Forecast of hourly average wind speed with ARMA models in Navarre. Solar Energy, 2005. **79**(1): p. 65-77.
37) Costa, A., A. Crespo, J. Navarro, G. Lizcano, H. Madsen, and E. Feitosa, A review on the young history of the wind power short-term prediction. Renewable and Sustainable Energy Reviews, 2007. **12**(6): p. 1725-1744.
38) Li, X., Y. Liu, and W. Xin. Wind speed prediction based on genetic neural network. in 2009 4th IEEE Conference on Industrial Electronics and Applications. 2009. Xi'an, China: IEEE.
39) Liu, H., H. Tian, X. Liang, and Y. Li, New wind speed forecasting approaches using fast ensemble empirical model decomposition, genetic algorithm, Mind Evolutionary Algorithm and Artificial Neural Networks. Renewable Energy, 2015. **83**: p. 1066-1075.
40) Liu, D., D. Niu, H. Wang, and L. Fan, Short-term wind speed forecasting using wavelet transform and support vector machines optimized by genetic algorithm. Renewable Energy, 2014. **62**: p. 592-597.
41) Wang, J., S. Qin, Q. Zhou, and H. Jiang, Medium-term wind speeds forecasting utilizing hybrid models for three different sites in Xinjiang, China. Renewable Energy, 2015. **76**: p. 91-101.
42) Foley, A.M., P.G. Leahy, A. Marvuglia, and E.J. McKeogh, Current methods and advances in forecasting of wind power generation. Renewable Energy, 2012. **37**(1): p. 1-8.
43) Ma, X., Y. Jin, and Q. Dong, A generalized dynamic fuzzy neural network based on singular spectrum analysis optimized by brain storm optimization for short-term wind speed forecasting. Applied Soft Computing 2017. **54**: p. 296-312.
44) Zameer, A., J. Arshad, A. Khan, and M.A.Z. Raja, Intelligent and robust prediction of short term wind power using genetic programming based ensemble of neural networks. Energy conversion and management, 2017. **134**: p. 361-372.
45) Shi, J., X. Qu, and S. Zeng, Short-term wind power generation forecasting: direct versus indirect ARIMA-based approaches. International Journal of Green Energy 2011. **8**(1): p. 100-112.
46) De Giorgi, M.G., A. Ficarella, and M. Tarantino, Error analysis of short term wind power prediction models. Applied Energy, 2011. **88**(4): p. 1298-1311.
47) Cadenas, E. and W. Rivera, Short term wind speed forecasting in La Venta, Oaxaca, México, using artificial neural networks. Renewable Energy, 2009. **34**(1): p. 274-278.
48) Ak, R., O. Fink, and E. Zio, Two machine learning approaches for short-term wind speed time-series prediction. IEEE transactions on neural networks and learning systems, 2016. **27**(8): p. 1734-1747.
49) Finamore, A.R., V. Calderaro, V. Galdi, A. Piccolo, and G. Conio. A wind speed forecasting model based on artificial neural network and meteorological data. in Environment and Electrical Engineering (EEEIC), 2016 IEEE 16th International Conference on. 2016. IEEE.
50) Li, S., Wind power prediction using recurrent multilayer perceptron neural networks. In IEEE Power Engineering Society General Meeting, 2003. **4**: p. 2325-2330.
51) Jung, S. and S.-D. Kwon, Weighted error functions in artificial neural networks for improved wind energy potential estimation. Applied Energy, 2013. **111**: p. 778-790.
52) Palomares-Salas, J.C., A. Agüera-Pérez, J.J.G. de la Rosa, and A. Moreno-Muñoz, A novel neural network method for wind speed forecasting using exogenous measurements from agriculture stations. Measurement, 2014. **55**: p. 295-304.
53) Noorollahi, Y., M.A. Jokar, and A. Kalhor, Using artificial neural networks for temporal and spatial wind speed forecasting in Iran. Energy Conversion and Management, 2016. **115**: p. 17-25.
54) Li, G. and J. Shi, On comparing three artificial neural networks for wind speed forecasting. Applied Energy, 2010. **87**: p. 2313-2320.
55) Xin, W., Y. Liu, and X. Li. Short-term forecasting of wind turbine power generation based on genetic neural network. in Intelligent Control and Automation (WCICA), 2010 8th World Congress on. 2010. IEEE.
56) Li, G., J. Shi, and J. Zhou, Bayesian adaptive combination of short-term wind speed forecasts from neural network models. Renewable Energy, 2011. **36**(1): p. 352-359.
57) Chang, G.W., H.J. Lu, Y.R. Chang, and Y.D. Lee, An improved neural network-based approach for short-term wind speed and power forecas. Renewable Energy, 2017. **105**: p. 301-311.
58) Shetty, R.P., A.P.S.P. Sathyabhama, and A.A. Rai. Optimized radial basis function neural network model for wind power prediction. in 2016 Second International Conference on Cognitive Computing and Information Processing (CCIP). 2016. IEEE.
59) Jafarian, M. and A.M. Ranjbar, Fuzzy modeling techniques and artificial neural networks to estimate annual energy output of a wind turbine. Renewable Energy, 2014. **35**(9): p. 2008-2014.
60) Barbounis, T.G., J.B. Theocharis, M.C. Alexiadis, and P.S. Dokopoulos, Long-term wind speed and power forecasting using local recurrent neural network models. IEEE Transactions on Energy Conversion, 2006. **21**(1): p. 273-284.
61) Barbounis, T.G. and J.B. Theocharis., Locally recurrent neural networks for long-term wind speed and power prediction. Neurocomputing, 2006. **69**(4-6): p. 466-496.

62) Wu, W., K. Chen, Y. Qiao, and Z. Lu. Probabilistic short-term wind power forecasting based on deep neural networks. in 2016 International Conference on Probabilistic Methods Applied to Power Systems (PMAPS). 2016. IEEE.
63) Liu, Z., W. Gao, Y.-H. Wan, and E. Muljadi. Wind power plant prediction by using neural networks. in IEEE Energy Conversion Congress and Exposition (ECCE). 2012. IEEE.
64) Doucoure, B., K. Agbossou, and A. Cardenas, Time series prediction using artificial wavelet neural network and multi-resolution analysis: Application to wind speed data. Renewable Energy, 2016. **92**: p. 202-211.
65) Bhaskar, K. and S.N. Singh, AWNN-assisted wind power forecasting using feed-forward neural network. IEEE Transactions on Sustainable Energy, 2012. **3**(2): p. 306-315.
66) Liu, H., H.-q. Tian, X.-f. Liang, and Y.-f. Li, Wind speed forecasting approach using secondary decomposition algorithm and Elman neural networks. Applied Energy, 2015. **157**: p. 183-194.
67) Ramirez-Rosado, I.J., L.A. Fernandez-Jimenez, C. Monteiro, J. Sousa, and R. Bessa, Comparison of two new short-term wind-power forecasting systems. Renewable Energy, 2009. **34**(7): p. 1848-1854.
68) Philippopoulos, K. and D. Deligiorgi, Application of artificial neural networks for the spatial estimation of wind speed in a coastal region with complex topograph. Renewable Energy, 2012. **38**(1): p. 75-82.
69) Sideratos, G. and N.D. Hatziargyriou, Wind power forecasting focused on extreme power system events. IEEE Transactions on Sustainable Energy, 2012. **3**(2): p. 445-454.
70) Miranda, M.S. and R.W. Dunn. One-hour-ahead wind speed prediction using a Bayesian methodology. in Power Engineering Society General Meeting, 2006. IEEE. 2006. IEEE.
71) Zhang, W., J. Wu, J. Wang, W. Zhao, and L. Shen, Performance analysis of four modified approaches for wind speed forecasting. Applied Energy, 2012. **99**: p. 324-333.
72) Mohandes, M.A., S. Rehman, and T.O. Halawani, A neural networks approach for wind speed prediction. Renewable Energy, 1998. **13**(3): p. 345-354.
73) Dowell, J. and P. Pinson, Very-short-term probabilistic wind power forecasts by sparse vector autoregression. IEEE Transactions on Smart Grid, 2016. **7**(2): p. 763-770.
74) Soder, L. Simulation of wind speed forecast errors for operation planning of multiarea power systems. in Probabilistic Methods Applied to Power Systems, 2004 International Conference on. 2004. IEEE.
75) Erdem, E. and J. Shi, ARMA based approaches for forecasting the tuple of wind speed and direction. Applied Energy, 2011. **88**(4): p. 1405-1414.
76) Liu, H., E. Erdem, and J. Shi, Comprehensive evaluation of ARMA–GARCH (-M) approaches for modeling the mean and volatility of wind speed. Applied Energy, 2011. **88**(3): p. 724-732.
77) Milligan, M., M. Schwartz, and Y.-h. Wan., Statistical wind power forecasting models: Results for US wind farms. No. NREL/CP-500-33956. National Renewable Energy Lab.(NREL), Golden, CO (United States), 2003.
78) Kulkarni, P.A., W. Hu, A.S. Dhoble, and P.M. Padole, Statistical wind prediction and fatigue analysis for horizontal-axis wind turbine composite material blade under dynamic loads. Advances in Mechanical Engineering, 2017. 9(9): p. 1687814017724088.
79) Damousis, I.G., M.C. Alexiadis, J.B. Theocharis, and P.S. Dokopoulos, A fuzzy model for wind speed prediction and power generation in wind parks using spatial correlation. IEEE Transactions on Energy Conversion, 2004. **19**(2): p. 352-361.
80) Madadi, S., M. Nazari-Heris, B. Mohammadi-Ivatloo, and S. Tohidi, Implementation of Genetic-Algorithm-Based Forecasting Model to Power System Problems. In Handbook of Research on Predictive Modeling and Optimization Methods in Science and Engineering, pp. 140-155. IGI Global, 2018.
81) Cornejo-Bueno, L., C. Camacho-Gómez, A. Aybar-Ruiz, L. Prieto, and S. Salcedo-Sanz, Feature Selection with a Grouping Genetic Algorithm–Extreme Learning Machine Approach for Wind Power Prediction. In Conference of the Spanish Association for Artificial Intelligence, pp. 373-382. Springer, Cham, 2016.
82) Salcedo-Sanz, S., E.G. Ortiz-Garcı, Á.M. Pérez-Bellido, A. Portilla-Figueras, and L. Prieto, Short term wind speed prediction based on evolutionary support vector regression algorithms. Expert Systems with Applications, 2011. **38**(4): p. 4052-4057.
83) Lydia, M., A.I. Selvakumar, S.S. Kumar, and G.E.P. Kumar, Advanced algorithms for wind turbine power curve modeling. IEEE Transactions on Sustainable Energy, 2013. **4**(3): p. 827-835.
84) Kariniotakis, G., E. Nogaret and G. Stavrakakis. A fuzzy logic and a neural network based wind power forecasting model. in EU Wind Energy Conference. 1996.
85) Hong, Y.-Y., H.-L. Chang, and C.-S. Chiu, Hour-ahead wind power and speed forecasting using simultaneous perturbation stochastic approximation (SPSA) algorithm and neural network with fuzzy inputs. Energy, 2010. **35**(9): p. 3870-3876.
86) Blonbou, R., Very short-term wind power forecasting with neural networks and adaptive Bayesian learning. Renewable Energy, 2011. **36**(3): p. 1118-1124.
87) Wang, S., N. Zhang, L. Wu, and Y. Wang, Wind speed forecasting based on the hybrid ensemble empirical mode decomposition and GA-BP neural network method. Renewable Energy, 2016. **94**: p. 629-636.
88) Salcedo-Sanz, S., A.M. Perez-Bellido, E.G. Ortiz-García, A. Portilla-Figueras, L. Prieto, and F. Correoso, Accurate short-term wind speed prediction by exploiting diversity in input data using banks of artificial neural networks. Neurocomputing, 2009. **72**(4-6): p. 1336-1341.
89) Peng, H., F. Liu, and X. Yang, A hybrid strategy of short term wind power prediction. Renewable Energy, 2013. **50**: p. 590-595.
90) Shi, J., J. Guo, and S. Zheng, Evaluation of hybrid forecasting approaches for wind speed and power generation time series. Renewable and Sustainable Energy Reviews, 2012. **16**(5): p. 3471-3480.
91) Supeni, E.E., J.A. Epaarachchi, M.M. Islam, and K.T. Lau. Development of artificial neural network model in predicting performance of the smart wind turbine blade. in Proceedings of the 3rd Malaysian Postgraduate Conference (MPC 2013), pp. 233-242. Education Malaysia. 2013.
92) Hu, W., D. Park, and D. Choi, Structural optimization procedure of a composite wind turbine blade for reducing both material cost and blade weight. Engineering Optimization, 2012. **45**(12): p. 1469-1487.
93) Rubio, J.d.J., Analytic neural network model of a wind turbine. Soft Computing, 2015. **19**(12): p. 3455-3463.
94) Lee, J.-C., S.-C. Shin, and S.-Y. Kim, An optimal design of wind turbine and ship structure based on neuro-response surface method. International Journal of Naval Architecture and Ocean Engineering, 2015. **7**(4): p. 750-769.
95) Chan, C.M., H.L. Bai, and D.Q. He, Blade shape optimization of the Savonius wind turbine using a genetic algorithmm. Applied Energy, 2018. **213**: p. 148-157.
96) Tchakoua, P., R. Wamkeue, M. Ouhrouche, F. Slaoui-Hasnaoui, T.A. Tameghe, and G. Ekemb, Wind turbine condition monitoring: State-of-the-art review, new trends, and future challenges. Energies, 2014. **7**(4): p. 2595-2630.
97) Wilkinson, M., B. Darnell, T.v. Delft, and k. Harman, Comparison of methods for wind turbine condition monitoring with SCADA data. IET Renew Power Gener, 2014. **8**: p. 390-397.
98) Marugán, A.P., F.P.G. Márquez, and J.M.P. Pérez, Optimal Maintenance Management of Offshore Wind Farms. Energies, 2016. **9**(1): p. 1-20.
99) Li, Z., Z. Ma, Y. Liu, W. Teng, and R. Jiang, Crack Fault Detection for a Gearbox Using Discrete Wavelet Transform and an Adaptive Resonance Theory Neural Network. Journal of Mechanical Engineering 2015. **61**(1): p. 63-73.
100) Kusiak, A. and A. Verma, Analyzing bearing faults in wind turbines; A data-mining approach. Renewable Energy, 2012. **48**: p. 110-116.

INDEX

C

D

E

F

K

L

M

N

O

P

R

S

T

U

V

W

X

Z

www.ingramcontent.com/pod-product-compliance
Lightning Source LLC
LaVergne TN
LVHW062148200926
848484LV00007B/1884